T0292021

Dimensions of Uncertainty in Communication Engineering

Dimensions of Uncertainty in Communication Engineering

Ezio Biglieri

ACADEMIC PRESS

An imprint of Elsevier

Academic Press is an imprint of Elsevier
125 London Wall, London EC2Y 5AS, United Kingdom
525 B Street, Suite 1650, San Diego, CA 92101, United States
50 Hampshire Street, 5th Floor, Cambridge, MA 02139, United States
The Boulevard, Langford Lane, Kidlington, Oxford OX5 1GB, United Kingdom

Notices

Knowledge and best practice in this field are constantly changing. As new research and experience broaden our understanding, changes in research methods, professional practices, or medical treatment may become necessary.

Practitioners and researchers must always rely on their own experience and knowledge in evaluating and using any information, methods, compounds, or experiments described herein. In using such information or methods they should be mindful of their own safety and the safety of others, including parties for whom they have a professional responsibility.

To the fullest extent of the law, neither the Publisher nor the authors, contributors, or editors, assume any liability for any injury and/or damage to persons or property as a matter of products liability, negligence or otherwise, or from any use or operation of any methods, products, instructions, or ideas contained in the material herein.

ISBN: 978-0-323-99275-6

For information on all Academic Press publications
visit our website at https://www.elsevier.com/books-and-journals

Publisher: Mara Conner
Acquisitions Editor: Tim Pitts
Editorial Project Manager: Jai Marie Jose
Production Project Manager: Manju Paramasivam
Designer: Vicky Pearson Esser

Typeset by VTeX

Working together
to grow libraries in
developing countries

www.elsevier.com • www.bookaid.org

Omne meum, nihil meum, *'tis all mine, and none mine.*
As a good housewife out of divers fleeces weaves one piece of cloth,
a bee gathers wax and honey out of many flowers, and makes a new bundle of all,
Floriferis ut apes in saltibus omnia libant,
I have laboriously collected this cento out of divers writers,
and that sine injuria, I have wronged no authors, but given every man his own.

Robert Burton, "The Anatomy of Melancholy"

Contents

Preface

As we know, there are known knowns; there are things we know we know.
We also know there are unknowns;
that is to say, we know there are some things we do not know.
But there are also unknown unknowns—the ones we don't know we don't know.
Donald Rumsfeld, 2002

Knowing that you do not know is the best.
Not knowing that you do not know is an illness.
Lǎozǐ, Dàodé Jīng

It is common practice in communication engineering to face uncertainty. Specifically, unknown outcomes of experiments run under similar conditions have to be dealt with. Probability theory has long become the tool of choice in this discipline, which implicitly assumes that all forms of uncertainty can be treated in terms of a single dimension. One should, however, realize that uncertainty presents itself under multiple dimensions: specifically, "aleatory" (or "stochastic") uncertainty, caused by the randomness of system behavior, differs in a substantial way from "epistemic" uncertainty, which is caused by the imperfect knowledge of the factors affecting the behavior of a system. In other words, while aleatory uncertainty affects events that cannot be predicted, epistemic uncertainty is caused by imperfect information about the probabilistic model of aleatory uncertainty. Different types of information deficiency determine the type of the associated uncertainty: as observed in [143, p. 6], information may be incomplete, imprecise, fragmentary, unreliable, vague, or even contradictory. Epistemic uncertainty can be reduced by a deeper analysis, while aleatory uncertainty cannot. Probability theory is only sufficient once epistemic uncertainty is removed from the problem.

In wireless communication, a considerable amount of research activity has been spent in the search for accurate statistical models of the transmission channel. The model selected should satisfy the requirement of yielding a design that performs satisfactorily and provides reliable guidelines for decision making. A major problem there is caused by the fact that no single model (e.g., Rayleigh or Rice probability distributions modeling fading effects) can be accurate enough for a wide variety of channels appearing in practice. Considerable efforts have also been spent in the search for general classes of probability measures which are at the same time physically justified and flexible enough to fit a large mass of data gathered experimentally. Previous work in the area of wireline transmission was primarily focused, under the optimistic assumption that the channel model was perfectly known, on the uncertainties due to randomness. More recent research involving wireless channels has realized that the impact of epistemic uncertainty on performance analysis could not be neglected.

The notion of distinguishing epistemic from aleatory uncertainty is far from new. Original probability theory was rooted in the analysis of games of chance (Gerolamo Cardano, Blaise Pascal, and Pierre De Fermat), and hence was based on purely aleatory uncertainty. Later, Pascal presented his famous wager, observing that, since either God exists or not, one can either wager for God or against Him, with widely different outcomes. This was a decision made under purely epistemic uncertainty, due to the missing knowledge about the actual existence of God.

Purely aleatory uncertainty assumes full confidence in the probability distribution model chosen, and implies "decisions under risk." Purely epistemic uncertainty is naturally measured by confidence, or by subjective knowledge [88, p. 9]. In [263], the distinction between epistemic and aleatory uncertainty is made from the viewpoint of financial investing, where an agent is expected to simultaneously seek maximum expected returns and minimum return variability. Here, distinct investment strategies reflect the relative weights assigned to epistemic and aleatory uncertainty. Since epistemic uncertainty results from incomplete knowledge, viewing the market's nature as purely epistemic will make an investor try to reduce uncertainty by seeking the maximum available amount of information before making investment choices. Investors who view the market as purely aleatory, and hence inherently unpredictable, will try to reduce variability using strategies like asset diversification (see also our discussion of decisions under risk and agent attitudes in Chapter 7). More generally, one might find himself in a situation where aleatory and epistemic uncertainty coexist. Consider the example of an urn A containing 50 blue balls and 50 red balls, and an urn B containing 100 red and blue balls in an unknown proportion. Consider a bet in which an agent draws a ball from one of the urns, and receives a prize if the ball is red. Draws from urn A are affected by pure aleatory uncertainty, while draws from urn B are affected by a mixture of aleatory and epistemic uncertainty.

Goal of this book

Mathematical tools useful for dealing with epistemic uncertainty are the subject of the present monograph. My leitmotiv here is that one should (i) avoid performance analyses relying on unwarranted statistical assumptions, and at the same time (ii) evaluate the cost of the inaccuracy of a physical model. Citing verbatim from [81, p. 112], "it is better to have a correct analysis that honestly distinguishes between variability and incertitude than an analysis that depends on unjustified assumptions and wishful thinking. (⋯) If the price of a correct assessment is broad uncertainty as a recognition or admission of limitations in our scientific knowledge, then we must pay that price."

The "law of decreasing credibility," formulated by C.F. Manski [166, p. 1] states that "the credibility of inference decreases with the strength of the assumptions maintained," and brings explicitly into the argument the reliability of conclusions, or performance evaluations, derived from assumptions lacking a strong support. One should keep in mind Bertrand Russell's quip about unjustified assumptions: "The method of 'postulating' what we want has many advantages; they are the same as the advantages of theft over honest toil" [215, p. 71], and the celebrated phrase used by Isaac Newton in his essay "General Scholium" and appended to the second (1713) edition of the Principia: "Hypotheses non fingo" (Latin for "I do not feign hypotheses," or "I contrive no hypotheses") [185, p. 589]. For example, Chapter 6 will describe how wrong assumptions about the dependence structure of random variables may heavily affect the accuracy of system performance. Thus, one should accept that, if calculations are known to be affected by epistemic uncertainty, at least an analysis of the sensitivity of the final results to the inaccurate assumptions is called for. One way of proceeding, as followed here, is to derive bounds to system performance that reflect the amount and the quality of epistemic uncertainty. Wide bounds should guide the search for ways to reduce uncertainty, while tight bounds indicate that the influence of epistemic uncertainty is limited.

Citing verbatim from [143, p. 9], "Our choice of uncertainty theory for dealing with each given problem should be determined solely by the nature of the problem. The chosen theory should allow us to express fully our ignorance and, at the same time, it should not allow us to ignore any available information." Since theories of epistemic uncertainty were developed across different fields, they took various forms and resulted into a fragmented terminology. Several mathematical structures are available: these include interval analysis, possibility theory, imprecise probability theory, random-set theory, fuzzy measure theory [278], Dempster–Shafer theory [229], and evidence theory [114, p. 605] (relations among these theories, and some equivalence results, are discussed, for example, in [8,163,207,272]). While a few of these techniques will be studied, it is beyond the scope of this book to pursue an extensive comparison of their merits. One related topic that will not be covered in this book is *Robust Statistics*, which carries an intimate relationship with uncertainty theories. As observed in [119, p. 1], simplifications are often justified by appealing to a sort of continuity principle, which assumes that a minor error in mathematical models causes a small error in performance prediction. Robust Statistics theory collects procedures aimed at reducing the sensitivity of a system behavior to errors caused by epistemic uncertainty, like modeling errors.

Although my selection of material reflects its focus on communication engineering, the techniques I describe are also of interest in broader contexts involving statistics and computational probability theory. The reader will realize that those techniques reflect research work done in disparate areas, sometimes having very little in common. Given the importance to other applications that many topics covered by this book have, extensive commonalities with areas like transportation theory and financial mathematics are often pointed at through a number of references to research done in those areas.

Each chapter is devoted to a topic which could be easily expanded to book length. To avoid writing an *omnium-gatherum*, only the essential aspects of each topic are highlighted, sometimes at the price of loss of rigor. Inevitable omissions and meager coverage of certain important topics, caused by space constraints, are partly compensated by the generous use of bibliographical references to previous related work, which can be used for deepening the treatment, widening the range of its applications, and giving the reader a view of how different contributions are related both in time and context. Some references, placed in the "Further reading" section of bibliography, are not cited in the text but could be useful for additional reading. The "Sources and Parerga" section at the end of each chapter is intended as a warehouse storing the pieces that were left out of the construction of the main text.

Audience

The material here does not demand any great mathematical knowledge. It only assumes a firm grasp of the fundamental concepts of probability, information, and communication theories as presented in senior-level college courses. Elementary notions of optimization theory may be useful to follow some arguments in a few sections of Chapters 1 to 3. Although readers would probably benefit the most by reading this book from cover to cover, I tried to keep each chapter standing on its own, so that individual chapters could be read separately, with no need to work through the entire text. The reader will realize that, due to the variety of topics covered, the introduction of proofs of each individual statement would require an unreasonable expansion of the mathematical background needed to appreciate their details. Thus, although I loathe (especially in textbooks) the expression "it can be proved that . . . ," I could not avoid using it often, lest the size of the treatment was expanded beyond control. In exchange, I provide abundant references to papers and journals where the interested reader can find the details that were omitted, or delve deeper (after all, as Joseph Conrad wrote in 1897 in a letter to Cunninghame Graham, "One writes only half the book; the other half is with the reader").

This book is meant for researchers and scientists working in communication engineering and for graduate students at any level. It may also be useful as a supplementary textbook in courses devoted to communication engineering or information theory.

Tabula gratulatoria

I have been developing the material included in this book for about a decade. Much of it includes work I have done along the years with several coauthors, from whom I have learned a good deal about the topics covered here. Among them I want to thank Nabil Alrajeh, Daniele Angelosante, Emanuele Grossi, I-Wei Lai, Marco Lops, Giorgio Taricco, Emanuele Viterbo, Cheng-An Yang, and especially Kung Yao. I also thank Anthony Ephremides and Sergio Verdú for a friendship rooted in scholarship and truth.

Ezio Biglieri
Universitat Pompeu Fabra,
Barcelona, Spain
December 2021

Notations and symbols

1. The set of real numbers is denoted by $\mathbb{R}$; $\mathbb{R}^+$ and $\mathbb{N}$ denote the sets of nonnegative real numbers and nonnegative integers, respectively; $\overline{\mathbb{R}}$ denotes the extended real line $\mathbb{R} \cup \{-\infty, \infty\}$.
2. log denotes the natural logarithm.
3. $\Re$ denotes the real part of a complex number.
4. $\mathsf{var}(X)$ denotes the variance of the random variable X, and $\mathsf{cov}(X, Y)$ the covariance of the pair of random variables X, Y.
5. If A is a set, A^c denotes its complement and $|A|$ its cardinality; $\mathbb{1}_A$ denotes its indicator (or characteristic) function, viz., $\mathbb{1}_A(x) = 1$ if $x \in A$, and $\mathbb{1}_A(x) = 0$ if $x \notin A$; 2^A denotes the power set of A, i.e., the family of all subsets of A. If A and B are subsets of X, $A - B$ denotes their difference $\{x \in X \mid x \in A, x \notin B\}$.
6. The notation $[a, b]$ denotes a closed interval, and (a, b) an open interval; $\mathbb{I}$ denotes the interval $[0, 1]$.
7. The symbol $\sim$ denotes asymptotic equality, $\approx$ approximate equality (in numerical value), and $\triangleq$ equality by definition.
8. Lower-case boldface symbols denote vectors, and capital boldface symbols denote matrices. In particular, $\mathbf{I}$ denotes the identity matrix, and $\mathbf{0}$ the matrix or vector, all of whose entries are zero. The superscript T denotes the transpose of a vector or matrix; $\mathbf{A} \geqslant \mathbf{0}$ means that matrix $\mathbf{A}$ is nonnegative definite.
9. The support $\mathcal{X}$ of a random variable X is a set such that $X \in \mathcal{X}$ with probability one; $\mathcal{X}$ is also called the support set of the probability measure, of the probability density function, and of the cumulative distribution function of X.
10. We write $X \sim \mathcal{N}(\mu, \sigma^2)$ to indicate that the random variable X has a Gaussian distribution with mean μ and variance σ^2, $\mathbf{X} \sim \mathcal{N}(\boldsymbol{\mu}, \mathbf{R})$ to indicate that $\mathbf{X}$ is a Gaussian random vector with mean $\boldsymbol{\mu}$ and covariance matrix $\mathbf{R}$, with $\mathcal{N}_c$ indicating that X is a complex random variable. We write $X \sim \mathcal{U}[a, b]$ to indicate that X has a uniform distribution with support set $[a, b]$, and $X \sim \mathcal{G}(\alpha, \beta)$ to indicate that X has a Gamma distribution with parameters α, β.
11. The symbol $\mathbb{P}$ denotes probability measure, and $\mathbb{E}$ expectation.
12. The cumulative distribution function associated with $\mathbb{P}$ is denoted by F, and the probability density function, assumed to exist, is denoted by f. If convenient, when we refer to the probability measure of a random variable X, we specify its cumulative distribution function and probability density function writing $F_X(x)$ and $f_X(x)$, respectively. The function $F^{\leftarrow}$ denotes the quasi-inverse of the cumulative distribution function F.
13. $\delta_{i,j}$ denotes the Kronecker delta function, taking value 1 if $i = j$ and 0 otherwise; $u(x)$ denotes the unit-step function, taking value 1 if $x > 0$, and 0 otherwise; $\mathsf{sgn}(x)$ denotes the *signum* function, taking value -1 if $x < 0$, 0 if $x = 0$, and 1 if $x > 0$.
14. $\langle X \rangle$ denotes the empirical mean of the random variable X.
15. $x_n = O[g(n)]$ means that a positive constant M exists such that $|x_n| \leqslant M \cdot |g(n)|$ for all sufficiently large n.

Acronyms and abbreviations

a.e.	Almost everywhere
AIC	Akaike information criterion
a.s.	Almost surely
CDF	Cumulative distribution function
CR	Cognitive radio
DM	Decision maker
DS	Dempster–Shafer
EUT	Expected utility theory
GQR	Gauss quadrature rule
KKT	Karush–Kuhn–Tucker
KL	Kullback–Leibler
LHS	Left-hand side
LR	Likelihood ratio
MDL	Minimum description length
MGF	Moment-generating function
MIMO	Multiple-input, multiple-output
ML	Maximum likelihood
pdf	Probability density function
POD	Positive orthant dependent
PQD	Positive quadrant dependent
PR	Principal representation
QR	Quadrature rule
RHS	Right-hand side
ROC	Receiver operating characteristic
RV	Random variable
SDP	Semidefinite program
SIRP	Spherically invariant random process
SoN	State of Nature

Model selection

1

In this initial chapter we show how channel modeling can be done when epistemic uncertainty may play a major role in the prediction of system performance. Although most of the tools described here can be adapted to different situations, we shall focus our attention on the derivation, based on incomplete knowledge available, of the probability density function (pdf) of the random fading attenuation process affecting wireless communications. We proceed as follows:

① We first describe how a pdf of a given family, whose form is known but whose parameters are unspecified, can be fitted to experimental data: Rayleigh, Rice, and Nakagami-*m* fading models.

② If only incomplete information about a random variable (RV) is available, and in particular its pdf has an unknown form, we describe how the *maximum-entropy method* can be used to determine a pdf consistent with what is known about the RV.

③ Assuming a given class of fading models in which the true model, or at least a good approximation to it, is believed to lie, we show how the Akaike Information Criterion can be used to make the best model choice.

1.1 Parametric models

Here we consider the selection of a class of parametric models: although in common practice neither the model nor its parameters are perfectly known, we may accept to proceed under the assumption that the model structure is known and correct, while only its parameters need be determined. The specification of such parametric class carries more epistemic effort than the estimation of its parameters. In particular, models selected in a candidate set should make as much physical sense as possible. Moreover, a model should be selected with an eye on its practicality/usefulness (as an adage goes, "all models are wrong, but some are useful" [41, p. 20]). The ideal model should be simple enough to be mathematically convenient, thus satisfying an implicit "law of parsimony," even at the cost of some inexactness, and hence should be chosen with the smallest number of parameters sufficient to represent the data in an adequate way. In any case, as suggested for example in [8, p. 144], models "should not be taken too seriously," i.e., they should be understood as crude approximations to reality, and thus the choice of a certain class of models is often strongly influenced by mathematical convenience. Nonetheless, the assumption that the "true model" generating the available experimental data lies within the set of candidate models is generally implicit, although sometimes unwarranted.

Dimensions of Uncertainty in Communication Engineering. https://doi.org/10.1016/B978-0-32-399275-6.00009-5

Let the model parameters be summarized by vector $\boldsymbol{\theta}$. Four cases categorize the situation in which a model depending on $\boldsymbol{\theta}$ can be selected [41, p. 14]. What is relevant may be:

(i)   Only the given model structure, and not the value of $\boldsymbol{\theta}$.
(ii) The given model structure, plus the theoretically best value of $\boldsymbol{\theta}$.
(iii) The given model structure, plus the best estimate $\hat{\boldsymbol{\theta}}$ fitting the data.
(iv) The model structure obtained by fitting experimental data.

1.2   Wireless channel models

It is generally assumed that the signal transmitted over a wireless channel is additively perturbed by zero-mean white Gaussian noise. Moreover, the received power in a wireless channel is affected by attenuations, statistically independent of the noise, caused by various phenomena that are usually characterized as the combination of three effects:

(i)   *Path loss*, due to the fact that the power received by an antenna decreases as the distance from the transmitter increases.
(ii) *Shadowing loss*, due to the absorption of the radiated signal by scattering structures located between transmitter and receiver.
(iii) *Fading loss*, occurring as the combination of two effects: *multipath propagation* and *Doppler frequency shift*.

Due to the complicated nature of wireless channels, their precise model is either unknown or too complex to be handled. Here we make the common assumption that the receiver is able to compensate for the phase shift introduced by the channel, so that only the fading envelope need modeling. An important special fading model is *slow fading*, which is approximately constant over a transmitted-symbol time, so that the fading random process can be modeled by a single RV R with pdf $f_R(r), 0 \leqslant r < \infty$, describing the envelope

$$R(t) \triangleq \sqrt{X_{\mathrm{R}}^2(t) + X_{\mathrm{I}}^2(t)} \tag{1.1}$$

of the stationary complex fading process $X_{\mathrm{R}}(t) + jX_{\mathrm{I}}(t)$.

Several fading-envelope models have been proposed (for a summary, see, e.g., [238, Table 2.2, p. 21]). The most useful among these include the following:

1.2.1   Rayleigh fading

The Rayleigh pdf is

$$f_R(r) = \frac{r}{\sigma^2} e^{-r^2/2\sigma^2}, \qquad r \geqslant 0, \tag{1.2}$$

and the corresponding cumulative distribution function (CDF) is

$$F_R(r) = 1 - e^{-r^2/2\sigma^2}, \qquad r \geqslant 0. \tag{1.3}$$

This model has a single parameter $\sigma^2 = \mathbb{E}[R^2]/2$, the average power of the envelope. The "normalized" form of the pdf

$$f_R(r) = 2re^{-r^2}, \qquad r \geqslant 0, \tag{1.4}$$

which has $\mathbb{E}[R^2] = 1$, is often used.

1.2.2 Rice fading

If the propagation medium includes a major fixed path in addition to several "scatter" paths, the Rice fading envelope model is often used. This pdf has two parameters, σ and v, and is given by

$$f_R(r) = \frac{r}{\sigma^2} \exp\left(-\frac{r^2 + v^2}{2\sigma^2}\right) I_0\left(\frac{rv}{\sigma^2}\right), \qquad r \geqslant 0, \tag{1.5}$$

where $I_0(\cdot)$ denotes the modified Bessel function of the first kind with order zero. The average power of the envelope is $\mathbb{E}[R^2] = 2\sigma^2 + v^2$, where v^2 is the power of the fixed-path (line-of-sight) component, and $2\sigma^2$ the power of the scatter component. The CDF can be written in the form

$$F_R(r) = 1 - Q_1\left(\frac{v}{\sigma}, \frac{r}{\sigma}\right), \qquad r \geqslant 0, \tag{1.6}$$

where $Q_1(\cdot, \cdot)$ is called *Marcum Q-function* [237, p. 12]. Introducing the *shape parameter* $K \triangleq v^2/2\sigma^2$, the pdf can also be written as

$$p_R(r) = 2r(1 + K)\exp\left[-(1 + K)r^2 - K\right] I_0\left(2r\sqrt{K(1 + K)}\right), \qquad r \geqslant 0. \tag{1.7}$$

Since $I_0(0) = 1$, as $K \to 0$, the Rice pdf tends to the Rayleigh pdf (1.4). As the power of the scatter component tends to zero, and hence $K \to \infty$, the Rice pdf tends to a Gaussian pdf with mean value tending to infinity.

1.2.3 Nakagami-m fading

This pdf [180] depends on two parameters, m and Ω, where Ω denotes the average power of the envelope and

$$m \triangleq \frac{\Omega^2}{\mathbb{E}\left[(R^2 - \Omega^2)^2\right]} \tag{1.8}$$

is called the *shape*, or *fading figure*. The expression of the pdf is

$$f(r) = \frac{2m^m}{\Gamma(m)\Omega^m} r^{2m-1} \exp\left(-\frac{m}{\Omega} r^2\right), \qquad m \geqslant 1/2, \ \Omega > 0, \ r \geqslant 0, \qquad (1.9)$$

which yields $\mathbb{E}[R^2] = \Omega$.

Fig. 1.1 shows this pdf for $\Omega = 1$ and some values of m. For $m = 0.5$, it is a one-sided Gaussian density, and for $m = 1$ a Rayleigh density.

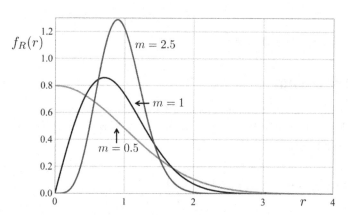

Figure 1.1 Nakagami-m pdf for $\Omega = 1$ and $m = 0.5$, $m = 1$, and $m = 2.5$.

1.3   Parameter estimation

Once an envelope pdf model is chosen, and it depends on one or more parameters, these should be estimated from observed data. We examine here a situation in which we know the parametric form of the pdf, but not the value of its parameters (this is to be contrasted to the case where the pdf is entirely determined by the data without any model assumption). So we assume a parameter vector $\boldsymbol{\theta}$, which is unknown but fixed (i.e., not a RV) and belongs to a known set Θ. We write $f_{\boldsymbol{\theta}}(x)$ for the corresponding pdf. Assume that the real data vector $\mathbf{x} = (x_1, x_2, \dots, x_n)$ has been observed, whose components are identically distributed samples taken from the random process under investigation. The maximum-likelihood (ML) parameter estimation procedure consists of choosing the parameter value $\hat{\boldsymbol{\theta}}$ that makes the observed data most probable. The *likelihood of* $\boldsymbol{\theta}$ with respect to $\mathbf{x}$ is given by the joint pdf of the data vector,

$$L(\boldsymbol{\theta}) \triangleq f_{\boldsymbol{\theta}}(\mathbf{x}). \qquad (1.10)$$

When the samples are statistically independent, we have, with obvious notations,

$$L(\boldsymbol{\theta}) = \prod_{i=1}^{n} f_{\boldsymbol{\theta}}(x_i). \qquad (1.11)$$

The ML estimate of θ is then

$$\hat{\theta} \triangleq \arg\max_{\theta} L(\theta), \tag{1.12}$$

as θ ranges in the parameter space Θ. If the function $\hat{\theta} : \mathbb{R}^n \mapsto \Theta$ is measurable, then it is called the *maximum-likelihood estimator* of θ.

When (1.11) holds, it may be easier to maximize the logarithm of $L(\theta)$, which yields

$$
\begin{aligned}
\hat{\theta} &= \arg\max_{\theta} \log L(\theta) \\
&= \arg\max_{\theta} \log \prod_{i=1}^{n} f_{\theta}(x_i) \\
&= \arg\max_{\theta} \sum_{1=1}^{n} \log f_{\theta}(x_i).
\end{aligned}
\tag{1.13}
$$

Further details about ML estimates can be found, for example, in [137, Chapter 7], [235, Chapter 4].

Example 1.1. A central role in communication theory is played by the Gaussian (or *normal*) distribution. If the RV X has a Gaussian pdf, mean μ, and variance σ^2, i.e.,

$$f_X(x) = \frac{1}{\sqrt{2\pi\sigma^2}} \exp\left(-\frac{1}{2\sigma^2}(x-\mu)^2\right), \tag{1.14}$$

we write $X \sim \mathcal{N}(\mu, \sigma^2)$. $\diamond$

Example 1.2. For n independent normally distributed observations $\mathbf{x}$ with components $x_i \sim \mathcal{N}(\mu, \sigma^2)$, $1 \leqslant i \leqslant n$, the likelihood function is

$$L(\mu, \sigma) = \frac{1}{(2\pi\sigma^2)^{n/2}} \exp\left(-\frac{1}{2\sigma^2} \sum_{i=1}^{n}(x_i-\mu)^2\right), \tag{1.15}$$

and its logarithm is

$$\log[L(\mu, \sigma)] = -\frac{n}{2}\log(2\pi\sigma^2) - \frac{1}{2\sigma^2}\sum_{i=1}^{n}(x_i-\mu)^2. \tag{1.16}$$

Taking the derivative of (1.16) with respect to μ and equating it to zero, we obtain the equation whose solution is $\hat{\mu}$, the ML estimate of μ,

$$\frac{\partial}{\partial\mu}\log[L(\mu, \sigma^2)] = \frac{1}{2\sigma^2}2n\left(\langle\mathbf{x}\rangle - \mu\right) = 0, \tag{1.17}$$

where $\langle \mathbf{x} \rangle$ is the sample mean

$$\langle \mathbf{x} \rangle \triangleq \frac{1}{n} \sum_{i=1}^{n} x_i. \tag{1.18}$$

Thus,

$$\hat{\mu} = \langle \mathbf{x} \rangle. \tag{1.19}$$

This is a maximum of $\log[L(\mu, \sigma^2)]$, as the second derivative is strictly negative. Similarly, taking the derivative of (1.16) with respect to σ and equating it to zero, we obtain $\hat{\sigma}$:

$$\frac{\partial}{\partial \sigma} \log[L(\mu, \sigma)] = -\frac{n}{\sigma} + \frac{1}{\sigma^3} \sum_{i=1}^{n} (x_i - \mu)^2 = 0, \tag{1.20}$$

whose solution is

$$\hat{\sigma}^2 = \frac{1}{n} \sum_{i=1}^{n} (x_i - \mu)^2. \tag{1.21}$$

Replacing $\hat{\mu}$ for μ in (1.21), we finally obtain

$$\begin{aligned} \hat{\sigma}^2 &= \frac{1}{n} \sum_{i=1}^{n} (x_i - \mu)^2 \\ &= \frac{1}{n} \sum_{i=1}^{n} x_i^2 - \langle \mathbf{x} \rangle^2. \end{aligned} \tag{1.22}$$

Direct calculations show that the expected value of $\hat{\mu}$ is equal to the true value μ, which is expressed by saying that the ML estimate $\hat{\mu}$ is *unbiased*. On the contrary,

$$\mathbb{E}[\hat{\sigma}^2] = \frac{n-1}{n} \sigma^2, \tag{1.23}$$

showing that $\hat{\sigma}$ is a *biased* estimate. $\diamond$

Example 1.3. The Rayleigh pdf in its unnormalized form (1.2) depends on parameter σ^2. We have

$$\log f_{\theta}(\mathbf{r}) = -n \log(\sigma^2) + \sum_{i=1}^{n} \log(r_i) - \frac{1}{2\sigma^2} \sum_{i=1}^{n} r_i^2 = 0. \tag{1.24}$$

Equating to zero the derivative taken with respect to σ^2, the ML estimate of σ^2 is obtained as the solution of

$$-\frac{n}{\sigma^2} + \frac{1}{2\sigma^4} \sum_{i=1}^{n} r_i^2 = 0, \tag{1.25}$$

i.e.,

$$\hat{\sigma}^2 = \frac{1}{2n} \sum_{i=1}^{n} r_i^2, \tag{1.26}$$

which is unbiased. Notice that the ML estimate of σ, given by the square root of (1.26), is biased. $\diamond$

Example 1.4. The Nakagami-m pdf (1.9) depends on two parameters, m and Ω. We have, with $\boldsymbol{\theta} = (m, \Omega)$ and $r_i \geq 0$ for all i,

$$\begin{aligned} \log f_{\boldsymbol{\theta}}(\mathbf{r}) &= \log \left[\prod_{i=1}^{n} \frac{2m^m}{\Gamma(m)\Omega^m} r_i^{2m-1} \exp(-mr_i^2/\Omega) \right] \\ &= n \log \left[\frac{2m^m}{\Gamma(m)\Omega^m} \right] + (2m-1) \sum_{i=1}^{n} \log(r_i) - \frac{m}{\Omega} \sum_{i=1}^{n} r_i^2. \end{aligned} \tag{1.27}$$

If Ω is known, taking the partial derivative of (1.27) with respect to m and equating it to zero, we obtain the ML estimate $\hat{m}$ as the solution of

$$-\psi(m) + \log(m) = \frac{1}{n\Omega} \sum_{i=1}^{n} r_i^2 - 1 + \log(\Omega) - \frac{1}{n} \sum_{i=1}^{n} \log(r_i^2), \tag{1.28}$$

where $\psi(m) \triangleq \Gamma'(m)/\Gamma(m)$ denotes the digamma function. We may solve (1.28) for m after replacing the average power Ω with its ML estimate $\hat{\Omega} = (1/n) \sum_{i=1}^{n} r_i^2$. Approximate solutions of (1.28) are described in [44,281].

   Simpler (non-ML) estimation methods use the sample estimates of the kth moment $\mu_k \triangleq \mathbb{E}[R^k]$, viz.,

$$\hat{\mu}_k \triangleq \frac{1}{N} \sum_{n=1}^{N} r_n^k. \tag{1.29}$$

Since for the Nakagami-m pdf direct calculation yields

$$\mu_k = \frac{\Gamma(m+k/2)}{\Gamma(m)m^{k/2}} \Omega^{k/2}, \tag{1.30}$$

a moment-based estimate $\hat{m}$ of m may be obtained, for any given k, as the solution of [1]

$$\frac{\Gamma(\hat{m}+k/2)}{\Gamma(\hat{m})\hat{m}^{k/2}} = \frac{\mu_k}{\hat{\mu}_2^{k/2}}, \tag{1.31}$$

obtained by identifying Ω with $\hat{\mu}_2 \triangleq (1/n) \sum_{i=1}^{n} r_i^2$. For even k, the above is an algebraic equation. For $k = 4$, we obtain the estimate

$$\hat{m} = \frac{\hat{\mu}_2^2}{\hat{\mu}_4 - \hat{\mu}_2^2}. \tag{1.32}$$

In [45] an estimator based on $\hat{\mu}_1$, $\hat{\mu}_2$, and $\hat{\mu}_3$ is derived as

$$\hat{m} = \frac{\hat{\mu}_1 \hat{\mu}_2}{2(\hat{\mu}_3 - \hat{\mu}_1 \hat{\mu}_2)}. \tag{1.33}$$

A slightly different approach, advocated in [174], rewrites $f_{\boldsymbol{\theta}}(\mathbf{r})$ after defining $\sigma \triangleq \Omega/m$ as

$$f_{\boldsymbol{\theta}}(\mathbf{r}) = \left(\frac{2}{\Gamma(m)\sigma^m}\right)^n \prod_{i=1}^{n} r_i^{(2m-1)} \exp\left(-\frac{1}{\sigma} \sum_{i=1}^{n} r_i^2\right) \tag{1.34}$$

and equates to zero the partial derivatives of the logarithm, which yields

$$\frac{1}{\sigma^2}\left(-nm\sigma + \sum_{i=1}^{n} r_i^2\right) = 0, \tag{1.35a}$$

$$-n\psi(m) - n\log\sigma + \sum_{i=1}^{n} \log r_i^2 = 0. \tag{1.35b}$$

Eq. (1.35a) has the solution $\sigma = (\sum_i r_i^2)/nm$, and (1.35b) can be solved numerically.
A comparison of various parameter estimators is made in [281]. $\diamond$

Remark 1.1. In the calculations above, we have implicitly used the "invariance property" of maximum-likelihood estimators [137, Theorem 7.2], which states that, if $g(\cdot)$ is a one-to-one function, then $\widehat{g(\theta)} = \hat{g}(\theta)$. If the number of model parameters is larger than two, *Stein paradox* [76,77] should be taken into account. It states that, when $\boldsymbol{\theta}$ has three or more components, and the quality of the estimator is evaluated in terms of its expected mean-square error, then combined estimators may be more accurate than any method that separately estimates the individual components of $\boldsymbol{\theta}$.

1.3.1   A word of caution

Assume that the pdf of an RV X, modeled using estimates of its parameters, is used to evaluate the probability of an event much rarer than the occurrence of the values taken by the samples used for the estimate. Thus, a model generated using frequent, and hence "typical," events is used to compute the probability of events occurring very seldom. In this situation, errors in parameter estimation may be magnified whenever the pdf resulting from the estimate is used to predict rare events [13].

Example 1.5. Consider, for example, the estimate of the parameter a of the exponential CDF $F(x) = 1 - e^{-ax}$, $x \geqslant 0$. The estimated value $\hat{a}$ may be used to compute the approximate probability $\epsilon \triangleq \mathbb{P}[X > T] = e^{-\hat{a}T}$ that a threshold T is exceeded. The fractional change of $\mathbb{P}[X > T]$ due to the error in the estimate of a is given by

$$
\begin{aligned}
\eta &\triangleq \frac{e^{-aT} - e^{-\hat{a}T}}{e^{-\hat{a}T}} \\
&= e^{-(a-\hat{a})T} - 1 \\
&= \epsilon^{-(1-a/\hat{a})} - 1.
\end{aligned}
\tag{1.36}
$$

(See Fig. 1.2.)   ◇

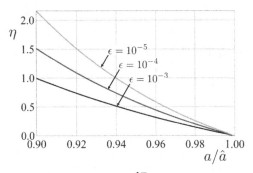

Figure 1.2 Relative error in the estimate $\epsilon = e^{-\hat{a}T}$ caused by replacing the true value a with its estimate $\hat{a}$.

1.4 Differential entropy and Kullback–Leibler divergence

In the rest of this chapter we shall be using two concepts derived from information theory and statistics, namely *differential entropy* and *Kullback–Leibler (KL) divergence*. We first provide their definitions and basic properties.

1.4.1 Differential entropy

The differential entropy of a continuous RV X with support set $\mathcal{X}$ and pdf $f_X(x)$ is defined as [55, Chapter 8]

$$
\mathsf{h}(X) \triangleq - \int_{\mathcal{X}} f_X(x) \log[f_X(x)] \, dx
\tag{1.37}
$$

and measured in *nats*. Sometimes this is conveniently denoted by $\mathsf{h}(f)$, which stresses the fact that $\mathsf{h}(X)$ depends on X only through its pdf. As a function of f_X, the differential entropy is concave $\cap$ over a convex $\cup$ set [55, p. 409].

Example 1.6. Direct calculation shows that an RV $X \sim \mathcal{N}(0, \sigma^2)$ has entropy

$$h(X) = \frac{1}{2} \log(2\pi e \sigma^2). \tag{1.38}$$

$\diamond$

The definition of differential entropy can be extended to random vectors $\mathbf{X}$ with joint pdf $f(\mathbf{x})$, yielding the "joint differential entropy"

$$h(f) \triangleq \int_{\mathcal{X}} f(\mathbf{x}) \log[f(\mathbf{x})] \, d\mathbf{x}. \tag{1.39}$$

Example 1.7. The multivariate pdf of a Gaussian random n-vector $\mathbf{X}$ with mean $\boldsymbol{\mu}$ and covariance matrix $\mathbf{R} \triangleq \mathbb{E}[(\mathbf{X} - \boldsymbol{\mu})(\mathbf{X} - \boldsymbol{\mu})^{\mathsf{T}}]$, which we denote $\mathbf{X} \sim \mathcal{N}(\boldsymbol{\mu}, \mathbf{R})$, is

$$f(\mathbf{x}) = \frac{1}{\sqrt{(2\pi)^n \det(\mathbf{R})}} \exp\left\{ -\frac{1}{2} (\mathbf{x} - \boldsymbol{\mu})^{\mathsf{T}} \mathbf{R}^{-1} (\mathbf{x} - \boldsymbol{\mu}) \right\}, \tag{1.40}$$

where $\det(\mathbf{R})$ denotes the determinant of $\mathbf{R}$. In this case, direct calculation yields [55, p. 250]

$$h(f) = \frac{1}{2} \log\left[(2\pi e)^n \det(\mathbf{R}) \right]. \tag{1.41}$$

$\diamond$

Some properties of differential entropy are the following [55, p. 253–254]:

 (i) $h(X + c) = h(X)$, i.e., translations do not affect the differential entropy.
 (ii) $h(aX) = h(X) + \log|a|$.
 (iii) If $\mathbf{X}$ is a random vector and $\mathbf{A}$ a nonrandom square matrix, $h(\mathbf{AX}) = h(\mathbf{X}) + \log(|\det(\mathbf{A})|)$.

1.4.2 Kullback–Leibler divergence

The *Kullback–Leibler (KL) divergence* is the relative (differential) entropy, or cross-entropy, between two RVs with pdfs f and g. It is defined as [55, p. 251]

$$D(f \parallel g) \triangleq \int_{\mathcal{X}} f(x) \log \frac{f(x)}{g(x)} \, dx. \tag{1.42}$$

This is finite only if the support set $\mathcal{X}$ of f is contained in the support set of g.

Kullback–Leibler divergence $D(f \parallel g)$ is always nonnegative, and zero if and only if $f = g$ almost everywhere (a.e.). To prove the latter property, we need Jensen inequality, which we state in the form of a lemma [55, pp. 25–27]:

Lemma 1.1. *Let $c(x)$ be a function convex $\cup$ over interval (a, b), i.e., such that*

$$c(\lambda x_1 + (1 - \lambda) x_2) \leqslant \lambda c(x_1) + (1 - \lambda) c(x_2) \tag{1.43}$$

for every $x_1, x_2 \in (a,b)$ and $\lambda \in [0,1]$. Then for any RV X the following inequality holds:

$$c[\mathbb{E}[X]] \leqslant \mathbb{E}[c(X)]. \tag{1.44}$$

If $c(x)$ is concave $\cap$, i.e., (1.43) holds with $\geqslant$, then inequality (1.44) is reversed. Equality in (1.44) holds if and only if c is an affine function on the support set of X [165].

Denoting by $\mathcal{X}$ the support set of X, and using Jensen inequality, we have [55, p. 252–253]

$$
\begin{aligned}
-D(f \parallel g) &= \int_{\mathcal{X}} f(x) \log \frac{g(x)}{f(x)} \, dx \\
&\leqslant \log \int_{\mathcal{X}} f(x) \frac{g(x)}{f(x)} \, dx \\
&= \log \int_{\mathcal{X}} g(x) \, dx \\
&\leqslant \log(1) = 0,
\end{aligned}
\tag{1.45}
$$

with equality if and only if $f = g$ almost everywhere, which yields an equality in the second line of (1.45).

Remark 1.2. Although sometimes referred to as the KL *distance*, $D(f \parallel g)$ is not technically a distance—it is not symmetric, as generally $D(f \parallel g) \neq D(g \parallel f)$, and it does not satisfy the triangle inequality. It could rather be interpreted as a "directed distance" between two pdf models f and g. A discussion of this point can be found in [123, Section IV]. In our context, KL divergence measures the inefficiency of assuming that the pdf is g when the true pdf is f [55, p. 19], or the information lost when g is used to approximate f [41, p. 51].

Example 1.8. Let $X \sim \mathcal{N}(\mu_X, \sigma_X^2)$ and $Y \sim \mathcal{N}(\mu_Y, \sigma_Y^2)$ be two Gaussian RVs with pdfs f_X, f_Y, respectively. Direct calculation yields

$$D(f_X \parallel f_Y) = \log \frac{\sigma_Y}{\sigma_X} + \frac{(\sigma_X^2 - \sigma_Y^2) + (\mu_X - \mu_Y)^2}{2\sigma_Y^2}. \tag{1.46}$$

$\diamond$

Example 1.9. Two numerical examples of the values taken on by KL divergence are shown in Figs. 1.3–1.4. Fig. 1.3 shows, vs. m, the values of the divergence between the normalized versions of Rayleigh density g and Nakagami-m density f as in Fig. 1.1. Fig. 1.4 shows, vs. the shape parameter K, the values of the divergence between the normalized Rice density (1.7), denoted by f_K, and the Rayleigh density, denoted by f_0. $\diamond$

Remark 1.3. KL divergence, besides yielding an intuitive information-theoretical interpretation, is a tool widely used to compare pdfs when computational tractability is sought. Yet, it would seem more natural to use, instead of KL divergence, a quantity

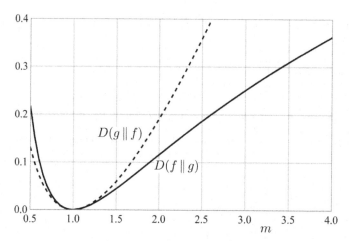

Figure 1.3 KL divergence between RVs with pdfs f (Nakagami-m) and g (Rayleigh) vs. m.

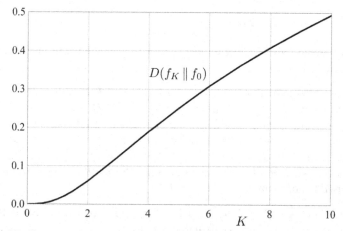

Figure 1.4 KL divergence between RVs with pdfs f_K (normalized Rice) and f_0 (Rayleigh) vs. K.

with the properties of a distance, which would a satisfactory geometric interpretation. Two such quantities are the *total variation distance*, defined as the L_1-distance between two pdfs f and g with common support set $\mathcal{X}$ [194, p. 60]:

$$d_{\text{TV}}(f, g) \triangleq \int_{\mathcal{X}} |f(x) - g(x)| \, dx \tag{1.47}$$

(which satisfies the property $0 \leqslant d_{\text{TV}}(f, g) \leqslant 1$), and *Hellinger distance*, defined as $1/2$ of the L_2-distance between the square roots of the pdfs [194, p. 61], [251, p. 83]:

$$d_{\text{H}}^2(f, g) \triangleq \frac{1}{2} \int_{\mathcal{X}} \left(\sqrt{f(x)} - \sqrt{g(x)} \right)^2 dx = 1 - \int_{\mathcal{X}} \sqrt{f(x)g(x)} \, dx. \tag{1.48}$$

The property $0 \leqslant d_H(f, g) \leqslant 1$ is derived from Cauchy–Schwarz inequality. In particular, $d_H(f, g) = 0$ when $\sqrt{f} = \sqrt{g}$ a.e., and $d_H(f, g) = 1$ when $fg = 0$ a.e.

The *Bhattacharyya distance*, taking values in $\mathbb{R}^+$, is defined as [134]

$$d_B(f, g) \triangleq -\log \rho_B(f, g), \tag{1.49}$$

where $\rho_B(f, g)$ is the *Bhattacharyya coefficient*, lying between 0 and 1,

$$\rho_B(f, g) \triangleq \int_{\mathcal{X}} \sqrt{f(x)g(x)}\, dx. \tag{1.50}$$

Notice that d_B does not obey the triangle inequality [134], and hence should be more appropriately classified as a *semimetric*. Bhattacharyya and Hellinger distances are related by

$$d_H(f, g) = \sqrt{1 - \rho_B(f, g)}. \tag{1.51}$$

The total variation distance is related to KL divergence by *Pinsker inequality* [251, p. 88]

$$d_{TV}(f, g) \leqslant \sqrt{2D(f \parallel g)} \tag{1.52}$$

and to Hellinger distance by the inequalities [194, p. 61], [251, p. 86]

$$2d_H^2(f, g) \leqslant d_{TV}(f, g) \leqslant 2d_H(f, g)\sqrt{2 - d_H^2(f, g)} \leqslant 2\sqrt{2}d_H(f, g). \tag{1.53}$$

1.5 KKT optimality conditions

Consider the standard form of a scalar optimization problem [38, §5.5.3]

$$\underset{x}{\text{minimize}} \ f(x) \tag{1.54a}$$

$$\text{subject to } a_i(x) \leqslant 0, \quad i = 1, \ldots, m, \tag{1.54b}$$

$$b_j(x) = 0, \quad j = 1, \ldots, p, \tag{1.54c}$$

whose optimal solution, i.e., the minimum constrained value of $f(x)$, will be denoted by $p^\star \triangleq f(x^\star)$. This problem may be *convex*, which occurs if functions $f(x)$ and $a_i(x)$, $i = 1, \ldots, m$, are convex, and $b_j(x)$, $j = 1, \ldots, p$, are affine.

The *Lagrangian functional* associated with (1.54) is

$$L(x, \boldsymbol{\lambda}, \boldsymbol{v}) \triangleq f(x) + \sum_{i=1}^{m} \lambda_i a_i(x) + \sum_{j=1}^{p} v_j b_j(x), \tag{1.55}$$

where λ_i and v_j, the components of vectors $\boldsymbol{\lambda}$ and $\boldsymbol{v}$, are the *Lagrange multipliers*.

The *Lagrange dual function* is the minimum value of $L(x, \lambda, \nu)$ over x,

$$\Lambda(\lambda, \nu) \triangleq \inf_x L(x, \lambda, \nu), \qquad (1.56)$$

and is concave. This function yields lower bounds on $p^\star$: in fact, for any λ all of whose components are nonnegative (we denote this writing $\lambda \geqslant 0$) and any ν, we have

$$\Lambda(\lambda, \nu) \leqslant p^\star. \qquad (1.57)$$

The best lower bound obtainable on $p^\star$ in the form (1.57) is found as the solution of the *dual problem* [38, p. 223]

maximize $\Lambda(\lambda, \nu)$ $\qquad\qquad$ (1.58a)

subject to $\lambda \geqslant 0$. $\qquad\qquad$ (1.58b)

The optimal λ, ν are denoted by $\lambda^\star$, $\nu^\star$. Problem (1.58) is convex even when (1.54) is not. Denoting by $d^\star$ the optimal value of the Lagrange dual problem, we have

$$d^\star \leqslant p^\star. \qquad (1.59)$$

The difference $p^\star - d^\star$ is referred to as the *optimal duality gap*. The important special case

$$d^\star = p^\star \qquad (1.60)$$

which allows the solution to the "primal" problem (1.54) to be obtained by solving (1.58), occurs when *Slater condition* is satisfied, i.e., an x exists such that $a_i(x) < 0, i = 1, \ldots, m$, and $b_j(x) = 0, j = 1, \ldots, p$ [38, p. 226].

The Karush–Kuhn–Tucker (KKT) optimality conditions are

$$\frac{d}{dx} \left\{ f(x) + \sum_{i=1}^{m} \lambda_i a_i(x) + \sum_{j=1}^{p} \nu_j b_j(x) \right\} = 0, \qquad (1.61a)$$

$$a_i(x) \leqslant 0, \qquad i = 1, \ldots, m, \qquad (1.61b)$$

$$b_j(x) = 0, \qquad j = 1, \ldots, p, \qquad (1.61c)$$

$$\lambda_i \geqslant 0, \qquad i = 1, \ldots, m, \qquad (1.61d)$$

$$\lambda_i a_i(x) = 0, \qquad i = 1, \ldots, m. \qquad (1.61e)$$

If the duality gap is zero, the KKT conditions are necessary and sufficient for the optimal point of primal and dual problems.

In the special case of no inequality constraints in (1.54), the KKT conditions become *Lagrange conditions*.

1.6   Choosing the best model: the maximum-entropy principle

Assume that, because of epistemic uncertainty, the model pdf we are looking for is only known to belong to a given class of pdfs consistent with our state of knowledge and among which we want to choose the "best" one. The *maximum-entropy principle* states that, if a random-variable model is to be built from incomplete information, the one with the largest differential entropy should be chosen among those consistent with prior information. It is argued (see, e.g., [127,233]) that the maximum-entropy principle is the correct method of inference when incomplete information is available about a certain model.

This principle is justified intuitively by observing that the maximum-entropy pdf "is the least biased estimate possible on the given information," and hence "uniquely determined as the one which is maximally noncommittal with regard to missing information" [127, pp. 620–623]. A rigorous justification of the maximum-entropy principle is exhibited in [233], where it is proved that the only pdf satisfying the given constraints and some consistency axioms is the one satisfying the maximum-entropy principle. The consistency axioms are based on the fundamental principle that if a problem can be solved in more than one way, then the results should be consistent. Thus, "maximizing any function but entropy will lead to inconsistencies unless that function and entropy have identical maxima" [233, p. 27].

To find the maximum-entropy pdf when the constraints are expressed by known values of some moments of a RV X, we solve the optimization problem

$$\underset{f}{\text{maximize}} \ \ h(f) \tag{1.62a}$$

$$\text{subject to} \int f(x)\,dx = 1, \tag{1.62b}$$

$$\int h_i(x)f(x)\,dx = \mu_i, \qquad i \in \mathfrak{I}, \tag{1.62c}$$

where $f(x) \geqslant 0$ with $f(x) = 0$ outside a support set $\mathfrak{X}$, and the moment constraints are defined through the functions $h_i(x)$, with i in the finite index set $\mathfrak{I}$. A general solution is obtained as follows [55, p. 409 ff.]: form the Lagrangian functional

$$L(f) = -\int f(x)\log f(x)\,dx + \lambda_0 \int f(x)\,dx + \sum_{i \in \mathfrak{I}} \lambda_i \int h_i(x)f(x)\,dx \tag{1.63}$$

and equate to zero its functional derivative

$$\frac{dL}{df} = -\log f(x) - 1 + \lambda_0 + \sum_{i \in \mathfrak{I}} \lambda_i h_i(x) = 0. \tag{1.64}$$

The solution of this equation is

$$f^\star(x) = \exp\left\{\lambda_0 - 1 + \sum_{i \in \mathcal{I}} \lambda_i h_i(x)\right\}, \qquad x \in \mathcal{X}, \tag{1.65}$$

with Lagrange multipliers λ_i chosen so that the constraints in (1.62) are satisfied. To prove that (1.65) actually yields a maximum of relative entropy, define $f^\star(x)$ as in (1.65) with the constraints satisfied. We have, with $g(x)$ a pdf satisfying the constraints (1.62b)–(1.62c) and the integral extended to $\mathcal{X}$,

$$\mathsf{h}(g) = -\int g(x) \log[g(x)]\,dx \tag{1.66a}$$

$$= -\int g(x) \log\left[\frac{g(x)}{f^\star(x)} f^\star(x)\right] dx \tag{1.66b}$$

$$= -D(g \parallel f^\star) - \int g(x) \log[f^\star(x)]\,dx \tag{1.66c}$$

$$\leqslant -\int g(x) \log[f^\star(x)]\,dx \quad \text{(KL divergence is nonnegative)} \tag{1.66d}$$

$$= -\int g(x) \left(\lambda_0 - 1 + \sum_{i \in \mathcal{I}} \lambda_i h_i(x)\right) dx \quad \text{(from the definition of } f^\star\text{)}$$
$$\tag{1.66e}$$

$$= -\int f^\star(x) \left(\lambda_0 - 1 + \sum_{i \in \mathcal{I}} \lambda_i h_i(x)\right) dx \quad (f^\star, g \text{ satisfy the constraints)}$$
$$\tag{1.66f}$$

$$= -\int f^\star(x) \log[f^\star(x)]\,dx \tag{1.66g}$$

$$= \mathsf{h}(f^\star). \tag{1.66h}$$

Equality $\mathsf{h}(g) = \mathsf{h}(f^\star)$ holds if and only if (1.66d) holds with equality, i.e., if $D(g \parallel f^\star) = 0$, which implies $g(x) = f^\star(x)$ a.e.

Remark 1.4. As observed in [55, p. 411], the same proof holds, *mutatis mutandis*, for discrete and vector random variables.

Example 1.10. With no constraints other than that of a compact support set $\mathcal{X} = [a, b]$, the pdf maximizing the differential entropy is uniform in $[a, b]$.   ◇

Example 1.11. Let the real random n-vector $\mathbf{X}$ have zero mean and covariance matrix $\mathbf{R} = \mathbb{E}[\mathbf{X}\mathbf{X}^\mathsf{T}]$. Its differential entropy is maximized if $\mathbf{X} \sim \mathcal{N}(\mathbf{0}, \mathbf{R})$ [55, p. 254]. To prove this, let $g(\mathbf{x})$ denote any pdf satisfying the covariance constraint and

$$f(\mathbf{x}) = \frac{1}{\sqrt{(2\pi)^n \det(\mathbf{R})}} \exp\left\{-\frac{1}{2}\mathbf{x}^\mathsf{T}\mathbf{R}^{-1}\mathbf{x}\right\}. \tag{1.67}$$

Then we have

$$0 \leqslant D(g \parallel f) \tag{1.68a}$$

$$= -\mathsf{h}(g) - \int g(\mathbf{x}) \log[f(\mathbf{x})] d\mathbf{x} \tag{1.68b}$$

$$= -\mathsf{h}(g) - \int f(\mathbf{x}) \log[f(\mathbf{x})] d\mathbf{x} \tag{1.68c}$$

$$= -\mathsf{h}(g) + \mathsf{h}(f), \tag{1.68d}$$

which proves that (1.67) maximizes the entropy given a covariance constraint. The equality (1.68c) follows from the fact that $\log[f(\mathbf{x})]$ is a quadratic form, whose moments are the same for f and g. $\diamond$

Example 1.12. Given a stationary discrete-time random process $\mathbf{X} = (X_1, \ldots, X_n)$, whose components may not be independent and equally distributed, its *differential entropy rate* is defined as $(1/n)\mathsf{h}(\mathbf{X})$, with $n \to \infty$, when the limit exists. The empirical discrete distribution of the eigenvalues of the covariance matrix of $\mathbf{X}$ tends to a limit which is the power density spectrum of the process. Now, assume that $\mathbf{X}$ is stationary Gaussian, obtained by sampling a continuous process at a rate that satisfies Shannon sampling theorem [230]. Let its power density spectrum $S_{\mathbf{X}}(v)$ have support set $[-1/2, 1/2]$. A result due to Kolmogorov shows that the differential entropy rate of $\mathbf{X}$ is given by [55, p. 416 ff.]

$$\mathsf{h} = \frac{1}{2} \log(2\pi e) + \frac{1}{2} \int_{-1/2}^{1/2} \log[S_{\mathbf{X}}(v)] dv. \tag{1.69}$$

If we assume that our state of knowledge of process $\mathbf{X}$ is restricted to the values of some power moments of $S_{\mathbf{X}}(v)$,

$$\mu_k \triangleq \int_{-1/2}^{1/2} v^k S_{\mathbf{X}}(v) dv, \qquad k \in \mathcal{K}, \tag{1.70}$$

with $\mathcal{K}$ a finite index set, we may search for the maximum-entropy estimate of $S_{\mathbf{X}}(v)$, the one which maximizes $\int_{-1/2}^{1/2} \log[S_{\mathbf{X}}(v)] dv$ under the constraints (1.70). Here μ_0 turns out to be the total average power of $\mathbf{X}$, μ_1 is an indicator of the symmetry of the power density spectrum, and μ_2 characterizes the root-mean-square spread of the spectrum [242], defined as $D \triangleq \sqrt{\mu_2/\mu_0 - (\mu_1/\mu_0)^2}$. Taking the functional derivative of

$$L(S_{\mathbf{X}}) = \int_{-1/2}^{1/2} \log[S_{\mathbf{X}}(v)] dv + \sum_{k \in \mathcal{K}} \lambda_k \left(\mu_k - \int_{-1/2}^{1/2} v^k S_{\mathbf{X}}(v) dv \right) \tag{1.71}$$

with respect to $S_{\mathbf{X}}$ and equating it to zero, we obtain

$$S_{\mathbf{X}}(v) = \left(\sum_{k \in \mathcal{K}} \lambda_k v^k \right)^{-1}, \tag{1.72}$$

where the Lagrange multipliers λ_k can be found as the solutions of

$$\int_{-1/2}^{1/2} v^k \left(\sum_{k \in \mathcal{K}} \lambda_k v^k \right)^{-1} dv = \mu_k, \qquad k \in \mathcal{K}. \tag{1.73}$$

It is shown in [242] that for $\mu_0 = 1$, $\mu_1 = 0$, and $\mu_2 = 1/8$ the maximum-entropy power spectrum is close to the "Jakes spectrum"

$$S_X(v) = \frac{2}{\pi \sqrt{1 - 4v^2}}, \qquad |v| < 1/2. \tag{1.74}$$

$\diamond$

Example 1.13. Consider multiple-input, multiple-output (MIMO) transmission with n_T transmitting and n_R receiving antennas. The mathematical model of the relation between the transmitted vector $\mathbf{x}$ and the received vector $\mathbf{y}$ is (see, for example, [26, 113])

$$\mathbf{y} = \sqrt{\frac{\mathsf{snr}}{n_T}} \mathbf{H} \mathbf{x} + \mathbf{n}, \tag{1.75}$$

where snr denotes the received signal-to-noise ratio, $\mathbf{H}$ the random channel transfer matrix with components h_{ij}, $i = 1, \ldots, n_R$, $j = 1, \ldots, n_T$, n_T the number of transmit antennas, n_R the number of receive antennas, and $\mathbf{n}$ the additive random noise vector $\sim \mathcal{N}(\mathbf{0}, \mathbf{I})$. We want to obtain the "best" model of $\mathbf{H}$ under the only knowledge of the average energy $\mathcal{E}$ carried by the channel, that is, if $\mathbb{E}[\mathbf{H}] = \mathbf{0}$,

$$\mathcal{E} \triangleq \frac{1}{n_R n_T} \mathbb{E} \left(\sum_{i=1}^{n_R} \sum_{j=1}^{n_T} |h_{ij}|^2 \right). \tag{1.76}$$

More precisely, since the components of $\mathbf{H}$ are RVs, what model pdf $f(\mathbf{H})$ should be chosen for the channel? One may select the pdf complying with constraint (1.76) and maximizing the differential entropy [61]. To do this, form the Lagrangian functional

$$L(f) = - \int f(\mathbf{H}) \log f(\mathbf{H}) d\mathbf{H}$$
$$+ \beta \left[1 - \int f(\mathbf{H}) d\mathbf{H} \right] + \gamma \sum_{i=1}^{n_R} \sum_{j=1}^{n_T} \left[\mathcal{E} - \int |h_{ij}|^2 f(\mathbf{H}) d\mathbf{H} \right] \tag{1.77}$$

and compute the functional derivative of $L(f)$ with respect to f. Equating this to zero, we obtain

$$\frac{dL(f)}{df} = -1 - \log f(\mathbf{H}) - \beta - \gamma \sum_{i=1}^{n_R} \sum_{j=1}^{n_T} |h_{ij}|^2 = 0, \tag{1.78}$$

which yields

$$f(\mathbf{H}) = \exp\left(-1 - \beta - \gamma \sum_{i=1}^{n_R} \sum_{j=1}^{n_T} |h_{ij}|^2\right)$$

$$= e^{-1-\beta} \prod_{i=1}^{n_R} \prod_{j=1}^{n_T} \exp\left(-\gamma |h_{ij}|^2\right) \tag{1.79}$$

$$= \prod_{i=1}^{n_R} \prod_{j=1}^{n_T} \exp\left[-\left(\gamma |h_{ij}|^2 - (1+\beta)/(n_r n_T)\right)\right].$$

We can immediately observe that, with the only constraint involving average energy, the maximum-entropy principle yields a model matrix $\mathbf{H}$ having Gaussian entries which are independent and identically distributed. For a full derivation of $f(\mathbf{H})$, coefficients β and γ need be computed, which requires solving the equations

$$\int \sum_{i=1}^{n_R} \sum_{j=1}^{n_T} |h_{ij}|^2 f(\mathbf{H}) \, d\mathbf{H} = n_T n_R \mathcal{E}, \tag{1.80a}$$

$$\int f(\mathbf{H}) \, d\mathbf{H} = 1. \tag{1.80b}$$

The result is [61, p. 1670]

$$f(\mathbf{H}) = \frac{1}{(\pi \mathcal{E})^{n_T n_R}} \exp\left\{-\frac{1}{\mathcal{E}} \sum_{i=1}^{n_R} \sum_{j=1}^{n_T} |h_{ij}|^2\right\}. \tag{1.81}$$

$\diamond$

1.6.1 Maximizing entropy with order-statistics constraints

In [208], a procedure is described to find the distribution of an RV X maximizing the differential entropy under the constraints of known expected values taken with respect to order statistics. The optimization problem (1.62) has constraint (1.62c) changed into

$$\int x f_{(k)}(x) \, dx = \mu_k, \qquad k \in \{1, \ldots n\}, \tag{1.82}$$

where $f_{(k)}(x)$ denotes the pdf of the kth RV in the ordered set $X_{(1)} \leqslant X_{(2)} \leqslant \cdots \leqslant X_{(n)}$ of n independent RVs having common pdf $f(x)$ and CDF $F(x)$. The corresponding CDF is given by [58, p. 9]

$$F_{(k)}(x) = \sum_{i=k}^{n} \binom{n}{i} [F(x)]^i [1 - F(x)]^{n-i}, \quad k = 1, \ldots, n, \tag{1.83}$$

and is obtained by observing that $F_{(k)}(x)$ is the probability that at least k of the X_i are less than or equal to x. The pdf $f_{(k)}(x)$ is obtained as

$$f_{(k)}(x) = k \binom{n}{k} [F(x)]^{k-1} [1 - F(x)]^{n-k} f(x), \quad k = 1, \ldots, n. \tag{1.84}$$

The values of μ_i can be estimated by observing N values of X, ordering them in an increasing order

$$x_{(1)} \leqslant x_{(2)} \leqslant \cdots \leqslant x_{(N)}$$

and computing the empirical means $\langle \mu_k \rangle$ of n ordered segments of N/n values each. In [208] it is suggested that this procedure can find application to statistical signal processing in a nonstationary environment. Exact solutions for small values of n are provided, and simulation results show the merits of this approach.

1.6.2 Spherically invariant processes

In [33,39], the search for a statistical model of fading in wireless communications is restricted to the realm of *spherically invariant random processes* (SIRPs). Among the reasons motivating this restriction are a good deal of generality, flexibility, and the fact that, under suitable constraints, SIRPs maximize differential entropy, which is a property satisfied by the ubiquitous Gaussian distribution when the constraint is that of a bounded variance. A real random vector $\mathbf{X}$ is spherically invariant if there exists a function $g(\cdot)$ and a positive definite matrix $\mathbf{R}$ such that its pdf $f_{\mathbf{X}}(\mathbf{x})$ can be written in the form

$$f_{\mathbf{X}}(\mathbf{x}) = g(\mathbf{x}^{\mathsf{T}} \mathbf{R} \mathbf{x}). \tag{1.85}$$

An SIRP is completely characterized by its mean value, its covariance function, and its univariate pdf. This first-order pdf can be either prescribed or obtained from experimental data [39]. The connection between SIRP RVs and Gaussian RVs is made explicit by the following property: if the complex process $X_R(t) + jX_I(t)$ is an SIRP, then the pdf of its envelope $R(t) = \sqrt{X_R^2(t) + X_I^2(t)}$ is the same as that of a product of two independent nonnegative RVs the form [33, p. 5528], [275]

$$R = VX, \tag{1.86}$$

where X is the envelope of a complex Gaussian process, so that X has a Rayleigh or Rice pdf. In other terms, the envelope of an SIRP process can be modeled as that of a Rayleigh or Rice process with a random variance. The pdf of R satisfies

$$f_R(r) = \int_0^\infty (1/v) f_X(r/v) f_V(v) \, dv, \qquad 0 < r < \infty, \tag{1.87}$$

which is called *Mellin convolution* of pdfs f_X and f_V. This shows how f_R is a *mixture* (or *compound*) pdf resulting from the two densities f_X and f_V.

Convolution (1.87) can be transformed into a product if Mellin transforms are used. The Mellin transform [42,78] of the univariate pdf $f_X(x)$ with support set $\mathbb{R}^+$ is

$$\tilde{f}_X(s) \triangleq \mathbb{E}[X^{s-1}] = \int_0^\infty x^{s-1} f_X(x) \, dx. \tag{1.88}$$

The integral above is defined for any Lebesgue-integrable function, and converges in a suitable vertical strip of the complex plane. The inverse Mellin transform yields

$$f_X(x) = \frac{1}{2\pi j} \int_{c-j\infty}^{c+j\infty} x^{-s} \tilde{f}_X(s) \, ds, \qquad c \in \mathbb{R}, x \in \mathbb{R}^+. \tag{1.89}$$

Raising both sides of (1.86) to power $s - 1$, and taking expectations after using the independence of V and R, we obtain from (1.88)

$$\tilde{f}_R(s) = \tilde{f}_V(s) \tilde{f}_X(s), \tag{1.90}$$

or, equivalently,

$$\tilde{f}_V(s) = \frac{\tilde{f}_R(s)}{\tilde{f}_X(s)}, \tag{1.91}$$

so that $f_V(x)$ can be computed as the inverse Mellin transform of the ratio $\tilde{f}_R(s)/\tilde{f}_X(s)$.

Example 1.14. (Nakagami-m fading) Consider the Nakagami-m pdf

$$f_R(x) = \frac{2m^m}{\Gamma(m)\Omega^m} x^{2m-1} \exp\left(-\frac{m}{\Omega}x^2\right). \tag{1.92}$$

With X a Rayleigh fading envelope, we obtain from direct calculation

$$f_V(v) = \frac{2m^m}{\Gamma(m)\Gamma(1-m)\Omega^m} v^{2m-1} \left(1 - \frac{m}{\Omega}v^2\right)^{-m}, \qquad v \in \left(0, \sqrt{\Omega/m}\right). \tag{1.93}$$

$\diamond$

Remark 1.5. One should carefully examine the conditions under which (1.86) is valid. In fact, the SIRP process modeling a fading envelope R cannot be ergodic— otherwise it would be Gaussian. Actually, for an ergodic process the channel would "reveal" its probability distribution to a single realization, which is not consistent with the assumption that every realization of the random process is affected by a value taken by the RV V. Suitable assumptions of deep interleaving should be added, and $V(t)$ should be slowly varying [33, p. 5532].

1.7   Choosing the best model in a set: Akaike Information Criterion

In this section we suppose that, after observing a signal received at the output of a channel, one wants to choose a statistical model for the observations. If it occurs that similar channels had previously been modeled using one of common pdfs, it would be tempting to insert a few of these in a model pool to be used with the available data, and look for the model that best captures the features present in the observed data and which are deemed important. To do this, one should assume that one model can be found in the pool which is the most appropriate to fit the available data (notice that this is just an assumption, as it is possible that none of the models under scrutiny can provide a perfect, or even good, fit to the data).

For general pdfs, a number of techniques have been advocated to select the best model in a set of candidates. We describe here the *Akaike Information Criterion* (AIC) [3,41]. This criterion provides an estimate of the Kullback–Leibler (KL) information loss due to the use of an inaccurate model for the pdf which actually generates the observed data, so that the model with the best AIC value approximates the true (underlying) pdf with the minimum loss of information among the candidate models. We observe again that, while the AIC criterion is useful to select the best model among a set, even that relatively best model might be poor in an absolute sense [41, p. 62]. Thus, every effort should be made to ensure that the set of candidate models is well founded.

The idea underlying the AIC is the minimization of the information we lose if the candidate pdf g is used in lieu of the true pdf f. As f is not known, this loss cannot be computed exactly. However, we can compare the amount of information lost if f is used instead of g_1 against the amount lost if f is used instead of g_2. This creates a comparison between models g_1 and g_2. With this procedure, we learn nothing about the absolute quality of a model, just the quality relative to other models in a pool.

Denote by g_θ the candidate pdf, and assume that g depends on a real vector parameter $\theta \in \Theta \subseteq \mathbb{R}^K$. Based on the observation $\mathbf{x} = (x_1, \dots, x_n)$ of the vector RV $\mathbf{X}$, we form the model pool

$$\mathcal{M} \triangleq \{g_\theta(\mathbf{x}) \mid \theta \in \Theta\}. \tag{1.94}$$

If f is the "true" pdf, to compare f against g_θ, we use the KL divergence

$$\begin{aligned} D(f \parallel g_\theta) &= \int f(\mathbf{x}) \log[f(\mathbf{x})]\, d\mathbf{x} - \int f(\mathbf{x}) \log[g_\theta(\mathbf{x})]\, d\mathbf{x} \\ &= \mathbb{E}_f\big[\log[f(\mathbf{X})]\big] - \mathbb{E}_f\big[\log[g_\theta(\mathbf{X})]\big]. \end{aligned} \tag{1.95}$$

The first expectation in (1.95) is a constant depending only on the true pdf, and hence unknown. Writing

$$D(f \parallel g_\theta) = C - \mathbb{E}_f\big[\log[g_\theta(\mathbf{X})]\big], \tag{1.96}$$

we may interpret $D(f \parallel g_\theta) - C$ as a relative distance between f and g_θ, and use this as the cost function for the comparison among models.

In principle, we may assume that the best approximating model g_θ in $\mathcal{M}$ is that with the minimum KL divergence value as $\theta \in \Theta$,

$$g_{\theta_0} \triangleq \arg\min_{\theta \in \Theta} D(f \parallel g_\theta). \tag{1.97}$$

If we were able to compute the value θ_0, we would characterize the perfect model as that achieving $D(f \parallel g_\theta) = 0$, and hence evaluate the relative goodness of a model on the basis of the value of $D(f \parallel g_{\theta_0})$. However, since θ_0 cannot be computed, the model selection criterion should be modified accordingly. The idea here is to settle for the minimization of the *expected* KL value. Focusing on the difference $D(f \parallel g_\theta) - C$, we evaluate an approximation of $\mathbb{E}_f[\log[g_\theta(\mathbf{X})]]$. Observe that the maximum of this expectation is achieved when $\theta = \hat{\theta}$, which is the ML estimate of θ based on the observation of $\mathbf{x}$, and hence an approximation of θ_0. AIC consists of selecting, among competing model pdfs g_θ, the one yielding the minimum value of

$$\text{AIC} \triangleq -2\log[g_{\hat{\theta}}(\mathbf{x})] + 2K, \tag{1.98}$$

where the "bias correction term" $2K$ is chosen as twice the number K of estimable parameters in the approximating pdf. This choice makes AIC an asymptotically unbiased estimate of the expected value of the KL divergence between the fitted model and the unknown pdf generating the observed data.

We see from (1.98) that AIC rewards goodness of fit (as assessed by the likelihood function), but it also includes a penalty that is an increasing function of the number of estimated parameters. When the number K of parameters in θ is increased in the approximating model, the first term on the RHS of (1.98) decreases, while the second increases. This fact reflects the tradeoff between overfitting and underfitting [41, p. 62]. The parameter K penalizes model complexity, and hence prevents favoring models that are too complex. For small sample sizes, the following criterion, called AIC$_c$, was developed as a correction of AIC to avoid overfitting:

$$\text{AIC}_c \triangleq \text{AIC} + \frac{2K^2 + 2K}{n - K - 1}. \tag{1.99}$$

This is essentially AIC with an extra penalty term for the number of parameters. Note that, as $n \to \infty$, this extra term tends to 0, so that AIC$_c$ converges to AIC.

Example 1.15. AIC$_c$ is used in [87] to select one among Rayleigh, lognormal, and Nakagami-m pdfs to model a body-area propagation channel, and in [54] to select among lognormal, Gamma, and Rice pdfs to model an indoor wearable active Radio Frequency Identification channel. In [225], AIC is used to discuss a model for ultrawideband channels based on Rayleigh and Rice pdfs.   ◇

1.8 Choosing the best model in a set: minimum description length criterion

A point of view differing from AIC was taken by Schwarz [226] and Rissanen [209]. In [226], using a Bayesian approach, it is assumed that each model in a class is assigned an a priori probability, and the model is selected which approximately yields the maximum a posteriori probability. In [209], the best model is the one that yields the shortest description of the observed data. The idea of this "Minimum Description Length" (MDL) criterion is to characterize a probability model based on the shortest length of a computer program describing an event. The information-theoretic interpretation of this concept, connected to Shannon source coding theorem [55, §5.4], refers to the minimum length of a string of digits necessary to encode the observed data. As observed in [104, p. xxv], "MDL is based on the following insight: any regularity in the data can be used to *compress* the data, i.e., to describe it using fewer symbols than the number of symbols needed to describe the data literally. The more regularities there are, the more the data can be compressed."

As discussed in [268], in the large-sample limit Schwarz's and Rissanen's approaches yield the same MDL criterion

$$\text{MDL} \triangleq -\log[g_{\hat{\theta}}(\mathbf{x})] + \frac{1}{2} K \log(n). \tag{1.100}$$

Example 1.16. The MDL criterion (1.100) is chosen in [249] to select a fading channel model based on measurements taken in a suburban environment. The class of models includes Rayleigh, Rice, and Nakagami-m pdfs. $\diamond$

Sources and parerga

1. Book [41] is entirely devoted to the use of information-theoretic approaches in the analysis of empirical data.
2. Total-variation and Hellinger distances are sometimes defined in a way slightly different from (1.47) and (1.48), due to the inclusion of a factor 2.
3. Additional properties of distances between pdfs and inequalities connecting these can be found in [251, Section 2.4].
4. The "law of parsimony" is summarized by the dictum attributed to William of Occam (1288–1347) that "Non sunt multiplicanda entia sine necessitate" (Entities are not to be multiplied without necessity), or "What is unnecessary should be shaved away" (the "Occam razor," in Latin "novacula Occami").
5. Since the MDL criterion chooses, among models that fit the data equally well, the one allowing for the shortest description of the data, and hence the simplest, it can be viewed as an implementation of the law of parsimony [104, p. 29].
6. Reference [152] contains an information-theoretic analysis of communication systems in which both transmitter and receiver have uncertain knowledge of the

probability law governing the channel. The capacities of various channel models are derived.

7. It should be observed that typical maximum-entropy pdfs have exponentially small tails, whereas many practical problems involve heavy-tailed pdfs [246, p. 298].

8. If a scalar Gaussian model is used, it should be kept in mind that the light tails of its pdf do not allow enough weight to be assigned to extreme events. With a multivariate Gaussian model, the joint tails of the pdf may cause the joint probability of extreme outcomes to be underestimated. Moreover, the Gaussian assumption induces a strong symmetry in the model. Bruno de Finetti, cited in [74, p. 189], commented on

> *the unjustified and harmful habit of considering the Gaussian distribution in too exclusive a way, as if it represented the rule in almost all the cases arising in probability and in statistics, and as if each non-Gaussian distribution constituted an exceptional or irregular case (even the name of "normal distribution" may contribute to such an impression, and it would therefore perhaps be preferable to abandon it).*

9. The 1965 Nobel Prize in Literature was awarded to Mikhail Aleksandrovič Šolokhov, the author of the epic novel *And Quiet Flows the Don*. In 1974 some literary critics, including Aleksandr Isaevič Solženicyn (1970 Nobel Prize in Literature), claimed that the book was plagiarized from work of Fëdor Dmitrievič Kryukov, an anti-bolshevik author who died in 1920. A statistical analysis was carried on using AIC, and led to the conclusion that Šolokhov was the likely author [49, p. 7 ff.]. Other examples of application of AIC, described in [49], include football matches, exponential decay of beer froth, and rat teratology data.

10. The review paper [111] contains a tutorial presentation of the approaches in [209, 226]. A detailed treatment of the MDL criterion can be found in the book [104].

Performance bounds from epistemic uncertainty

2

While in Chapter 1 we were focused on the choice of a model, here we examine goal-oriented optimization, i.e., the search for probability measures that optimize an assigned performance parameter. The simple, yet fairly general, system performance index we consider here, and denote by η, is the expected value of a known function $h(\cdot) \geqslant 0$, computed with respect to the probability measure $\mathbb{P}$ of a random variable (RV) X,

$$\eta \triangleq \mathbb{E}_{\mathbb{P}}[h(X)]. \tag{2.1}$$

An important special case of function h is the indicator function $\mathbb{1}_{\mathcal{S}}$ of a given set $\mathcal{S}$, which yields $\eta = \mathbb{P}[X \in \mathcal{S}]$.

When $\mathbb{P}$ is known, η can be computed exactly. Here we are interested in a situation where epistemic uncertainty makes $\mathbb{P}$ only partially known, in which case we may settle for the computation of an approximation, or, better, of upper and lower bounds on η, consistent with the partial knowledge we have about $\mathbb{P}$. Whenever available, we are especially interested in *sharp* bounds, i.e., bounds that cannot be further tightened unless additional knowledge is gathered. A special case, which will be dealt with in Chapter 3, is the derivation of sharp upper and lower bounds on η when X is only known through some of its moments.

2.1 Model robustness

We start considering model robustness, i.e., the problem of determining to what extent a divergence of the (unknown) true probability density function (pdf) from a selected pdf may degrade the performance of a system. Specifically, we examine how performance varies as the channel model runs through an *uncertainty set* which surrounds the nominal model and whose size is measured using a suitably defined distance from the nominal probability distribution. This analysis applies, for example, to a situation where a signal is transmitted over a random channel whose model is derived from measurements, and hence is known with limited accuracy.

As we did in Chapter 1, to measure the difference between two pdfs f_0 and f, Kullback–Leibler (KL) divergence is used. The solution of an optimization problem (a) yields the pdf which has a given KL divergence from the nominal one, say f_0, and exhibits the worst value of η, and (b) provides the range of values taken by η in (2.1) when a pdf f is used in lieu of f_0. Fig. 2.1 provides a qualitative illustration of this concept.

Dimensions of Uncertainty in Communication Engineering. https://doi.org/10.1016/B978-0-32-399275-6.00010-1

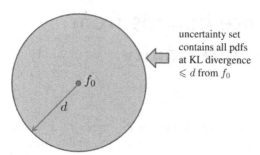

Figure 2.1 Illustration of the search for the worst pdf within an uncertainty set of pdfs f having a KL divergence $\leqslant d$ from the nominal distribution f_0.

The general problem to be solved is the optimization

$$\underset{f}{\text{maximize}} \quad \mathbb{E}_f[h]$$
$$\text{subject to} \quad D(f \parallel f_0) \leqslant d, \tag{2.2}$$

i.e., the search for the pdf whose KL divergence is within d of the known nominal pdf f_0, and which yields the maximum (i.e., worst) value of the cost function $\eta = \mathbb{E}_f[h]$. Explicitly, we should solve

$$\underset{f}{\text{maximize}} \int h(x) f(x)\, dx \tag{2.3a}$$

$$\text{subject to} \int f(x) \log \frac{f(x)}{f_0(x)}\, dx \leqslant d, \tag{2.3b}$$

$$\int f(x)\, dx = 1. \tag{2.3c}$$

(Condition $f(x) \geqslant 0$ should be added, unless automatically satisfied by the solution of (2.3).)

The solution to (2.3) is derived in [123, Theorem 1]. It is obtained by observing that the objective of (2.3) is linear and the constraint set is convex, so that (2.3) is a convex optimization problem. Since Slater condition (see Section 1.5 *supra* or [38, p. 244]) is satisfied, strong duality holds, and hence the Karush–Kuhn–Tucker (KKT) conditions are sufficient for optimality. The Lagrangian L of (2.3) is

$$L = \int h(x) f(x)\, dx - s\left(\int f(x) \log \frac{f(x)}{f_0(x)}\, dx - d\right) - \mu\left(\int f(x)\, dx - 1\right). \tag{2.4}$$

Taking the functional derivative of L with respect to f, the KKT conditions (1.61) are

$$h(x) - s\left(\log \frac{f(x)}{f_0(x)} + 1\right) - \mu = 0, \tag{2.5a}$$

$$\int f(x)\,dx = 1, \tag{2.5b}$$

$$s \geqslant 0, \tag{2.5c}$$

$$s\left(\int \log \frac{f(x)}{f_0(x)} f(x)\,dx - d\right) = 0. \tag{2.5d}$$

For $s > 0$, the maximum is achieved at the boundary (see (2.5d)). In this case, from (2.5a) we have

$$f(x) = f_0(x) e^{h(x)/s} e^{-\mu/s - 1} \tag{2.6}$$

and, using the normalization condition (2.5b), we obtain the optimizing $f(x)$ in the form

$$f^\star(x) = f_0(x) \frac{e^{h(x)/s^\star}}{\xi(s^\star)}, \tag{2.7}$$

where

$$\xi(s) \triangleq \int e^{h(x)/s} f_0(x)\,dx. \tag{2.8}$$

Inserting (2.6) into (2.5d), we obtain the equation yielding $s^\star$,

$$\frac{\xi^\bullet(s)}{\xi(s)} - s \log \xi(s) = ds, \tag{2.9}$$

where

$$\xi^\bullet(s) \triangleq \frac{d\xi(s)}{d(1/s)} = \int h(x) e^{h(x)/s} f_0(x)\,dx. \tag{2.10}$$

The resulting maximum value of $\mathbb{E}_f[h]$, denoted $\eta_{\max}$, is

$$\eta_{\max} \triangleq \int h(x) f^\star(x)\,dx = \frac{\xi^\bullet(s^\star)}{\xi(s^\star)}. \tag{2.11}$$

Remark 2.1. We might also proceed using duality. Since the duality gap of (2.3) is zero [123], the solution to (2.3) is also obtained by minimizing with respect to $s \geqslant 0$ the convex dual function

$$L(s) \triangleq \max_f L(f, s) \tag{2.12}$$

with the constraint (2.3c), and with

$$L(f, s) \triangleq \int h(x) f(x)\,dx + s\left(\int f(x) \log \frac{f(x)}{f_0(x)}\,dx - d\right). \tag{2.13}$$

This yields

$$\eta_{max} = \min_{s \geq 0} \left[s \log \xi(s) + sd \right], \tag{2.14}$$

where the function of s in square brackets is convex. Equating to zero the derivative of this function taken with respect to $1/s$, we obtain again (2.9).

Remark 2.2. Define $\eta_0 \triangleq \int h(x) f_0(x) \, dx$, that is, the expected value of h taken with respect to the nominal pdf f_0. Observe that, since as $s \to \infty$ we have $\xi(s) \to 1$ and $\xi^{\bullet}(s) \to \eta_0$, (2.9) has $s^* \to \infty$ for $d \to 0$, so that $d = 0$ yields

$$\eta_{max} = \eta_0$$

as it should, because $d = 0$ implies $D(f \| f_0) = 0$ and hence $f = f_0$.

Remark 2.3. The procedure outlined shows how the problem of determining the worst pdf can be reduced to the scalar problem (2.14). Its solution can be obtained by using standard numerical algorithms.

Example 2.1 (Error probability). Here we examine the error probability p of uncoded binary antipodal modulation with equally likely signals having a common energy $\mathcal{E}$. The channel is affected by ergodic fading with envelope amplitude R and additive white Gaussian noise with power spectral density $N_0/2$. If the receiver has perfect channel-state information, i.e., knows the value taken on by R, we have [23, Chapter 4]

$$\mathsf{p} = \mathbb{P}\left[\sqrt{R^2 \mathcal{E}} + n < 0 \right] \tag{2.15}$$

$$= \mathbb{E}_R Q\left(R\sqrt{2\,\mathsf{snr}} \right), \tag{2.16}$$

where $Q(\cdot)$ denotes the Gaussian tail function

$$Q(x) \triangleq \frac{1}{\sqrt{2\pi}} \int_x^{\infty} e^{-z^2/2} \, dz = \frac{1}{2} \mathrm{erfc}\left(\frac{x}{\sqrt{2}} \right) \tag{2.17}$$

and $\mathsf{snr} \triangleq \mathcal{E}/N_0$ is the signal-to-noise ratio. The expectation is carried with respect to the RV R, whose pdf is only partially known. With

$$h(x) = Q\left(x\sqrt{2\,\mathsf{snr}} \right) \tag{2.18}$$

and f_0 the Rayleigh pdf, we obtain

$$\mathsf{p}_0 = \frac{1}{2}\left[1 - \sqrt{\frac{\mathsf{snr}}{1 + \mathsf{snr}}} \right]. \tag{2.19}$$

The "worst" pdf f^* at $\mathsf{snr} = 10$ dB is shown in Fig. 2.2 for $d = 0.1$ and $d = 1$. The

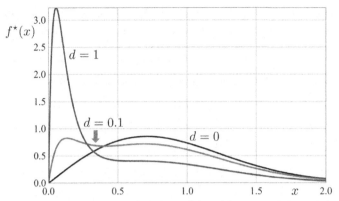

Figure 2.2 "Worst" probability density functions $f^\star(x)$ with snr $= 10$ dB and f_0 the Rayleigh pdf for various values of d.

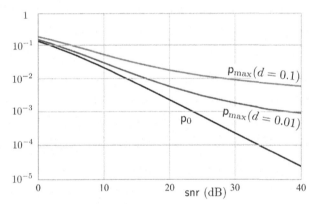

Figure 2.3 p_0 and p_{max} vs. snr for binary antipodal error probability with $d = 0.1$ and $d = 0.01$.

resulting variation of p_{max}, with $d = D(f_K \| f_0)$ and f_K the Rice density with parameter K, is illustrated in Fig. 2.3. This depicts the behavior of p_0 and p_{max} vs. snr for $d = 0.1$ and $d = 0.01$. It is seen that at large values of snr the curve slope (known as "diversity") decreases, showing how the performance loss is mainly due to model uncertainty, while for small snr it is approximately independent of d, reflecting the fact that the model uncertainty is "masked" by the additive noise effect. Stated in different terms, at small snr the performance is dictated by aleatory uncertainty, while at large snr it is affected principally by epistemic uncertainty.   ◇

Example 2.2 (Ergodic channel capacity). Error probability, as computed in Example 2.1, is the appropriate parameter for the evaluation of the quality of a transmission channel when no error-control code is used in it. For a coded ergodic channel, i.e., one "displaying all its features" along a codeword, (Shannon) ergodic capacity can be used instead, as it describes the quality of a channel in which a powerful (i.e., capacity-approaching) code is used.

In a channel with ergodic fading whose envelope R has pdf f_R, under the assumption that the receiver has perfect knowledge of the value taken by the fading amplitude R, the capacity is given by [23, p. 92 ff.], [31]

$$C(f) = \int_0^\infty \log(1 + r^2 \mathsf{snr}) f_R(r) \, dr \qquad \text{nat/dimension pair.} \tag{2.20}$$

For example, if f_R is the Rayleigh pdf (1.4), we obtain

$$C_0 = -\exp(1/\mathsf{snr}) \operatorname{Ei}(-1/\mathsf{snr}), \tag{2.21}$$

where $\operatorname{Ei}(\cdot)$ denotes the exponential integral function

$$\operatorname{Ei}(x) \triangleq \int_{-\infty}^x \frac{e^t}{t} \, dt = -\int_{-x}^\infty \frac{e^{-t}}{t} \, dt. \tag{2.22}$$

To find the pdf at KL divergence within d, we define

$$h(x) = \log(1 + x^2 \mathsf{snr}) \tag{2.23}$$

so that

$$\xi(s) = \int_0^\infty (1 + x^2 \mathsf{snr})^{1/s} f_0(x) \, dx. \tag{2.24}$$

The pdf that minimizes $C(f)$ in (2.20) is given by (2.7). We obtain

$$C_{\min} = \frac{\xi^\bullet(s^\star)}{\xi(s^\star)}. \tag{2.25}$$

Using the dual approach, we may also obtain $s^\star$ as the value of s that maximizes $s \log \xi(s) + sd$ for $s \leqslant 0$, namely

$$C_{\min} = \max_{s \leqslant 0} \left[s \log \xi(s) + sd \right]. \tag{2.26}$$

(Notice the difference between (2.26) and (2.14), due to the fact that, unlike in the former example, we are looking for the pdf that *minimizes* the objective function.)

Fig. 2.4 shows the behavior of $C_{\min}$ vs. snr for f_0 the Rayleigh density and $d = 0.1$; C_0 is the capacity (2.21), while

$$C_{\mathrm{AWGN}} = \log(1 + \mathsf{snr}) \tag{2.27}$$

is the capacity of the (unfaded) AWGN channel. $\qquad\qquad\qquad\qquad\qquad \diamond$

Example 2.3 (Ergodic channel capacity). A number of wireless channels cannot be modeled as ergodic (see, e.g., [23, Chapter 4], [31]). An important family of non-ergodic channels consists of block-fading, or "quasi static," channels, whose fading attenuations remain fixed during the transmission of a codeword, and change, randomly and independently, from codeword to codeword [23, Section 4.3]. In this case

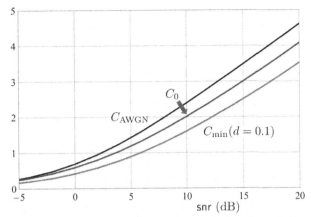

Figure 2.4 Ergodic capacities (in nat/dimension pair) of a fading channel: C_{AWGN}, C_0, and C_{min} vs. snr for f_0 the Rayleigh density and $d = 0.1$.

it has been argued that the *information-outage probability* p_{out} should be used to evaluate the quality of a coded channel. This is the probability that the transmission rate ρ, measured in nats per dimension pair, exceeds the instantaneous mutual information of the channel. Any error-control code approaching p_{out} should have a word-error probability that, as the code length increases, becomes independent of the code length, and hence p_{out} provides a fundamental lower bound on coded word-error rates for sufficiently long code words. The combined calculation of error probability and information-outage probability allows channel quality to be evaluated under the separate assumptions that no code or a powerful code is used.

When the "channel state information" available at the transmitter is the pdf of the fading envelope R, and the input data distribution is assigned, the information-outage probability is defined by

$$\mathsf{p}_{\text{out}} \triangleq \mathbb{P}\left[\log(1 + R^2\mathsf{snr}) < \rho\right]. \tag{2.28}$$

We may rewrite p_{out} in the form

$$\mathsf{p}_{\text{out}} = \mathbb{P}[R \in \mathcal{I}], \tag{2.29}$$

where $\mathcal{I}$ denotes the interval $\left(0, \sqrt{(e^\rho - 1)/\mathsf{snr}}\right)$. In our context, $h(x)$ is the indicator function of $\mathcal{I}$,

$$h(x) = \mathbb{1}_{\mathcal{I}}(x) \triangleq \begin{cases} 1, & x \in \mathcal{I}, \\ 0, & \text{otherwise} \end{cases} \tag{2.30}$$

so that

$$\mathsf{p}_{\text{out}} = \mathbb{E}_R[h(X)]. \tag{2.31}$$

We have explicitly, from definition (2.8),

$$\xi(s) = e^{1/s} \int_{\mathcal{I}} f_0(x)\, dx + \int_{\mathcal{I}^c} f_0(x)\, dx \tag{2.32}$$

$$= 1 + \mathsf{p}_{\text{out},0}(e^{1/s} - 1), \tag{2.33}$$

where $\mathcal{I}^c$ denotes the complement of $\mathcal{I}$, and

$$\mathsf{p}_{\text{out},0} \triangleq \int_{\mathcal{I}} f_0(x)\, dx \tag{2.34}$$

is the nominal outage probability. Moreover, we have

$$\xi^\star(s) = \mathsf{p}_{\text{out},0} e^{1/s} \tag{2.35}$$

so that, from (2.11),

$$\mathsf{p}_{\text{out,max}} = \frac{\mathsf{p}_{\text{out},0} e^{1/s^\star}}{1 + \mathsf{p}_{\text{out},0}(e^{1/s^\star} - 1)}, \tag{2.36}$$

where $s^\star$ is the solution of

$$-\log\left(1 + \mathsf{p}_{\text{out},0}(e^{1/s} - 1)\right) + \frac{\mathsf{p}_{\text{out},0} e^{1/s}}{s(1 + \mathsf{p}_0(e^{1/s} - 1))} = d. \tag{2.37}$$

The dual formulation yields [123]

$$\mathsf{p}_{\text{out,max}} = \min_{s \geq 0}\left[s \log(1 + \mathsf{p}_{\text{out},0}(e^{1/s} - 1)) + sd \right]. \tag{2.38}$$

Simple calculations based on (2.36) show that $\mathsf{p}_{\text{out,max}} = 1$ if and only if $\mathsf{p}_{\text{out},0} = 1$ or $s^\star = 0$, and $\mathsf{p}_{\text{out,max}} = 0$ if and only if $\mathsf{p}_{\text{out},0} = 0$. Moreover, $\mathsf{p}_{\text{out,max}} \geqslant \mathsf{p}_{\text{out},0}$, with equality if and only if $\mathsf{p}_{\text{out},0} = 0$ or $\mathsf{p}_{\text{out},0} = 1$ or $d = 0$. Since the first derivative of the LHS of (2.37) is negative, it reaches its maximum value $\log(1/\mathsf{p}_{\text{out},0})$ at $s = 0$. There we have $\mathsf{p}_{\text{out,max}} = \mathsf{p}_{\text{out},0} = 1$. When $d > \log(1/\mathsf{p}_{\text{out},0})$, we have $\mathsf{p}_{\text{out,max}} = 1$. Fig. 2.5 shows the behavior of $\mathsf{p}_{\text{out,max}}$ vs. the value d of the KL divergence (observe how the nominal pdf f_0 affects $\mathsf{p}_{\text{out,max}}$ only via the nominal outage probability $\mathsf{p}_{\text{out},0}$).

Fig. 2.6 shows $\mathsf{p}_{\text{out,max}}$ and $\mathsf{p}_{\text{out},0}$ for two values of snr and f_0 the Rayleigh pdf.

◇

2.2 Performance optimization with divergence constraints

In this section we examine the following problem: Given a nominal pdf $f_0(x)$ of the RV X yielding the performance-index value η_0, evaluate the range of variation of η

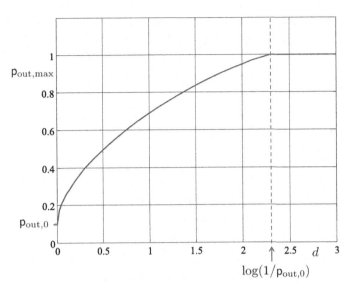

Figure 2.5 $\mathsf{p}_{out,max}$ vs. d. Here $\mathsf{p}_{out,0} = 0.1$.

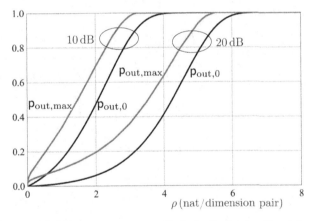

Figure 2.6 Outage probability $\mathsf{p}_{out,max}$ vs. ρ with f_0 the Rayleigh pdf and $d = 0.1$.

when $f_0(x)$ is replaced by a pdf $f(x)$, ranging in a set $\mathcal{D}$ of alternative pdfs dictated by epistemic uncertainty and whose KL divergence $D(f \| f_0)$ from $f_0(x)$ is exactly d. To do this, we compute the bounds

$$\underline{\eta} \triangleq \inf_{g \in \mathcal{D}} \mathbb{E}_g[h], \tag{2.39a}$$

$$\overline{\eta} \triangleq \sup_{g \in \mathcal{D}} \mathbb{E}_g[h], \tag{2.39b}$$

with $\mathcal{D} \triangleq \{g \mid D(g \| f_0) = d\}$.

The calculations involved in (2.39) can be reduced to the optimization of a single real parameter. The proof is based on two variational formulations of KL divergence, which can be used to derive upper and lower bounds on system performance.

Theorem 2.1. *Let f and g be two pdfs with support set $\mathcal{X}$, and $k(x)$ a bounded real function. The following equalities hold* [47], [73, *Proposition 1.4.2*], [157]:

$$-\log \int_{\mathcal{X}} e^{-k(x)} g(x)\, dx = \inf_{f}\left[D(f \parallel g) + \int_{\mathcal{X}} k(x) f(x)\, dx \right], \tag{2.40a}$$

$$\log \int_{\mathcal{X}} e^{k(x)} g(x)\, dx = \sup_{f}\left[-D(f \parallel g) + \int_{\mathcal{X}} k(x) f(x)\, dx \right], \tag{2.40b}$$

with inf *and* sup *achieved if and only if $f(x) = f^{\star}(x)$, with*

$$f^{\star}(x) \triangleq g(x)\frac{e^{-k(x)}}{\xi}, \qquad \xi \triangleq \int_{\mathcal{X}} e^{-k(x)} g(x)\, dx. \tag{2.41}$$

Proof. To prove (2.40a), consider the following chain of equalities:

$$
\begin{aligned}
D(f \parallel g) + \int_{\mathcal{X}} k(x) f(x)\, dx &= \int_{\mathcal{X}} \log\left(\frac{f(x)}{g(x)}\right) f(x)\, dx + \int_{\mathcal{X}} k(x) f(x)\, dx \\
&= \int_{\mathcal{X}} \log\left(\frac{f(x)}{f^{\star}(x)}\right) f(x)\, dx \\
&\quad + \int_{\mathcal{X}} \log\left(\frac{f^{\star}(x)}{g(x)}\right) f(x)\, dx + \int_{\mathcal{X}} k(x) f(x)\, dx \\
&= D(f \parallel f^{\star}) - \int_{\mathcal{X}} \left(\log \int_{\mathcal{X}} e^{-k(x)} g(x)\, dx\right) f(x)\, dx \\
&= D(f \parallel f^{\star}) - \log \int_{\mathcal{X}} e^{-k(x)} g(x)\, dx.
\end{aligned}
\tag{2.42}
$$

Since $D(f \parallel f^{\star}) \geq 0$, with equality if and only if $f(x) = f^{\star}(x)$, we obtain (2.40a). To prove (2.40b), we proceed in a similar way:

$$
\begin{aligned}
&-D(f \parallel g) + \int_{\mathcal{X}} k(x) f(x)\, dx \\
&= -\int_{\mathcal{X}} \log\left(\frac{f(x)}{g(x)}\right) f(x)\, dx + \int_{\mathcal{X}} k(x) f(x)\, dx \\
&= -\int_{\mathcal{X}} \log\left(\frac{f(x)}{f^{\star}(x)}\right) f(x)\, dx \\
&\quad - \int_{\mathcal{X}} \log\left(\frac{f^{\star}(x)}{g(x)}\right) f(x)\, dx + \int_{\mathcal{X}} k(x) f(x)\, dx \tag{2.43} \\
&= -D(f \parallel f^{\star}) + \int_{\mathcal{X}} \left(\log \int_{\mathcal{X}} e^{-k(x)} g(x)\, dx\right) f(x)\, dx
\end{aligned}
$$

$$= -D(f \parallel f^\star) + \log \int_{\mathcal{X}} e^{k(x)} g(x) \, dx.$$

Since $D(f \parallel f^\star) \geqslant 0$, with equality if and only if $f(x) = f^\star(x)$, we obtain the inequality

$$\log \int_{\mathcal{X}} e^{k(x)} g(x) \, dx \geqslant -D(f \parallel f^\star) + \int_{\mathcal{X}} k(x) f(x) \, dx, \qquad (2.44)$$

and consequently (2.40b).   □

Eqs. (2.40) lead to the following bounds:

$$\int_{\mathcal{X}} k(x) f(x) \, dx \geqslant -\log \int e^{-k(x)} g(x) \, dx - D(f \parallel g), \qquad (2.45a)$$

$$\int_{\mathcal{X}} k(x) f(x) \, dx \leqslant \log \int e^{k(x)} g(x) \, dx + D(f \parallel g). \qquad (2.45b)$$

Choose now $k(x) = c\,h(x)$, with $c \in (0, \infty)$. Since the bounds (2.45) are valid for all c, we may summarize them in the form [157]

$$\sup_{c \in (0, \infty)} \left[\Theta^-(c) - \frac{1}{c} D^\star(f \parallel g) \right] \leqslant \mathbb{E}_f[h(X)] \leqslant \inf_{c \in (0, \infty)} \left[\Theta^+(c) + \frac{1}{c} D^\star(f \parallel g) \right], \qquad (2.46)$$

where

$$D^\star(f \parallel g) \triangleq \sup_f D(f \parallel g),$$

$$\Theta^-(c) \triangleq -\frac{1}{c} \log \int_{\mathcal{X}} e^{-c\,h(x)} g(x) \, dx, \qquad (2.47)$$

$$\Theta^+(c) \triangleq \frac{1}{c} \log \int_{\mathcal{X}} e^{c\,h(x)} g(x) \, dx,$$

and it is assumed that $\Theta^-(c)$, $\Theta^+(c)$ are finite. In (2.46) the lower bound may be negative and the upper bound greater than one, so their numerical values should be constrained within the interval $[0, 1]$.

Bounds (2.46) provide the sought performance range, with calculations reduced to the optimization of a single real parameter.

Remark 2.4. Direct calculation yields

$$\lim_{c \to 0^+} \Theta^-(c) = \lim_{c \to 0^+} \Theta^+(c) = \eta_0. \qquad (2.48)$$

Example 2.4 (Bounds on failure probability). As an application of bounds (2.46), consider the evaluation of the "failure probability"

$$\mathsf{p}_F = \mathbb{P}(X \leqslant T) = \int_{-\infty}^{T} f_X(x) \, dx \qquad (2.49)$$

when full knowledge of the pdf of X is unavailable. This performance index can be dealt with by defining $h(x)$ as the indicator function of interval $(-\infty, T]$. We first examine a situation in which a closed-form expression of p_F is available, so that the quality of bounds (2.46) can be exactly assessed. With $X \sim \mathcal{N}(\mu, \sigma^2)$, we have

$$p_F = 1 - Q\left(\frac{T - \mu}{\sigma}\right), \tag{2.50}$$

where the Gaussian tail function $Q(\cdot)$ was defined in (2.17). Choose $f_X(x)$ within a family $\mathcal{F}$ of pdfs whose parameters are only known to lie within a known interval: we write $X \sim \mathcal{N}([\underline{\mu}, \overline{\mu}], [\underline{\sigma}^2, \overline{\sigma}^2])$. With $T = -1$, $\underline{\mu} = -0.2$, $\overline{\mu} = 0.2$, $\underline{\sigma} = 0.9$, and $\overline{\sigma} = 1.1$, the minimum and maximum of (2.50) are

$$\min_{\mathcal{F}}\left(1 - Q\left(\frac{T - \mu}{\sigma}\right)\right) = 0.0912, \tag{2.51a}$$

$$\max_{\mathcal{F}}\left(1 - Q\left(\frac{T - \mu}{\sigma}\right)\right) = 0.2335, \tag{2.51b}$$

corresponding to $\mu = 0.2$, $\sigma = 0.9$ and $\mu = -0.2$, $\sigma = 1.1$, respectively. To apply bounds (2.46), we choose $f_0 \in \mathcal{F}$ as a Gaussian pdf with $\mu_0 = 0$ and $\sigma_0 = 1$. The KL divergence between f and f_0 is, from (1.46),

$$D(f \parallel f_0) = \log\frac{\sigma_0}{\sigma} + \frac{(\sigma^2 - \sigma_0^2) + (\mu - \mu_0)^2}{2\sigma_0^2}, \tag{2.52}$$

and we obtain

$$\Theta^-(c) = -\frac{1}{c}\log\left[e^{-c}\left(1 - Q\left(\frac{T - \mu}{\sigma}\right)\right) + Q\left(\frac{T - \mu}{\sigma}\right)\right],$$

$$\Theta^+(c) = \frac{1}{c}\log\left[e^c\left(1 - Q\left(\frac{T - \mu}{\sigma}\right)\right) + Q\left(\frac{T - \mu}{\sigma}\right)\right]. \tag{2.53}$$

We have $D^* = 0.03036$. With $T = -1$, we obtain the behavior of $\Theta^+(c) + (1/c)D^*$ and $\Theta^-(c) - (1/c)D^*$ shown in Fig. 2.7. This yields lower and upper bounds on p_F equal to 0.076 and 0.255, respectively, while the nominal probability is $p_{F0} = 0.159$.

$$\diamond$$

2.3 Scenarios of uncertainty

Here we examine the general problem of bounding the expectation $\mathbb{E}_{\mathbb{P}}[h(X)]$ under the assumption that, in addition to the pdf of X, also the function h is affected by epistemic uncertainties. To model the uncertainty affecting the problem, we define a set $\mathcal{A}$ of scenarios including the admissible pairs $(h, \mathbb{P})$,

$$\mathcal{A} \triangleq (\mathcal{H}, \mathcal{P}). \tag{2.54}$$

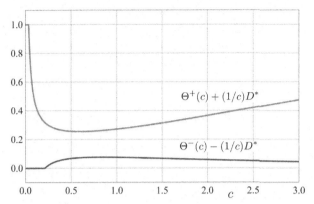

Figure 2.7 $\Theta^-(c) - (1/c)D^*$ and $\Theta^+(c) + (1/c)D^*$ (constrained within the interval $[0, 1]$) vs. c in Example 2.4.

The available information we have about the system is that any element of sets $\mathcal{H}$ and $\mathcal{P}$ is allowed to be the true h and $\mathbb{P}$, respectively, so that $\mathcal{A}$ represents an intermediate level of knowledge between known $\mathbb{P}$ and h and a complete lack of information about them.

With this formulation, our ultimate goal is the search for the two bounds,

$$\underline{\eta} \triangleq \inf_{(H,\mathbb{Q})\in\mathcal{A}} \mathbb{E}_{\mathbb{Q}}[H(X)], \tag{2.55a}$$

$$\overline{\eta} \triangleq \sup_{(H,\mathbb{Q})\in\mathcal{A}} \mathbb{E}_{\mathbb{Q}}[H(X)], \tag{2.55b}$$

where X has a known support set $\mathcal{X}$. These bounds are as tight as allowed by the constraint $(H, \mathbb{Q}) \in \mathcal{A}$ which reflects the information available.

Although the optimization problems involved in (2.55) are generally nonconvex and infinite-dimensional, we shall see later that in several important instances their solution can be simplified, and even given a closed form. Otherwise, one might settle for the search of bounds looser than $\underline{\eta}$ and $\overline{\eta}$, obtained, for example, by relaxing the constraints reflected by $\mathcal{A}$.

The above can be generalized to vector RVs $\mathbf{X}$, and hence to functions $h(\mathbf{x})$.

2.3.1 Defining the constraints

Several classes of possible constraints on h and $\mathbb{P}$ may be defined, leading to different scenarios and to corresponding families of bounds.

2.3.1.1 Constraints on $\mathbb{P}$

In scalar problems, one may assume that any probability measure with support set $\mathcal{X}$ is admissible. For n-dimensional problems, a common assumption is that the n components of the vector RV $\mathbf{X}$ are statistically independent, so that $\mathbb{P} = \bigotimes_{i=1}^{n} \mathbb{P}_i$ and

$\mathcal{X} = \bigotimes_{i=1}^{n} \mathcal{X}_i$, where $\mathbb{P}_i$ and $\mathcal{X}_i$ denote the probability measure and the support set of the ith component of $\mathbf{X}$, respectively. Specific constraints may include a number of known moments, for example, known expected values of X or of $h(X)$, or their bounds.

2.3.1.2 Constraints on h

In general, it will be implicitly assumed that h is measurable. Moreover, one may define a parametric or nonparametric class $\mathcal{H}$ of functions defined on $\mathcal{X}$. For one-dimensional random variables, one may choose bounded functions, i.e., those satisfying

$$\sup_{x,x' \in \mathcal{X}} |h(x) - h(x')| \leqslant D < \infty. \tag{2.56}$$

A useful generalization of (2.56) to functions with n variables can be obtained as follows: define the componentwise *oscillations* of h as the maximum variation of h in its ith argument:

$$\mathrm{osc}_i(h) \triangleq$$

$$\sup_{(x_1,\dots,x_n) \in \mathcal{X}} \sup_{x_i' \in \mathcal{X}_i} \left| h(x_1, \dots, x_{i-1}, x_i, x_{i+1}, \dots, x_n) - h(x_1, \dots, x_{i-1}, x_i', x_{i+1}, \dots, x_n) \right|$$

for $i = 1, \dots, n$. Oscillations $\mathrm{osc}_i(h)$ describe how a change in the value of x_i affects the value taken by function h, and hence quantify parameter sensitivities. The *diameter* of h is defined as

$$D_h \triangleq \sqrt{\sum_{i=1}^{n} \mathrm{osc}_i^2(h)}, \tag{2.57}$$

which, for $n = 1$, reduces to definition (2.56).

A function h is said to have the *bounded-difference property* if for every i, $1 \leqslant i \leqslant n$, the oscillation of h is bounded above by $D_i < \infty$. The condition

$$\mathrm{osc}_i(h) \leqslant D_i < \infty \tag{2.58}$$

is also called *McDiarmid constraint*.

With vector random variables, one might restrict attention to functions that are sums of single-variable functions,

$$h(x_i, \dots, x_n) = \sum_{i=1}^{n} h(x_i). \tag{2.59}$$

2.3.1.3 Presence of unknown unknowns

While the description above assumes a system whose behavior depends on the incompletely known RV X, one may add a further dimension of uncertainty by assuming that the system behavior is described by an RV having the form $k(X, Z)$, where Z accounts for *unknown unknowns*.

A further generalization arises in a scenario where it is not even known how many unknown unknowns there are. This corresponds to problems which may be treated using *random-set theory*, whose description is out of the scope of this book but can be found, for example, in [28,101,163,169,186].

2.4 Concentration-of-measure inequalities

Two classic "concentration-of-measure" inequalities, stated here in the form of theorems, will be used in the balance of this chapter.

Theorem 2.2 ((HOEFFDING INEQUALITIES) [37, pp. 34–35], [116, Theorem 1]). *Given n independent bounded random variables $X_1, \ldots, X_n$ with support sets $\mathcal{X}_i = [a_i, b_i]$ and common mean $\mu \triangleq \mathbb{E}[X_i]$, for any $T > 0$ the following holds:*

$$\mathbb{P}[\langle \mu \rangle - \mu \geqslant T] \leqslant \exp\left(-\frac{2n^2 T^2}{\sum_{i=1}^{n}(b_i - a_i)^2}\right), \tag{2.60a}$$

$$\mathbb{P}[\mu - \langle \mu \rangle \geqslant T] \leqslant \exp\left(-\frac{2n^2 T^2}{\sum_{i=1}^{n}(b_i - a_i)^2}\right), \tag{2.60b}$$

$$\mathbb{P}[|\mu - \langle \mu \rangle| \geqslant T] \leqslant 2\exp\left(-\frac{2n^2 T^2}{\sum_{i=1}^{n}(b_i - a_i)^2}\right), \tag{2.60c}$$

where $\langle \mu \rangle$ denotes the empirical mean

$$\langle \mu \rangle \triangleq \frac{1}{n}\sum_{i=1}^{n} X_i. \tag{2.61}$$

In particular, if $\mathcal{X}_i = [0, 1]$ for all i, inequalities (2.60) simplify to

$$\mathbb{P}[\langle \mu \rangle - \mu \geqslant T] \leqslant e^{-2nT^2}, \tag{2.62a}$$

$$\mathbb{P}[\mu - \langle \mu \rangle \geqslant T] \leqslant e^{-2nT^2}, \tag{2.62b}$$

$$\mathbb{P}[|\langle \mu \rangle - \mu| \geqslant T] \leqslant 2e^{-2nT^2}. \tag{2.62c}$$

Remark 2.5. Hoeffding inequalities show how the tails of the probability distributions on the LHS of (2.60)–(2.62) decay exponentially with T^2, thus behaving similarly to Gaussian distributions. This shows why these inequalities can be viewed as versions of the central limit theorem [258, p. 15].

Theorem 2.3 ((MCDIARMID, OR HOEFFDING–AZUMA, OR BOUNDED-DIFFERENCES INEQUALITIES) [37, p. 171], [171, Theorem 3.7]). *Consider independent random variables $X_1, \ldots, X_n$ and a real function $h(x_1, \ldots, x_n)$. Under condition (2.58),* *the following inequalities hold:*

$$\mathbb{P}[\mathbb{E}[h(X)] - h(X) \geqslant T] \leqslant \exp\left(-2\frac{T^2}{D_h^2}\right), \tag{2.63a}$$

$$\mathbb{P}[h(X) - \mathbb{E}[h(X)] \geqslant T] \leqslant \exp\left(-2\frac{T^2}{D_h^2}\right), \tag{2.63b}$$

$$\mathbb{P}\big[|h(X) - \mathbb{E}[h(X)]| \geqslant T\big] \leqslant 2\exp\left(-2\frac{T^2}{D_h^2}\right), \tag{2.63c}$$

where D_h is defined as in (2.57).

Remark 2.6. When $\mathcal{X}_i = [a_i, b_i]$ and $h(\mathbf{X}) = \sum_{i=1}^{n} X_i$, Hoeffding inequality

$$\mathbb{P}\left[\frac{1}{n}\sum_{i=1}^{n} X_i - \frac{1}{n}\sum_{i=1}^{n} \mathbb{E}[X_i] \geqslant T\right] \leqslant \exp\left(-2n\frac{T^2}{n^{-1}\sum_{i=1}^{n}(b_i - a_i)^2}\right) \tag{2.64}$$

turns out to be a special case of McDiarmid inequalities [159, p. 4607].

Remark 2.7. A different form of McDiarmid inequalities can be obtained as follows. Observe, for example, that (2.63b) can be written in the form

$$\mathbb{P}[h(X) \geqslant T + \mathbb{E}[h(X)]] \leqslant \exp\left(-2\frac{T^2}{D_h^2}\right) \tag{2.65}$$

so that, with $\tau \triangleq T + \mathbb{E}[h(X)]$,

$$\mathbb{P}[h(X) \geqslant \tau] \leqslant \exp\left(-2\frac{\big(\mathbb{E}[h(X)] - \tau\big)^2}{D_h^2}\right). \tag{2.66}$$

2.5 Some applications

We proceed by describing some specific uncertainty scenarios which yield explicit results.

2.5.1 $\mathbb{P}$ *unknown, h known*

If we only know that $\mathbb{P}$ is the probability measure of an RV X with support set $\mathcal{X}$, a simple (yet sometimes unacceptably complex) way to approximate η consists of bounding it empirically, i.e., using experimental testing or Monte Carlo methods [86]. Assume that we want to evaluate $\eta = \mathbb{P}[X \in \mathcal{S}]$, with $\mathcal{S} \subseteq \mathbb{R}$ a fixed set. If N independent observations of the realizations of X are available, say x_i, $i = 1, \ldots, N$, an unbiased estimator of η is

$$\hat{\eta} = \frac{1}{N} \sum_{i=1}^{N} \mathbb{1}_{\mathcal{S}}(x_i), \tag{2.67}$$

where $\mathbb{1}_{\mathcal{S}}$ denotes the indicator function of $\mathcal{S}$. By choosing a large enough value of N, Monte Carlo simulation can provide a good approximation to η which is simple to implement and does not require prior information about X.

2.5.2 A trivial bound

If $\mathcal{P}$ includes all the probability measures, i.e., any probability measure is admissible, and $\eta = \mathbb{P}[X \in \mathcal{S}]$, then $\eta = 1$ when $\mathcal{S} = \mathcal{X}$ and $\eta = 0$ when $\mathcal{S} \cap \mathcal{X} = \emptyset$. Otherwise, (2.55) yield the trivial results $\overline{\eta} = 1$ and $\underline{\eta} = 0$.

2.5.3 Some moments of X known, h known

If h is known, and the admissible probability measures are those yielding known values of some of its power moments

$$\mu_i \triangleq \mathbb{E}_{\mathbb{P}}[X^i], \qquad i \in \mathcal{I}, \tag{2.68}$$

then upper and lower bounds on $\mathbb{E}_{\mathbb{P}}[h(X)]$ can be obtained using moment-bound theory, to be described in Chapter 3.

2.5.4 McDiarmid constraint sets

Assume a random vector $\mathbf{X}$ with n independent components, and a function h constrained to satisfy McDiarmid constraints (2.58) and to have bounded expected value:

$$\mathsf{osc}_i(h) \leqslant c_i(h), \tag{2.69a}$$

$$\mathbb{E}_{\mathbb{P}}[h(\mathbf{X})] \leqslant 0. \tag{2.69b}$$

For $n = 1$, we have [188, Theorem 5.1]

$$\sup_{(H,\mathbb{Q}) \in \mathcal{A}} \mathbb{Q}\left[H(X) \geqslant T + \mathbb{E}_{\mathbb{Q}}[H(X)]\right] = \begin{cases} 0, & D \leqslant T, \\ 1 - \dfrac{T}{D}, & 0 \leqslant T \leqslant D, \end{cases} \tag{2.70}$$

with D as in (2.56).

For $n = 2$, we have [188, Theorem 5.2]

$$\sup_{(H,\mathbb{Q})\in\mathcal{A}} \mathbb{Q}\big[H(\mathbf{X}) \geqslant T + \mathbb{E}_{\mathbb{Q}}[H(\mathbf{X})]\big]$$

$$= \begin{cases} 0, & D_1 + D_2 \leqslant T, \\[2mm] \dfrac{(D_1 + D_2 - T)^2}{4D_1 D_2}, & |D_1 - D_2| \leqslant T \leqslant D_1 + D_2, \\[2mm] 1 - \dfrac{T}{\max(D_1, D_2)}, & 0 \leqslant T \leqslant |D_1 - D_2|, \end{cases} \qquad (2.71)$$

where D_1 and D_2 are defined in (2.58).

For $n = 3$, an explicit solution is shown in [188, Theorem 5.4]. Numerical solutions for $n > 3$ are available in [188, §5.1.3].

The results above should be compared with McDiarmid inequality, which states

$$\sup_{(H,\mathbb{Q})\in\mathcal{A}} \mathbb{Q}\big[H(\mathbf{X}) \geqslant T + \mathbb{E}_{\mathbb{Q}}[H(\mathbf{X})]\big] \leqslant \exp\left(-2\frac{T^2}{\sum_{i=1}^{n} D_i^2}\right). \qquad (2.72)$$

Example 2.5. Fig. 2.8 compares the optimum bound (2.71) with the bound obtained from McDiarmid inequality. Here $\mathbb{E}_{\mathbb{Q}}[H] = 0$ and $T = 0.75$. We may observe that

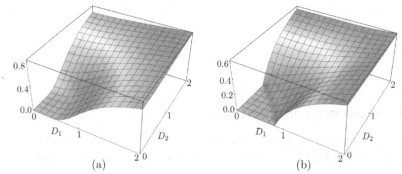

Figure 2.8 Upper bounds on $\mathbb{Q}[H(\mathbf{X}) \geqslant T]$: (a) McDiarmid bound, (b) optimum bound (2.71). Here $\mathbb{E}_{\mathbb{Q}}[H] = 0$, and $T = 0.75$.

if $D_1 \geqslant T + D_2$ (i.e., most of the uncertainty is contained in D_1), then a decrease in D_2 does not generate a decrease in bound (2.71), and hence to reduce the global uncertainty one should reduce D_1 [188, pp. 307–308].   ◇

2.5.5 Hoeffding constraint sets

Assume a random vector $\mathbf{X}$ with n independent components, and a function h constrained to be a sum of independent RVs, $h(\mathbf{X}) = X_1 + X_2 + \cdots + X_n$ with $\mathbb{E}_{\mathbb{P}}[h] \leqslant 0$.

With these constraints, Hoeffding inequality yields

$$\sup_{(H,\mathbb{Q})\in\mathcal{A}} \mathbb{Q}[H(\mathbf{X}) \geqslant T] \leqslant \exp\left(-2\frac{T^2}{\sum_{i=1}^{n} D_i^2}\right), \tag{2.73}$$

while the exact value of the upper bound is the same as in (2.71) [188, p. 311].

2.6 The certification problem

The bounds just described can be applied to the *certification problem*, where a decision has to be made as to whether a system is expected to perform within design specifications. Assume that the occurrence of event $\{Y \leqslant T\}$, with T a suitable threshold, is associated with the failure of a system whose performance is described by RV Y, which in turn is a (possibly uncertainly known) function of RV X. Ideal systems have $\mathbb{P}[Y \leqslant T] = 0$. However, when absolute guarantee of no failure cannot be attained, one may accept $\mathbb{P}[Y \leqslant T] \leqslant \varepsilon$, with ε called *certification tolerance* [159].

We may define as *performance margin* the difference between mean performance and threshold

$$m \triangleq \max(\mathbb{E}_{\mathbb{P}}[Y] - T, 0) \tag{2.74}$$

and as *uncertainty* the difference

$$u \triangleq \sup Y - \inf Y. \tag{2.75}$$

With these definitions, the *certification criterion*, or *confidence factor*, is the ratio m/u. If $\mathbb{E}_{\mathbb{P}}[Y]$ is unknown, one may replace it with an estimated average, which requires $\mathbb{P}[Y \leqslant T]$ to be determined to within a confidence interval which in turn should be included in a proper definition of margin (see [159]).

Example 2.6. Assume the calculation of $\eta = \mathbb{P}[h(X) \geqslant T]$, where T is a fixed threshold value and the expected value $\mathbb{E}[h(X)]$ is known and greater than T. McDiarmid inequality can be written in the form (2.66) as

$$\mathbb{P}[h(X) \geqslant T] \leqslant \exp\left(-2\frac{\left(\mathbb{E}[h(X)] - T\right)^2}{D_h^2}\right). \tag{2.76}$$

To obtain $\mathbb{P}[h(X) \geqslant T] < \varepsilon$, it is sufficient to choose

$$\frac{\left(\mathbb{E}[h(X)] - T\right)^2}{D_h^2} \geqslant \log\left(\sqrt{1/\varepsilon}\right). \tag{2.77}$$

In [159], D_h is called the *verification diameter*, and reflects epistemic uncertainty. The ratio between $(\mathbb{E}[h(Y)] - T)^2$ and D_h^2 is a confidence factor: if this ratio is sufficiently larger than one, the system may be deemed as reliable. $\diamond$

2.6.1   Empirical measures

Along with the failure event $\{Y \leqslant T\}$, consider the empirical failure probability

$$\langle \mathsf{p}_F \rangle \triangleq \frac{1}{n} \sum_{i=1}^{n} \delta_{y_i} \tag{2.78}$$

obtained by observing n independent values y_i taken on by Y, and setting $\delta_{y_i} = 1$ if $y_i \leqslant T$, and $\delta_{y_i} = 0$ otherwise. The expected value of $\langle \mathsf{p}_F \rangle$ is the failure probability p_F. Hoeffding inequality (2.62b) yields

$$\mathbb{P}\big[\mathsf{p}_F - \langle \mathsf{p}_F \rangle \geqslant \varepsilon \big] \leqslant e^{-2n\varepsilon^2}. \tag{2.79}$$

Defining

$$\varepsilon^{\star} \triangleq e^{-2n\varepsilon^2} \tag{2.80}$$

so that

$$\varepsilon = \left(\frac{1}{2n} \log \frac{1}{\varepsilon^{\star}} \right)^{1/2}, \tag{2.81}$$

we obtain that the event

$$\left\{ \mathsf{p}_F \leqslant \hat{\mathsf{p}}_F + \left(\frac{1}{2n} \log \frac{1}{\varepsilon^{\star}} \right)^{1/2} \right\} \tag{2.82}$$

occurs with probability $1 - \varepsilon^{\star}$. Hence, the inequality

$$\langle \mathsf{p}_F \rangle + \left(\frac{1}{2n} \log \frac{1}{\varepsilon^{\star}} \right)^{1/2} \leqslant \varepsilon \tag{2.83}$$

yields a conservative certification criterion [159, p. 4592]. This shows how the number n of samples of Y to be observed for certification is on the order of $\varepsilon^{-2} \log(1/\varepsilon^{\star})$.

2.6.2   Certification under epistemic uncertainty

When a system model is sought, to account for epistemic uncertainty, one may assume, as mentioned *supra*, that this uncertainty affects the nominal system function $h(X)$ by transforming it into the new function $k(X, Z)$ which includes, in addition to X, random *unknown unknowns* modeled by a new RV Z independent of X. In this case, we write

$$\begin{aligned}
\mathsf{p}_F &\triangleq \mathbb{P}[g(X, Z) \leqslant T] \\
&= \mathbb{P}[h(X) + \big(k(X, Z) - h(X)\big) \leqslant (T + \tau) - \tau] \\
&\leqslant \mathbb{P}[h(X) \leqslant T + \tau] + \mathbb{P}[k(X, Z) - h(X) \leqslant -\tau].
\end{aligned} \tag{2.84}$$

If we interpret $h(X) - k(X, Z)$ as the difference between predicted and measured system behavior, then $\mathbb{P}[k(X, Z) - h(X) \leqslant -\tau]$ is the probability that the difference between predicted and measured performance is more than τ, which may be regarded as a measure of modeling error, i.e., an indication of epistemic uncertainty [159, p. 4596]. We also observe that we cannot achieve $\mathsf{p}_F < \varepsilon$ unless $\mathbb{P}[k(X, Z) - h(X) \leqslant -\tau] < \varepsilon$. A conservative certification criterion is now

$$\mathbb{P}[h(X) \leqslant T + \tau] + \mathbb{P}[k(X, Z) - h(X) \leqslant -\tau] \leqslant \varepsilon, \tag{2.85}$$

where τ could be chosen to minimize the LHS of this inequality. Using previous results in this chapter, we can bound $\mathbb{P}[h(X) \leqslant T + \tau]$. Specifically, from McDiarmid inequality (2.63a), we have, assuming $\mathbb{E}[k - h] + \tau \geqslant 0$,

$$\mathbb{P}\big[k(X, Z) - h(X) \leqslant \mathbb{E}[k(X, Z) - h(X)] - \tau\big] \leqslant \exp\left(-2\frac{(\mathbb{E}[k - h] + \tau)^2}{D_{k-h}^2}\right), \tag{2.86}$$

where the diameter D_{k-h} is defined by

$$D_{k-h}^2 \triangleq \sup_{x, x', z} \big|[h(x) - k(x, z)] - [h(x') - k(x', z)]\big|^2 + \sup_{x, z, z'} |k(x, z) - k(x, z')|^2. \tag{2.87}$$

In practice, $\mathbb{E}[k - h]$ and $\mathbb{E}[h]$ must be estimated using empirical means. Reference [159] describes the steps necessary to introduce the estimation error in the calculations of the relevant bounds.

Sources and parerga

1. The "scenario of uncertainty," the set of probability measures expressing the uncertainty about the one to be used to find the best model or the performance bounds, is sometimes called "credal set" or "ambiguity set."
2. Concentration-of-measures inequalities describe how an RV deviates from its mean value. For example, Čebyšev inequality (3.7) can be viewed as a simple concentration inequality, very general but often too weak.
3. The "concentration of measure phenomenon" occurs with functions of a large number of variables, whereby if the local oscillations in each variable are small the functions are almost constant. In probability theory, concentration of measure occurs in laws of large numbers. Books [37,155] are extensive monographs on the subject. A succinct review of the subject can be found in [159].
4. Reference [159] contains a brief treatment of concentration inequalities applied to nonindependent RVs, to functions with unbounded oscillations, and to empirical processes defined by sampling.
5. The condition of independent RVs X_i in McDiarmid inequalities can be somewhat relaxed [171].

6. The case of multiple performance measures is dealt with in [159].

7. Conditions for the functions $\Theta^-(c)$ and $\Theta^+(c)$ in (2.47) to be differentiable and nonincreasing (resp., nondecreasing) and attaining a local minimum (resp., a local maximum) at a unique value of c are derived in [157, Proposition 3.7].

8. The results in Section 2.2 are generalized in [157], where bounds are derived which encompass the case of mixed aleatory and epistemic uncertainty. In [157] it is assumed that the performance index has the form

$$\eta = \int_{\mathcal{X}} \int_{\mathcal{Z}} h(x, z) f_X(x) f_Z(z) \, dx dz,$$

where the RV X has support set $\mathcal{X}$ and known pdf $f_X(x)$, while the RV Z has support set $\mathcal{Z}$ and incompletely known pdf $f_Z(z)$.

9. In [100], the quantity

$$\Xi(f \parallel g; h) \triangleq \inf_{c \in (0, \infty)} \left[\Theta^+(c) + \frac{1}{c} D(f \parallel g) \right] \tag{2.88}$$

is called *goal-oriented divergence*. It has the properties of a divergence, as $\Xi(f \parallel g; h) \geqslant 0$ and $\Xi(f \parallel g; h) = 0$ if and only if $f = g$ or h is a constant function. In [100], optimizations directly based on $\Xi(f \parallel g; h)$ are discussed.

10. In [188, p. 286] it is observed that bounds on $\mathsf{osc}_i(h)$ "are useful because they constitute a form of nonlinear sensitivity analysis and, combined with independence constraints, they lead to the concentration-of-measure phenomenon."

Moment bounds

3

In this chapter we derive bounds—or, if not easily available, approximations— to the value of performance index (2.1) under the assumption that the underlying probability measure $\mathbb{P}$ is unknown, but some of its moments are known (possibly within an uncertainty interval). We are especially interested in *sharp* upper and lower bounds, i.e., bounds which cannot be further improved unless epistemic uncertainty is reduced, or which can be achieved with equality by at least one probability measure consistent with the problem constraints.

Recalling the definitions introduced in Section 2.3, we start by dealing with cases where set $\mathcal{A}$, the uncertainty scenario, includes all the probability measures $\mathbb{P}$ of a real random variable (RV) X whose support set $\mathcal{X}$ and whose power moments

$$\mathbb{E}_{\mathbb{P}}[X^k], \qquad k \in \mathcal{K}, \tag{3.1}$$

are known, where $\mathcal{K}$ is an index set of positive integers. Moreover, we might be privy to some additional features of the unknown probability measure which would reduce the size of the uncertainty scenario. For example, one may want to restrict the RV X to be *unimodal*, i.e., to have only one peak in its pdf $f_X(x)$ at a certain value m of x, and to decrease as x moves away from m (in terms of the cumulative distribution function F_X, unimodality requires it to be convex $\cup$ on $(-\infty, m)$ and concave $\cap$ on $[m, \infty)$). Or, one may know that the underlying CDF $F_X(x)$ is continuous with bounded pdf. In the following we shall characterize an RV X interchangeably with its probability measure, its CDF $F_X(x)$, or its pdf $f_X(x)$.

3.1 Some classical results

Before proceeding further, we list a few examples of classical bounds obtained when the knowledge of $\mathbb{P}$ involves only a small number of scalar parameters associated with it.

Example 3.1 (Markov bound). If X is an RV with support set $\mathbb{R}^+$ and whose mean $\mu \triangleq \mathbb{E}_{\mathbb{P}}[X]$ is known, we have, for $T > 0$,

$$\mathbb{P}[X \geqslant T] \leqslant \frac{\mu}{T}. \tag{3.2}$$

(If $0 < T < \mu$, this inequality becomes trivial.) To prove (3.2), observe that

Dimensions of Uncertainty in Communication Engineering. https://doi.org/10.1016/B978-0-32-399275-6.00011-3

$$\mu \triangleq \int_0^\infty x f_X(x)\, dx$$

$$\geqslant \int_T^\infty x f_X(x)\, dx \tag{3.3}$$

$$\geqslant T \int_T^\infty f_X(x)\, dx$$

$$= T\, \mathbb{P}[X \geqslant T].$$

This bound is sharp: an RV X taking value T with probability 1 has $\mathbb{P}[X \geqslant T] = 1$ and $\mu = T$, which yields (3.2) with equality. $\diamond$

Example 3.2. Markov bound can be generalized by choosing an even nonnegative function ψ which is increasing on $(0, \infty)$. We obtain, for an RV X with support set $\mathbb{R}$,

$$\mathbb{E}_\mathbb{P}[\psi(X)] \triangleq \int_{-\infty}^\infty \psi(x) f_X(x)\, dx$$

$$= \int_{|x| \geqslant T} \psi(x)\, dF(x) + \int_{|x| < T} \psi(x) f_X(x)\, dx$$

$$\geqslant \int_{|x| \geqslant T} \psi(x) f_X(x)\, dx \tag{3.4}$$

$$\geqslant \psi(T) \int_{|x| \geqslant T} f_X(x)\, dx$$

$$= \psi(T)\, \mathbb{P}[|X| \geqslant T]$$

so that

$$\mathbb{P}[|X| \geqslant T] \leqslant \frac{\mathbb{E}_\mathbb{P}[\psi(X)]}{\psi(T)}. \tag{3.5}$$

Choosing, for example, $\psi = x^4$ we obtain

$$\mathbb{P}[|X| \geqslant T] \leqslant \frac{\mu_4}{T^4}. \tag{3.6}$$

$\diamond$

Example 3.3 (Čebyšev bound). If X is an RV with support set $\mathbb{R}$, mean μ, and variance $\sigma^2 \triangleq \mathbb{E}_\mathbb{P}[(X - \mu)^2]$, then, using Markov bound (3.5), we obtain, for $T > 0$,

$$\mathbb{P}[|X - \mu| \geqslant T] \leqslant \frac{\sigma^2}{T^2}. \tag{3.7}$$

To prove sharpness, choose a real number $\kappa \geqslant 1$, and consider an RV X taking values -1 with probability $1/2\kappa^2$, 0 with probability $1 - 1/\kappa^2$, and 1 with probability $1/2\kappa^2$. This RV has $\mu = 0$ and $\sigma^2 = 1/\kappa^2$, and hence $\mathbb{P}(|X| \geqslant 1) = 1/\kappa^2$, which satisfies (3.7) with equality. $\diamond$

Example 3.4 (Cantelli bound, or one-sided Čebyšev bound). Given an RV X with support set $\mathbb{R}$, mean μ, and variance σ^2, for any $T > 0$, the following inequality holds:

$$\mathbb{P}[X - \mu \geqslant T] \leqslant \frac{\sigma^2}{\sigma^2 + T^2}. \tag{3.8}$$

To prove it, observe that, for any real $u \geqslant 0$, defining $Y \triangleq X - \mu$, we have

$$\begin{aligned}
\mathbb{P}[Y \geqslant T] &= \mathbb{P}[Y + u \geqslant T + u] \\
&= \mathbb{P}[(Y + u)^2 \geqslant (T + u)^2] \\
&\leqslant \frac{\mathbb{E}_{\mathbb{P}}[(Y + u)^2]}{(T + u)^2} \quad \text{(from Markov inequality (3.5))} \\
&= \frac{\sigma^2 + u^2}{(T + u)^2}.
\end{aligned} \tag{3.9}$$

Now, choose the value of u that minimizes the bound. Differentiating, we obtain $u_{\mathrm{opt}} = \sigma^2/T$, which yields (3.8).

For $T < 0$, taking $s = -T > 0$ and proceeding as above, we obtain

$$\mathbb{P}[X - \mu < T] \leqslant \frac{\sigma^2}{\sigma^2 + s^2} = \frac{\sigma^2}{\sigma^2 + T^2}, \tag{3.10}$$

and hence, taking the complement,

$$\mathbb{P}[X - \mu \geqslant T] \geqslant 1 - \frac{\sigma^2}{\sigma^2 + T^2}. \tag{3.11}$$

To prove that (3.8) is sharp, consider the RV taking values $-\sigma^2$ with probability $1/(1 + \sigma^2)$ and 1 with probability $\sigma^2/(1 + \sigma^2)$. This RV has mean zero and variance σ^2, which yields $\mathbb{P}[X \geqslant 1] = \sigma^2/(1 + \sigma^2)$, and hence (3.8) with equality. $\diamond$

Other inequalities may involve distribution parameters other than power moments.

Example 3.5 (Distance between mean and median). For an RV X with mean μ and variance σ^2, if m denotes the median of X, i.e., $\mathbb{P}[X \geqslant m] \geqslant 1/2$ and $\mathbb{P}[X \leqslant m] \geqslant 1/2$, then

$$|m - \mu| \leqslant \sigma, \tag{3.12}$$

expressing the fact that the mean value of an RV is within one standard deviation of its median. A proof of (3.12) goes as follows [165]: Jensen inequality (1.44) yields

$$c[\mathbb{E}[Y]] \leqslant \mathbb{E}[c(Y)] \tag{3.13}$$

valid for c convex $\cup$. Using (3.13) with $c(y) = |y|$ and $Y = X - \mu$, we obtain

$$|\mathbb{E}[X - \mu]| \leqslant \mathbb{E}[|X - m|]. \tag{3.14}$$

Similarly, using (1.44) with $c(y) = y^2$ and $Y = |X - \mu|$, we have

$$|\mathbb{E}[X - \mu]| \leqslant \sqrt{\mathbb{E}[(X - \mu)^2]} = \sigma. \tag{3.15}$$

Next, observe that the median minimizes the mean absolute deviation, i.e., $\mathbb{E}[|X - a|]$ takes its minimum value when $a = m$. Using this property, (3.14)–(3.15), and Jensen inequality twice, we obtain

$$\begin{aligned}
|\mu - m| &\leqslant \left|\mathbb{E}[X - \mu]\right| \\
&\leqslant \mathbb{E}[|X - \mu|] \\
&\leqslant \sqrt{\mathbb{E}\left[(X - \mathbb{E}[X])^2\right]} \\
&= \sigma.
\end{aligned} \tag{3.16}$$

The equality $|\mu - m| = \sigma$ holds if X takes only two symmetric values [165]. $\diamond$

Remark 3.1. Some results are also available for situations in which moments are approximated using a set of samples from some population, as, for example, when μ and σ^2 are replaced by their sample equivalents [222,250]. This will be discussed in Chapter 5 within the context of Probability Boxes.

We now proceed as follows: after stating the general "moment problem," we study some cases in which the probability measure $\mathbb{P}$ is in principle known, but exact computations of $\eta \triangleq \mathbb{E}_{\mathbb{P}}[h(X)]$ are too complex, unless h is a power function. In this case, we show how bounds on η can be derived. We next state the conditions under which a given sequence of real numbers can be interpreted as a moment sequence, i.e., at least one probability measure exists such that the elements of this sequence can be viewed as moments generated by that measure. Next, we describe analytical and numerical methods allowing one to derive the tightest possible upper and lower bounds on the expected value of a known function of an RV known only through some of its moments. We first examine probability measures defined on a compact subset of $\mathbb{R}$, as this "scalar" case leads sometimes to closed-form solutions. Our first (and conceptually simplest) approach is based on a geometrical technique, which is practical only when the number of moments is very small. We next describe techniques based on quadrature rules, and finally describe some relatively recent results rooted in semidefinite programming. Finally, we study the moment problem on a subset of $\mathbb{R}^d$, $d > 1$. This has a higher complexity, leads to a number of yet unsolved problems, and its solution often calls for numerical techniques. Some applications will illustrate the solutions that can derived.

3.2 The general moment-bound problem

The general moment-bound problem, in its simplest version, assumes that a real random variable X exists which is inaccurately modeled, in the sense that its probability measure $\mathbb{P}$ (or, equivalently, its CDF $F_X(x)$ or its pdf $f_X(x)$) is known only through a (possibly small) number of its moments, i.e., of expected values of known functions of X. We look for tight, and possibly sharp, upper and lower bounds on the expected value of $\mathbb{E}_{\mathbb{P}}[\varphi(X)]$, with $\varphi(\cdot)$ a known function, under the constraint that the values of some moments of X are exactly known, as is the support set $\mathcal{X}$ of X. It makes sense to expect that, as more moments are known, the interval within which the exact value of $\mathbb{E}_{\mathbb{P}}[\varphi(X)]$ is confined becomes narrower. On the other hand, if moments are obtained through measurements, it may be expected that the accuracy of their measurement decreases as the moment order increases.

The (scalar) problem can be formulated in its generality as follows (see, for example, [240]): Let X be an RV with support set $\mathcal{X}$. We are given $N + 1$ measurable functions $h_i(x)$, $i = 0, \ldots, N$, defined on $\mathcal{X}$. The expectations of these functions, taken with respect to the probability measure of X and referred to as *moments* μ_i, are assumed to be known and finite, i.e., we know the values taken by

$$\mu_i \triangleq \mathbb{E}_{\mathbb{P}}[h_i] = \int_{\mathcal{X}} h_i(x) f_X(x) \, dx, \qquad i = 0, 1, \ldots, N, \tag{3.17}$$

but we do not know $f_X(x)$. Without any loss of generality, we assume from now on that the functions $h_i(x)$ are linearly independent: otherwise, some of the moments (3.17) could be derived from the others, and hence become redundant. Since f_X is a pdf, we take $h_0(x) = 1$ and $\mu_0 = 1$. (In the special case $h_i(x) = x^i$, μ_i is usually referred to as the ith *power moment*.) Given the real-valued measurable function $\varphi(x)$, we want to compute

$$\inf_{\mathbb{P} \in \mathcal{P}(\boldsymbol{\mu})} \mathbb{E}_{\mathbb{P}}[\varphi] \quad \text{and} \quad \sup_{\mathbb{P} \in \mathcal{P}(\boldsymbol{\mu})} \mathbb{E}_{\mathbb{P}}[\varphi], \tag{3.18}$$

where $\mathcal{P}(\boldsymbol{\mu})$ is the set of probability measures whose moments match $\boldsymbol{\mu} \triangleq (\mu_0, \mu_1, \ldots, \mu_N)$.

Since a solution to the fully general problem above is not available, to make it soluble, specific restrictions must often be considered. These consist of introducing in (3.18) some additional constraints.

Remark 3.2. In some cases, inf and sup in (3.18) can be replaced with min and max, respectively, but this may not be that relevant, since in most applications one is usually interested in the derivation of inequalities.

Example 3.6. Given N power moments $(\mu_1, \ldots, \mu_N)$ of an RV X with support set $\mathcal{X}$, we derive an upper bound to its differential entropy [240,280]. This turns out to be a special case of the maximum-entropy problem examined in Section 1.6. The problem to be solved is now

$$\underset{f}{\text{maximize}} \ \ \mathsf{h}(f)$$

$$\text{subject to} \int x^i f(x) \, dx = \mu_i, \qquad i = 0, 1, \ldots, N, \tag{3.19}$$

$$f(x) \geqslant 0.$$

Equating to zero the derivative of the Lagrangian functional

$$L(f) = -\int_{\mathcal{X}} f(x) \log[f(x)] \, dx + \sum_{i=0}^{N} \lambda_i \int_{\mathcal{X}} x^i f(x) \, dx, \tag{3.20}$$

we obtain the necessary condition for a stationary f,

$$-\log[f(x)] - 1 + \sum_{i=0}^{N} \lambda_i x^i = 0. \tag{3.21}$$

Since differential entropy is concave $\cap$ and the constraint set is convex $\cup$, the pdf obtained by solving (3.21), i.e.,

$$f^{\star}(x) = \exp\left(-1 + \sum_{i=0}^{N} \lambda_i x^i\right), \tag{3.22}$$

where the Lagrange coefficients are computed so that they satisfy the moment constraints, is entropy-maximizing. As the equations to be solved to obtain the λ_i are highly nonlinear, numerical procedures may be employed, as suggested in [240,280]. Some simple solutions are available for a small number of moments: if $\mathcal{X} = [a, b]$ and no moment constraints are given, then $f^{\star}$ is uniform; if $\mathcal{X} = [0, \infty)$ and μ_1 is known, then $f^{\star}$ is exponential; and if $\mathcal{X} = (-\infty, \infty)$ and μ_1, μ_2 are known, then $f^{\star}$ is Gaussian [230]. If $\mathcal{X} = (-\infty, \infty)$ and μ_1 is known, then there is no maximum-entropy pdf [55, p. 413]. In [68] it is shown that, for known μ_1, μ_2 and finite $\mathcal{X}$, the pdf maximizing the differential entropy exists and is finite. When $\mathcal{X} = [0, \infty)$, the maximum-entropy pdf exists if and only if $\mu_2 \leqslant 2\mu_1^2$. $\diamond$

When we consider the moment sequences forming the constraint set, we are faced with two separate problems. The first is the computation of moments of an RV X having a known structure (typically, a known function of one or more RVs whose probability measures are known, but whose explicit calculation is computationally too demanding). The second is the determination of conditions that must be satisfied by a real finite ("truncated") or infinite sequence for it to be a sequence of moments of an RV.

3.3   Calculation of moments

In some cases, direct calculation of $\mathbb{E}_{\mathbb{P}}[\varphi]$ may be hopelessly complicated, while certain moments can be obtained without too much effort. This occurs, for example, when X is a sum of a large number of independent RVs whose individual probability measures, and hence their power moments, are known,

$$X = \sum_{i=1}^{M} X_i. \tag{3.23}$$

A few standard techniques for the computation of power moments $\mathbb{E}[X^m]$ are listed below.

3.3.1 Using multinomial expansions

The multinomial equality

$$X^m = \sum_{k_1 + \cdots + k_M = m} \binom{m}{k_1, \ldots, k_M} \prod_{i=1}^{M} X_i^{k_i}, \tag{3.24}$$

where the *multinomial coefficients* are defined by

$$\binom{m}{k_1, \ldots, k_M} \triangleq \frac{m!}{k_1! \cdots k_M!}, \tag{3.25}$$

yields

$$\mathbb{E}[X^m] = \sum_{k_1 + \cdots + k_M = m} \binom{m}{k_1, \ldots, k_M} \prod_{i=1}^{M} \mathbb{E}[X_i^{k_i}]. \tag{3.26}$$

3.3.2 Using moment-generating functions

The *moment-generating function* (MGF) of X is defined as

$$M_X(t) \triangleq \mathbb{E}[e^{tX}] = \int_X e^{tx} f_X(x) \, dx, \tag{3.27}$$

which we assume existing in an open interval around $t = 0$. Expanding e^{tX} in a McLaurin series, we obtain

$$M_X(t) = \sum_{k=0}^{\infty} \frac{t^k}{k!} \mathbb{E}\left[X^k\right] \tag{3.28}$$

so that

$$\mathbb{E}[X^m] = \frac{d^m}{dt^m} M_X(t)\bigg|_{t=0}. \tag{3.29}$$

MGFs are especially useful in the case of a linear combination of independent RVs, as its MGF is the product of individual MGFs. Specifically, if $X \triangleq \sum_{i=1}^{M} a_i X_i$, we obtain, with obvious notations,

$$M_X(t) = \prod_{i=1}^{M} M_{X_i}(a_i t). \tag{3.30}$$

Remark 3.3. If some moments are not finite, then the MGF does not exist, but the finiteness of moments of all orders is not sufficient for existence. In general, if two RVs X_1, X_2 have the same MGF, their pdfs $f_{X_1}(x)$, $f_{X_2}(x)$ are equal for all values of x. However, this does not imply that two RVs with the same moments are identical: in fact, it may occur that all moments exist, while the MGF does not (see Example 3.13 *infra*).

Remark 3.4. The function $M_X(it)$, with $i \triangleq \sqrt{-1}$, is called the *characteristic function* of X, and may be viewed as the Fourier transform of the pdf of X; $M_X(-s)$ may also be viewed as the bilateral Laplace transform of f_X (see Example 3.12 *infra*).

3.3.3 Using recursive computations

A recursive method for computing numerically the moments of X in (3.23) runs as follows: Define the partial sums

$$\xi_m \triangleq \sum_{i=1}^{m} X_i \tag{3.31}$$

(so that $\xi_M = X$), and compute, recursively over $m = 1, \ldots, M$, the partial moments

$$\mu_k^{(m)} \triangleq \mathbb{E}[\xi_m^k] \tag{3.32}$$

by observing that

$$
\begin{aligned}
\mu_k^{(m+1)} &= \mathbb{E}[\xi_{m+1}^k] \\
&= \mathbb{E}\big[(\xi_m + X_{m+1})^k\big] \\
&= \sum_{j=0}^{k} \binom{k}{j} \mathbb{E}[\xi_m^j] \mathbb{E}[X_{m+1}^{k-j}] \\
&= \sum_{j=0}^{k} \binom{k}{j} \mu_j^{(m)} \mathbb{E}[X_{m+1}^{k-j}].
\end{aligned}
\tag{3.33}
$$

Example 3.7. If X_i are independent and uniformly distributed in $[-1, 1]$, their common moment generating function is

$$M_i(t) = \frac{\sinh(t)}{t}. \tag{3.34}$$

The RV $X = X_1 + X_2 + X_3 + X_4$ has MGF

$$M_X(t) = \left(\frac{\sinh(t)}{t}\right)^4 = 1 + \frac{2}{3}t^2 + \frac{1}{5}t^4 + \frac{34}{945}t^6 + \frac{62}{14175}t^8 + O(t^9), \tag{3.35}$$

which yields the moment sequence

$$\mu = (1, 0, 4/3, 0, 24/5, 0, 544/21, 0, 7936/45, \ldots). \tag{3.36}$$

◇

Example 3.8. Consider the "Čebyšev pulse," defined for $t \in \mathbb{R}$,

$$h(t) = 0.4023 \cos\big((2.839|t|/T) - 0.7553\big) \exp(-0.4587|t|/T)$$
$$+ 0.7162 \cos\big((1.176|t|/T) - 0.1602\big) \exp(-1.107|t|/T), \quad -\infty < t < \infty, \tag{3.37}$$

and its samples $h_i \triangleq h(iT)$ taken at multiples of T. Defining

$$X \triangleq \sum_{\substack{-30 \leqslant i \leqslant 30 \\ i \neq 0}} h_i X_i, \tag{3.38}$$

with each X_i taking values ± 1 with probability $1/2$, direct computation yields the MGFs $M_i(t) = \cosh(h_i t)$. Expanding in a McLaurin series the product (3.30), we obtain the moment sequence

$$\mu = (1, 0, 5.44461 \cdot 10^{-3}, 0, 8.30629 \cdot 10^{-5}, 0, 1.97265 \cdot 10^{-6}, 0, 6.12829 \cdot 10^{-6}, \ldots)$$

consistent with a result in [276]. ◇

3.4 Moments of unimodal pdfs

The constraint of a unimodal pdf can be dealt with using *Khinčin Characterization Theorem*. It states that an RV X which is unimodal about zero has the representation $X = UV$, where U and V are independent and $U \sim \mathcal{U}[0, 1]$ [80, p. 155], [128]). Thus, the power moments of X can be computed by observing that

$$\mathbb{E}[X^k] = \mathbb{E}[U^k] \cdot \mathbb{E}[V^k] = \frac{1}{k+1} \mathbb{E}[V^k]. \tag{3.39}$$

Example 3.9. We have, with obvious notations,

$$\mu_V \triangleq 2\mathbb{E}[X],$$
$$\sigma_V^2 \triangleq \mathbb{E}[V^2] - \big(\mathbb{E}[V]\big)^2 = 3\mathbb{E}[X^2] - 4\big(\mathbb{E}[X]\big)^2 = 3\sigma_X^2 - \mu_X^2. \tag{3.40}$$

◇

More generally, $\mathbb{E}_X[\varphi(X)]$ can be computed as follows: write

$$\mathbb{E}_X[\varphi(X)] = \mathbb{E}_V \mathbb{E}_{U|V}[\varphi(UV) \mid V]$$
$$= \mathbb{E}_V \left[\int_0^1 \varphi(uV) \, du \right] \tag{3.41}$$

so that, defining the function

$$\Phi(v) \triangleq \frac{1}{v} \int_0^v \varphi(y)\,dy, \tag{3.42}$$

we obtain

$$\mathbb{E}_X[\varphi(X)] = \mathbb{E}_V[\Phi(V)] \tag{3.43}$$

In conclusion, the problem of finding bounds on $\mathbb{E}_X[\varphi(X)]$, subject to X being unimodal with mode 0 and knowing some of its moments, is solved by finding bounds on $\mathbb{E}_V[\Phi(V)]$ subject to moment constraints on V.

3.4.1 Moment transfer

The technique just described for the calculation of moments of a unimodal RV can be viewed as a special case of the *moment transfer* technique described in [142, pp. 24–25]. It consists of transferring the original moment problem concerning an RV Y to a simpler problem involving another RV X which is connected to Y by the relation

$$f_Y(y) = \int K(y,x) f_X(x)\,dx, \tag{3.44}$$

with $K(y,x)$ a suitable kernel function. In this situation, we have

$$\begin{aligned}\mathbb{E}_Y[\varphi(Y)] &\triangleq \int \varphi(y) f_Y(y)\,dy \\ &= \iint \varphi(y) K(y,x) f_X(x)\,dx\,dy \\ &= \int \Phi(x) f_X(x)\,dx \\ &= \mathbb{E}_X[\Phi(X)],\end{aligned} \tag{3.45}$$

where

$$\Phi(x) \triangleq \int \varphi(y) K(y,x)\,dy. \tag{3.46}$$

Example 3.10. If $X = UV$ is unimodal as above, and V has support set $\mathbb{R}^+$, then, from the general expression (Mellin convolution (1.87)) for the pdf of the product of two independent RVs

$$f_X(x) = \int (1/v) f_U(x/v) f_V(v)\,dv \tag{3.47}$$

and the choice

$$K(u,v) = \frac{1}{v} f_U\left(\frac{u}{v}\right), \tag{3.48}$$

we obtain again

$$\mathbb{E}_X[\varphi(X)] = \mathbb{E}_V[\Phi(V)], \tag{3.49}$$

with Φ defined as in (3.42). $\diamondsuit$

3.5 When is a sequence a valid moment sequence?

To apply moment-bound theory to a sequence μ of real numbers, one should make sure that at least one $\mathbb{P}$ exists having μ as a valid moment sequence.

Example 3.11. When μ has a small number of components, simple conditions for μ to be a valid moment sequence can be obtained using elementary techniques. Consider, for example, an RV X with finite support set $[a, b]$. Taking the expected value in the inequalities $a \leqslant X \leqslant b$, valid with probability 1, we obtain $a \leqslant \mu_1 \leqslant b$, with an equality if X takes value a or b with probability 1. A condition involving the first two power moments can be obtained from the inequality $(b - X)(X - a) \geqslant 0$. Taking expectation, we obtain

$$\mu_2 \leqslant (a + b)\mu_1 - ab. \tag{3.50}$$

Subtracting μ_1^2 from both sides of (3.50), we obtain the condition involving the variance of X,

$$\sigma_X^2 \leqslant (b - \mu_1)(\mu_1 - a). \tag{3.51}$$

As a special case, a standardized RV (i.e., one with $\mu_1 = 0$ and $\sigma^2 = 1$) must satisfy the inequality $1 + ab \leqslant 0$. In this case, equality $1 + ab = 0$ holds for a discrete RV taking value a with probability $p = 1/(1 + a^2)$ and $-1/a$ with probability $1 - p$. Similar conditions involving μ_3 are derived in [120, §4.2]. $\diamondsuit$

To describe the results available, we classify the conditions under which they are obtained. First, we specify the support set $\mathcal{X}$ of the RV generating the moments. Next, we consider separately infinite and finite (or *truncated*) sequences of moments. The standard terminology here distinguishes three separate moment problems, viz.,

(i) *Hamburger moment problem*, $\mathcal{X} = \mathbb{R}$,
(ii) *Stieltjes moment problem*, $\mathcal{X} = \mathbb{R}^+$, and
(iii) *Hausdorff moment problem*, $\mathcal{X} = [a, b]$, a finite interval of the real line.

A probability measure *representing* a valid moment sequence μ is a measure $\mathbb{P}$ whose moments are the components of μ. If $\mathbb{P}$ is unique, then it is said to be *determined* by its moments.

3.5.1 Probability measures represented by moments

The three following theorems [153, Chapter 3] yield conditions for the existence of a representing probability measure $\mathbb{P}$ (on $\mathbb{R}$, $\mathbb{R}^+$, or $[a, b]$, respectively). These conditions are expressed in terms of definiteness of certain quadratic forms.

Theorem 3.1 yields conditions for the untruncated moment problem. Given the sequence $\boldsymbol{\mu}$, define the $(N+1) \times (N+1)$ "Hankel" matrices

$$\mathbf{H}_N(\boldsymbol{\mu})\big|_{i,j} \triangleq \mu_{i+j-2}, \qquad \mathbf{B}_N(\boldsymbol{\mu})\big|_{i,j} \triangleq \mu_{i+j-1}, \qquad \mathbf{C}_N(\boldsymbol{\mu})\big|_{i,j} \triangleq \mu_{i+j}, \quad (3.52)$$

for all integer pairs (i, j) with $1 \leqslant i, j \leqslant N+1$.

Then [153, Theorem 3.2]:

Theorem 3.1. *Let $\boldsymbol{\mu} \triangleq (\mu_k)_{k=0}^{\infty}$. Then*

(a) *$\boldsymbol{\mu}$ has a representing probability measure on $\mathbb{R}$ if and only if $\mathbf{H}_N(\boldsymbol{\mu})$ is nonnegative definite for all $N \in \mathbb{N}$.*

(b) *$\boldsymbol{\mu}$ has a representing probability measure on $\mathbb{R}^+$ if and only if $\mathbf{H}_N(\boldsymbol{\mu})$ and $\mathbf{B}_N(\boldsymbol{\mu})$ are nonnegative definite for all $N \in \mathbb{N}$.*

(c) *$\boldsymbol{\mu}$ has a representing probability measure on $[a, b]$ if and only if $\mathbf{H}_N(\boldsymbol{\mu})$ and $-\mathbf{C}_N(\boldsymbol{\mu}) + (a+b)\mathbf{B}_N(\boldsymbol{\mu}) - ab\mathbf{H}_N(\boldsymbol{\mu})$ are nonnegative definite for all $N \in \mathbb{N}$.*

The proof of this theorem, as well of those of Theorems 3.2 and 3.3 *infra*, can be found in [153, Chapter 3]. Here we describe only a central argument leading to the proof of Theorem 3.1(a). Consider the Hankel matrix $\mathbf{H}_N(\boldsymbol{\mu})$ associated with moments $\boldsymbol{\mu} = (\mu_i)_{i=0}^{2N}$,

$$\mathbf{H}_N(\boldsymbol{\mu}) \triangleq \begin{bmatrix} \mu_0 & \mu_1 & \cdots & \mu_N \\ \mu_1 & \mu_2 & \cdots & \mu_{N+1} \\ \vdots & \vdots & \ddots & \vdots \\ \mu_N & \mu_{N+1} & \cdots & \mu_{2N} \end{bmatrix}. \qquad (3.53)$$

This matrix is nonnegative definite. In fact, observe that, for any $\mathbf{y} \in \mathbb{R}^{N+1}$,

$$\mathbf{y}^{\mathsf{T}} \mathbf{H}_N(\boldsymbol{\mu}) \mathbf{y} = \sum_{i,j=0}^{N} y_i y_j \mathbb{E}[X^{i+j}] = \mathbb{E}\left[y_0 + y_1 X + \cdots + y_N X^N\right]^2 \geqslant 0. \quad (3.54)$$

The proof of the converse exhibits some intricacies that, as we shall see, make the multidimensional moment problem much more complex than its one-dimensional counterpart. It relies upon two basic results:

(i) [153, Theorem 2.5] Any real polynomial $p(x)$ of even degree is nonnegative if and only if it can be written as the sum of squares of other real polynomials $g_j(x)$, $j = 1, \ldots, r$, i.e.,

$$p(x) = \sum_{j=1}^{r} g_j^2(x). \qquad (3.55)$$

(ii) [153, Theorem 3.1] Given the infinite sequence of moments $\mu = (\mu_n)$ and a compact set $\mathcal{X} \subset \mathbb{R}$, a probability measure $\mathbb{P}$ on $\mathcal{X}$ exists such that $\mathbb{E}_{\mathbb{P}}[X^n] = \mu_n$ for all $n \in \mathbb{N}$ if and only if, for any real polynomial $f(x) \triangleq \sum_n y_n x^n$ nonnegative on $\mathcal{X}$, we have

$$\sum_n y_n \mu_n \geq 0. \tag{3.56}$$

Remark 3.5. The condition that the infinite sequence μ represent some measure on $\mathbb{R}$ (or the limit of some sequence of measures) can be expressed as a linear matrix inequality as in Theorem 3.1 together with the equality $\mu_0 = 1$. Using this fact, some moment problems can be cast as *semidefinite programs* [38, p. 168], as we shall illustrate later in this chapter.

3.5.1.1 The truncated moment problem

Theorems 3.2 and 3.3 *infra* show results for the truncated moment problem. The sequence μ may include an even or an odd number of moments, and even and odd cases are treated separately.

Write the $(N+1) \times (N+1)$ Hankel moment matrix $\mathbf{H}_N(\mu)$ in the partitioned form

$$\mathbf{H}_N(\mu) = [\mathbf{h}_0 \mid \mathbf{h}_1 \mid \cdots \mid \mathbf{h}_N], \tag{3.57}$$

where $\mathbf{h}_j$ are $(N+1) \times 1$ column matrices. The *Hankel rank* of μ, denoted by $\mathrm{rank}(\mu)$, is the smallest integer $1 \leq i \leq N$ such that $\mathbf{h}_i \in \mathrm{span}\{\mathbf{h}_0, \ldots, \mathbf{h}_{i-1}\}$. If $\mathbf{H}_N(\mu)$ is nonsingular, then its rank is $N+1$. For an $m \times n$ matrix $\mathbf{A}$, its range, denoted $\mathrm{range}[\mathbf{A}]$, is defined as the image span of $\mathbf{A}$, i.e., $\mathrm{range}(\mathbf{A}) = \{\mathbf{Ah} \mid \mathbf{h} \in \mathbb{R}^N\}$. We have [153, Theorems 3.3–3.4]:

Theorem 3.2. *Let $\mu \triangleq (\mu_i)_{i=0}^{2N}$. Then*

(a) μ *has a representing probability measure on $\mathbb{R}$ if and only if $\mathbf{H}_N(\mu)$ is nonnegative definite, and $\mathrm{rank}[\mathbf{H}_N(\mu)] = \mathrm{rank}[\mu]$.*

(b) μ *has a representing probability measure on $\mathbb{R}^+$ if and only if $\mathbf{H}_N(\mu)$ and $\mathbf{B}_{N-1}(\mu)$ are nonnegative definite, and the vector $(\mu_{N+1}, \ldots, \mu_{2N})$ is in $\mathrm{range}[\mathbf{B}_{N-1}(\mu)]$.*

(c) μ *has a representing probability measure on $[a,b]$ if and only if both $\mathbf{H}_N(\mu)$ and $(a+b)\mathbf{B}_{N-1}(\mu) - ab\mathbf{H}_{N-1}(\mu) - \mathbf{C}_{N-1}(\mu)$ are nonnegative definite.*

Theorem 3.3. *Let $\mu \triangleq (\mu_i)_{i=0}^{2N+1}$. Then*

(a) μ *has a representing probability measure on $\mathbb{R}$ if and only if $\mathbf{H}_N(\mu)$ is nonnegative definite, and $\mathbf{h}_{N+1} \in \mathrm{range}[\mathbf{H}_N(\mu)]$.*

(b) μ *has a representing probability measure on $\mathbb{R}^+$ if and only if $\mathbf{H}_N(\mu)$ and $\mathbf{B}_N(\mu)$ are nonnegative definite, and the vector $(\mu_{N+1}, \ldots, \mu_{2N+1})$ is in $\mathrm{range}[\mathbf{H}_N(\mu)]$.*

(c) μ *has a representing probability measure on $[a,b]$ if and only if $b\mathbf{H}_N(\mu) - \mathbf{B}_N(\mu)$ and $\mathbf{B}_N(\mu) - a\mathbf{H}_N(\mu)$ are nonnegative definite.*

3.5.1.2 Probability measures determined by moments

While the theorems above provide sufficient conditions to check whether a sequence of moments has a representing probability measure $\mathbb{P}$, additional conditions should be imposed to verify the uniqueness of $\mathbb{P}$, i.e., determinacy [153, Section 3.4]. In fact, if $\mathbb{P}$ is represented by its moment sequence, another probability measure $\mathbb{Q}$ may exist with the same moments as $\mathbb{P}$. If this is the case, the moment problem is called *indeterminate*.

Hausdorff moment problem is always determined. For Hamburger and Stieltjes moment problems, a number of checkable sufficient conditions for determinacy are available [158], for example, Carleman conditions:

(i) for the Hamburger moment problem,

$$\sum_{n=1}^{\infty} \mu_{2n}^{-1/2n} = \infty; \tag{3.58}$$

(ii) and for the Stieltjes moment problem,

$$\sum_{n=1}^{\infty} \mu_{n}^{-1/2n} = \infty. \tag{3.59}$$

Other sufficient conditions involve the growth rate of μ_n as $n \to \infty$:

(i) for the Hamburger moment problem,

$$\frac{\mu_{2(n+1)}}{\mu_{2n}} = O(n^2), \tag{3.60}$$

(ii) and for the Stieltjes moment problem,

$$\frac{\mu_{n+1}}{\mu_n} = O(n^2). \tag{3.61}$$

Converse criteria, in the form of sufficient conditions for indeterminacy, are listed in [158, Theorems 5–8].

Remark 3.6. A moment sequence may be determined for the Stieltjes problem, yet indeterminate for the Hamburger problem [4, p. 240], [158].

Example 3.12. From the definition (3.27) of MGF $M_X(t)$ it follows that, if $M_X(t) < \infty$ for t in some open interval about 0, then inverting $M_X(t)$ one can determine the pdf of X from its moments. A convenient way to do this consists of recognizing that $M_X(t)$ is related to the bilateral Laplace transform of the pdf $f_X(x)$,

$$\phi_X(s) \triangleq \int_{-\infty}^{\infty} e^{-sx} f_X(x)\, dx, \tag{3.62}$$

by the relation

$$M_X(s) = \phi_X(-s). \tag{3.63}$$

Thus, using the inverse Laplace transform, one can recover $f_X(x)$ from $M_X(s)$ as

$$f_X(x) = \frac{1}{2\pi j} \int_{-j\infty}^{j\infty} M_X(s) e^{-sx} \, ds. \tag{3.64}$$

Using this observation, we can find the pdf having moment sequence $\mu_n = n!$, $n = 0, 1, \ldots$, with $\mathcal{X} = \mathbb{R}^+$. This is determinate. In fact, using (3.61), we have

$$\frac{\mu_{n+1}}{\mu_n} = n + 1. \tag{3.65}$$

Its MGF (3.28) can be computed as

$$M(t) \triangleq \sum_{k-0}^{\infty} \frac{t^k}{k!} \mu_k = \frac{1}{1-t}, \qquad |t| < 1. \tag{3.66}$$

The equality

$$\int_0^\infty e^{(t-1)x} \, dx = \frac{1}{1-t} \tag{3.67}$$

shows that this sequence of moments determines the pdf $f_X(x) = e^{-x}$, $x \geqslant 0$. $\diamond$

Example 3.13. The lognormal pdf

$$f(x) = \frac{1}{\sqrt{2\pi}x} \exp[-(\log x)^2/2], \qquad x \in \mathbb{R}_+ \tag{3.68}$$

has moments $\mu_n = \exp(n^2/2)$ which do not satisfy condition (3.61), as

$$\frac{\mu_{n+1}}{\mu_n} = e^{n+1/2}. \tag{3.69}$$

It can be actually proved that an RV with lognormal density is not determined by its moments. In fact, different pdfs can be exhibited having the same moments [115] (see also [153, p. 52] and [158, §4]). $\diamond$

Remark 3.7. In the following a number of examples will show how, when the moment sequence is truncated, one can derive a family of pdfs generating that sequence.

3.6   Moments in a parallelepiped

In this section we follow [149, Chapter VI], to which the reader is referred for proofs and additional details. Consider two real vectors $\boldsymbol{\mu}'$ and $\boldsymbol{\mu}''$, with components $\mu'_1, \ldots, \mu'_N$ and $\mu''_1, \ldots, \mu''_N$, respectively. A partial ordering in $\mathbb{R}^N$, denoted $\trianglelefteq$, is defined as follows: $\boldsymbol{\mu}' \trianglelefteq \boldsymbol{\mu}''$ if and

$$(-1)^{N-K} \mu'_K \leqslant (-1)^{N-K} \mu''_K, \qquad K = 1, \ldots, N. \tag{3.70}$$

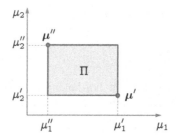

Figure 3.1 Locus Π of vectors $\boldsymbol{\mu}$ such that $\boldsymbol{\mu}' \trianglelefteq \boldsymbol{\mu} \trianglelefteq \boldsymbol{\mu}''$.

If in addition $\boldsymbol{\mu}' \neq \boldsymbol{\mu}''$, we write $\boldsymbol{\mu}' \lhd \boldsymbol{\mu}''$. We also define the parallelepiped Π as the set of vectors $\boldsymbol{\mu}$ which satisfies

$$\boldsymbol{\mu}' \trianglelefteq \boldsymbol{\mu} \trianglelefteq \boldsymbol{\mu}'' \tag{3.71}$$

and hence includes all $\boldsymbol{\mu}$ such that

$$(-1)^{N-K}\mu'_K \leqslant (-1)^{(N-K)}\mu_K \leqslant (-1)^{N-K}\mu''_K, \qquad K = 1, \ldots, N. \tag{3.72}$$

Fig. 3.1 shows $\boldsymbol{\mu}'$ and $\boldsymbol{\mu}''$ as the endpoints of the "oblique diagonal" of Π for $N = 2$. A key point in this theory of "moments in a parallelepiped" is that for $\mathcal{X} = [a, b]$, with $a > 0$, to verify that the entire parallelepiped is made of moment sequences, it is sufficient to verify that those endpoints are moment sequences: in other words, if $\boldsymbol{\mu}'$ and $\boldsymbol{\mu}''$ are moment sequences of an RV with support set $[a, b]$ and $a > 0$, then any $\boldsymbol{\mu}$ such that $\boldsymbol{\mu}' \trianglelefteq \boldsymbol{\mu} \trianglelefteq \boldsymbol{\mu}''$ is also a moment sequence.

Remark 3.8. To understand the significance of the result above, observe that in principle, to verify that all points in the parallelepiped Π are moment vectors, one should verify that the coordinates of all its vertices are moment vectors. However, the above shows that it is sufficient to verify that the two vertices of the oblique diagonal of Π have coordinates that are moment sequences.

Remark 3.9. The problem of finding upper and lower moment bounds on the probability $\mathbb{P}[Y \geqslant X]$, where X and Y are independent RVs with support set $\mathbb{R}^+$ and means and variances known within intervals, is solved in [8, Chapter 14].

3.7 Geometric bounds

A very general approach to moment bounds (see [70,71,141,142,149,277] and the references therein) is based on a geometric representation of the set of values that may be taken by the expectation $\mathbb{E}_\mathbb{P}[\varphi(X)]$ as $\mathbb{P}$ runs through a prescribed set $\mathcal{P}$ of probability measures, and makes it possible to derive sharp upper and lower bounds on the values of that expectation. The advantage of this elegant method is that, at least for a small number of moments, it provides bounds in a simple and immediate way. In addition, it endows the problem with a useful conceptual framework.

Recall that a *convex combination* of points $\mathbf{x}_1, \ldots, \mathbf{x}_m$ in $\mathbb{R}^N$ is a linear combination with nonnegative coefficients summing to 1, i.e., whose form is

$$\sum_{i=1}^{m} \lambda_i \mathbf{x}_i, \qquad \lambda_i \geqslant 0, \ \sum_{i=1}^{m} \lambda_i = 1. \tag{3.73}$$

The *convex hull* of a set $\mathcal{T} \subset \mathbb{R}^N$ is defined as the set of all convex combinations of all finite sets of points in $\mathcal{T}$.

Theorem 3.4 (CARATHÉODORY THEOREM, [258, p. 1]). *Every point in the convex hull of a set $\mathcal{T} \subset \mathbb{R}^N$ can be expressed as a convex combination of at most $N + 1$ points of $\mathcal{T}$.*

The *Nth moment space* $\mathcal{M}_N$ is defined as the set of points $\mathbf{m} \triangleq (m_1, \ldots, m_N)$ in $\mathbb{R}^N$ whose coordinates are the moments $\mu_1, \ldots, \mu_N$ of at least one probability measure, with $\mu_i \triangleq \mathbb{E}[h_i]$, and the functions h_i are linearly independent. Formally,

$$\mathcal{M}_N \triangleq \left\{ \mathbf{m} \in \mathbb{R}^N \ \middle| \ \mathbf{m} = \boldsymbol{\mu} \text{ for some } \mathbb{P} \in \mathcal{P} \right\}. \tag{3.74}$$

The set $\mathcal{M}_N$ has a nonempty interior because of the assumption of linear independence of functions h_i, and is closed, bounded, and convex. More specifically, $\mathcal{M}_N$ is a *cone*, i.e., if $\mathbf{m} \in \mathcal{M}_N$ then $\lambda \mathbf{m} \in \mathcal{M}_N$ for all $\lambda \geqslant 0$. The key property of $\mathcal{M}_N$ here is that it coincides with the convex hull of $\mathcal{C}_N$, where $\mathcal{C}_N$ is the curve traced out in N dimensions by $h_i(x)$, $i = 1, \ldots, N$, as x runs in $[a, b]$. This convex hull will be denoted by $\mathcal{H}_N$ (a heuristic proof of the properties of $\mathcal{M}_N$ can be found in [277]).

Example 3.14. A necessary and sufficient condition for a sequence $\boldsymbol{\mu}$ of numbers to be a moment sequence can be obtained by imposing the condition that the point $\boldsymbol{\mu} \in \mathbb{R}^N$ be in the moment space $\mathcal{M}_N$. Assume probability measures in $[0, 1]$, $N = 3$, and $h_1 = x$, $h_2 = x^2$, and $h_3 = x^2$. It is proved in [70] that the boundary of $\mathcal{H}_3$ consists of (a) the curve $\mathcal{C}_3$ traced out in three dimensions by x, x^2, and x^3 as x varies between 0 and 1, (b) the lines joining $(0, 0, 0)$ with every point of $\mathcal{C}_3$, and (c) the lines joining $(1, 1, 1)$ with every point of $\mathcal{C}_3$. With this characterization, necessary and sufficient conditions for $\mu_1 = \mathbb{E}[X]$, $\mu_2 = \mathbb{E}[X^2]$, and $\mu_3 = \mathbb{E}[X^3]$ to be moments can be derived as

$$0 \leqslant \mu_1 \leqslant 1, \tag{3.75a}$$

$$\mu_1^2 \leqslant \mu_2 \leqslant \mu_1, \tag{3.75b}$$

$$\frac{\mu_2^2}{\mu_1} \leqslant \mu_3 \leqslant \frac{\mu_2(1 - \mu_2) + \mu_1(\mu_2 - \mu_1)}{1 - \mu_1}. \tag{3.75c}$$

$\diamondsuit$

Using Carathéodory Theorem 3.4, we have that every $\mathbf{m} \in \mathcal{M}_N$ admits a representation

$$m_i = \sum_{j=1}^{N+1} w_j h_i(x_j), \qquad i = 1, \dots, N, \; w_j \geqslant 0, \tag{3.76}$$

which involves no more than $N + 1$ points x_j of $\mathcal{C}_N$. Thus each point in $\mathcal{M}_N$ can be represented by a probability measure whose finite support set is $\{x_1, \dots, x_{N+1}\}$, and these values are taken with probabilities $\{w_1, \dots, w_{N+1}\}$ [141, Theorem 1].

Remark 3.10. The above property shows that, for any truncated moment sequence $\boldsymbol{\mu} = (\mu_1, \dots, \mu_N)$, where

$$\mu_i = \mathbb{E}_{\mathbb{P}}[h_i(X)] \tag{3.77}$$

and $(h_1(x), \dots, h_N(x))$ defined over the closed interval $\mathcal{X} = [a, b]$, there is no loss of generality in the assumption that [6, p. 216]

$$\mathbb{E}_{\mathbb{P}}[h_i(X)] = \mathbb{E}_{\mathbb{P}_d}[h_i(X)], \qquad i = 1, \dots, N, \tag{3.78}$$

where $\mathbb{P}_d$ is a probability measure with discrete support set including at most $N + 1$ points. In other words, a representing probability measure for $\boldsymbol{\mu}$ exists whose support set is discrete and finite. This property will be extensively used later in this chapter to find explicit representing probability measures yielding upper and lower bounds on the expectation of a given function.

Remark 3.11. Later in this chapter (Section 3.9) we shall show that, for certain special choices of the functions h_i, the number of points involved in the representation of $\boldsymbol{\mu}$ can be approximately halved.

The properties of moment spaces can be used to find maximum and minimum values of $\mathbb{E}_{\mathbb{P}}[\varphi(X)]$ with respect to $\mathbb{P}$ when some moments $\mathbb{E}_{\mathbb{P}}[h_i(X)]$ are known. In fact, consider the moment space generated by the set of functions $\{h_1, \dots, h_{N-1}, \varphi\}$. The expected value μ_N of $\varphi(X)$ is one of the coordinates of $\mathcal{M}_N$. Thus, the smallest and largest values of this coordinate yield the minimum and maximum value of $\mathbb{E}_{\mathbb{P}}[\varphi(X)]$. For small values of N, an efficient graphical solution to this problem consists of plotting $\mathcal{C}_N$, completing the convex hull, and reading the values of its upper and lower envelope directly from the plot (see [277] for some simple applications). Although analytical solutions to the problem of finding upper and lower bounds on $\mathbb{E}[\varphi(X)]$ can be obtained with this geometric approach, these may be quite complex. In fact, it may happen that $\mathcal{H}_N$ decomposes into several different regions, each one with its own analytical formula [141]. A formal description of this procedure is in [6, p. 219] under the name "method of optimal distance": given the convex hull of the curve $\big(h_1(x), \dots, h_N(x)\big)$, the lower bound to $\mathbb{E}[h_N(X)]$ is given by the smallest Euclidean distance between vectors $(\mu_1, \dots, \mu_{N-1}, 0)$ and $(\mu_1, \dots, \mu_{N-1}, z) \in \mathcal{M}$. Similarly, the upper bound is given by their *largest* Euclidean distance.

3.7.1 Two-dimensional case

For $N = 2$, let $h_1(x)$ and $h_2(x)$ be two continuous functions defined over $[a, b]$. The moment space $\mathcal{M}_2$ is the set of the pairs

$$\left(\int_{[a,b]} h_1(x) f(x) \, dx, \int_{[a,b]} h_2(x) f(x) \, dx \right) \tag{3.79}$$

as $f(x)$ runs over all pdfs whose support set is $[a, b]$. This is the convex hull of the curve $\mathcal{C} \triangleq \{(h_1(x), h_2(x)) \mid x \in [a, b]\}$ in the two-dimensional Euclidean space $\mathbb{R}^2$.

Example 3.15. Given $h_1(x)$ and $h_2(x)$, the expected value of $h_2(X)$ can be identified with the second coordinate of $\mathcal{M}_2$. If the first coordinate is chosen as the known value of $\mathbb{E}[h_1(X)]$, upper and lower bounds on $\mathbb{E}[h_2(X)]$ are obtained by direct observation of the upper and lower envelopes of the curve $\mathcal{C} = (h_1(x), h_2(x))$. To illustrate this procedure, consider the simple case, illustrated in Fig. 3.2, in which $h_1(0) = 0$, and the curve $\mathcal{C}$ is convex $\cup$ for $x \in [0, x_0]$. The straight line bounding the convex hull of $\mathcal{C}$ has equation

$$\begin{aligned}
h_2(x) &= h_2(0) + \frac{h_1(x)}{h_1(x_0)} \big(h_2(x_0) - h_2(0) \big) \\
&= \left(1 - \frac{h_1(x)}{h_1(x_0)} \right) h_2(0) + \frac{h_1(x)}{h_1(x_0)} h_2(x_0),
\end{aligned} \tag{3.80}$$

which yields the upper bound

$$\mathbb{E}[h_2(X)] \leqslant \left(1 - \frac{\mathbb{E}[h_1(X)]}{h_1(x_0)} \right) h_2(0) + \frac{\mathbb{E}[h_1(X)]}{h_1(x_0)} h_2(x_0). \tag{3.81}$$

We may observe that this bound corresponds to choosing for X a discrete probability distribution having two masses, one at 0 and one at x_0, with probabilities $1 - \mathbb{E}[h_1(X)]/h_1(x_0)$ and $\mathbb{E}[h_1(X)]/h_1(x_0)$, respectively.

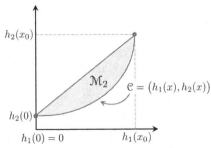

Figure 3.2 Derivation of sharp upper and lower bounds on $\mathbb{E}_{\mathbb{P}}[h_2(x)]$ assuming that $\mathbb{E}_{\mathbb{P}}[h_1(x)]$ is known.

The lower bound is obtained from the observation that the curve $\mathcal{C}$ yields itself the bound. Given $\mu = \mathbb{E}[h_1(X)]$ and the corresponding value $x_\mu = h_1^{-1}(\mu)$, we have

$$\mathbb{E}[h_2(X)] \geqslant h_2(x_\mu) \tag{3.82}$$

corresponding to a single-mass distribution located at x_μ. Thus, for X in $[a, b]$, $h_1(x) = x$, and $h_2(x)$ twice-differentiable with nonnegative second derivative in $[a, b]$, we obtain

$$h_2(\mu_1) \leqslant \mathbb{E}[h_2(X)] \leqslant h_2(a)p + h_2(b)(1 - p), \tag{3.83}$$

where

$$p \triangleq \frac{b - \mu_1}{b - a}. \tag{3.84}$$

A simple special case has $h_1(x) = x$, $h_2(x) = x^2$, and $x \in [0, 1]$. Equation (3.83) yields $\mu_1^2 \leqslant \mu_2 \leqslant \mu_1$. Given $\mu_1 \in [0, 1]$, the two limiting distributions have a single mass at μ_1, yielding $\mathbb{E}[X^2] = \mu_1^2$, and two masses at 0 and 1 with respective probabilities $1 - \mu_1$ and μ_1, yielding $\mathbb{E}[X^2] = \mu_1$. $\Diamond$

Example 3.16. This example deals with error probability in a digital communication system affected by interference [12, Chapter 7], [277]. Transmission is perturbed by white additive Gaussian noise $Z \sim \mathcal{N}(0, 1)$, and by a zero-mean interference denoted by Y, whose pdf is an even function but otherwise is unknown. A demodulation error occurs when $Y + Z$ takes a value outside of interval $[-T, T]$, where T is a known threshold depending on the demodulation scheme adopted. A usual assumption, the so-called "open-eye" condition $\max |Y| \leqslant T$, implies that no error is made when there is no noise. The probability of error has the expression

$$\begin{aligned}
\mathsf{p} &= \frac{1}{2}\big(\mathbb{P}[Y + Z > T] + \mathbb{P}[Y + Z < -T]\big) \\
&= \frac{1}{2}\mathbb{E}_Y\left[Q(T + |Y|) + Q(T - |Y|)\right],
\end{aligned} \tag{3.85}$$

where $Q(\cdot)$ denotes the Gaussian tail function (2.17). Assume that $\mu \triangleq \mathbb{E}\big[|Y|\big]$ is known. Choosing

$$h_1(x) = |x|, \qquad h_2(x) = \frac{1}{2}\big[Q(T + |x|) + Q(T - |x|)\big], \tag{3.86}$$

we can apply the approach taken in Example 3.15 to derive sharp upper and lower bounds on p. If $|x| \leqslant T$, the function $h_2(x)$ is convex $\cup$, so we have the bounds

$$h_2(\mu_1) \leqslant \mathsf{p} \leqslant \left(1 - \frac{\mu_1}{T}\right)h_2(0) + \frac{\mu}{T}h_2(T). \tag{3.87}$$

Consider next the case of interference having known variance σ^2. We have

$$h_1(x) = x^2, \qquad h_2(x) = \frac{1}{2}\big[Q(T + |x|) + Q(T - |x|)\big]/2. \tag{3.88}$$

The curve $\mathcal{C} = \big(h_1(x), h_2(x)\big)$ is convex $\cup$ for low-enough values of $|x|$, say $|x| < x_0$. In the range $(0, x_0^2)$ for the variance, we obtain the bound

$$h_2(\mu_2) \leqslant \mathsf{p} \leqslant \left(1 - \frac{\mu_2^2}{x_0^2}\right) h_2(0) + \frac{\mu_2^2}{x_0^2} h_2(x_0). \tag{3.89}$$

More general bounds, valid for a wider range of the variance, are tabulated in [170].

3.7.2 Application to spectrum sensing

Moment-bound theory can be applied to spectrum sensing, which is one of the major functions of interweaved cognitive radio [24,27]. Spectrum sensing detects and classifies "spectrum holes," i.e., regions of the frequency spectrum that are not occupied by primary users, and hence can be exploited opportunistically by secondary users. Upon reception of discrete signal $\mathbf{y}$, the spectrum sensor decides between the two hypotheses

$$
\begin{aligned}
&H_1 : \mathbf{y} \text{ contains the primary signal,}\\
&H_0 : \text{no primary signal is included in } \mathbf{y}.
\end{aligned}
\tag{3.90}
$$

By comparing the statistics Y, a suitable function of $\mathbf{y}$, against a threshold θ, the presence of the primary signal is detected when $Y > \theta$. The choice of Y depends on how much information about primary signal, interference, and noise is available to the sensor, and on the tolerable complexity of the calculations entailed in the decision process. In terms of performance, the two probabilities of interest here are the "false alarm" and "missed detection" probabilities, defined as

$$
\begin{aligned}
\mathsf{p}_{\mathrm{FA}} &\triangleq \mathbb{P}[Y > \theta \mid H_0],\\
\mathsf{p}_{\mathrm{MD}} &\triangleq \mathbb{P}[Y < \theta \mid H_1].
\end{aligned}
\tag{3.91}
$$

False alarms generate missed transmission opportunities, while missed detections cause transmission collisions.

The situation here may interpreted as one of epistemic uncertainty, as the decision is affected by an interference that is not modeled exactly. We assume in particular that only the amplitude range and the maximum power of interference are known. Moment-bound theory allows us to characterize probability distributions of the interference under given constraints, and to derive the maximum and minimum values that can be taken by p_{FA} and p_{D}.

3.7.2.1 Coherent sensing

Assume first that the primary signal is deterministic and known (for example, consisting of a known preamble), so that the signal observed at discrete time n has the form

$$y_n = \varepsilon x_n + i_n + z_n \tag{3.92}$$

where z_n are the complex noise samples, i_n the samples of the interfering signal, and ε takes value 1 if the primary signal x_n is present, and 0 otherwise. Here $n \in [0, N-1]$, the discrete sensing interval. We assume that the noise samples are independent, circularly symmetric, zero-mean complex Gaussian random variables with variance $\mathbb{E}[|z_n|^2] = \sigma_z^2$. In order to assess the presence of the primary-user signal, a sensor compares against threshold θ the maximum-likelihood statistics

$$
\begin{aligned}
Y &\triangleq \frac{1}{N} \sum_{n=0}^{N-1} \Re[y_n x_n^\star] \\
&= \frac{1}{N} \sum_{n=0}^{N-1} \Re[\varepsilon |x_n|^2 + i_n x_n^\star + z_n x_n^\star].
\end{aligned} \tag{3.93}
$$

The false-alarm and missed-detection probabilities take the form

$$
\begin{aligned}
\mathsf{p}_{\mathrm{FA}} &= \mathbb{P}\big[Y > \theta \mid \varepsilon = 0\big] \\
&= \mathbb{E}_{\mathrm{I}}\left[Q\left(\sqrt{N} \frac{\theta - I}{\sqrt{P\sigma_z^2}} \right) \right],
\end{aligned} \tag{3.94}
$$

$$
\begin{aligned}
\mathsf{p}_{\mathrm{MD}} &= \mathbb{P}\Big(Y < \theta \mid \varepsilon = 1\Big) \\
&= \mathbb{E}_{\mathrm{I}}\left[Q\left(\sqrt{N} \frac{P - \theta + I}{\sqrt{P\sigma_z^2}} \right) \right],
\end{aligned} \tag{3.95}
$$

where $\mathbb{E}_{\mathrm{I}}$ denotes expectation with respect to the probability distribution of the interference term

$$I \triangleq \frac{1}{N} \sum_{n=0}^{N-1} \Re[i_n x_n^*] \tag{3.96}$$

and P is the average signal power

$$P \triangleq \frac{1}{N} \sum_{n=0}^{N-1} |x_n|^2. \tag{3.97}$$

In the simple case where $N = 1$, $P = 1$, $I \in [-1, 1]$, and I has an even probability density function (which implies in particular that its mean value is zero), Fig. 3.3

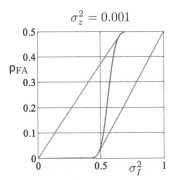

Figure 3.3 Coherent sensing and its probability of false alarm p_{FA}. Variation of p_{FA} uncertainty with σ_I^2 for $\theta = 0.75$ and noise power $\sigma_z^2 = 0.001$.

shows the convex hull of p_{FA} with $\theta = 0.75$ and noise power $\sigma_z^2 = 0.001$. The convex hull contains all possible values of p_{FA} for a given value of $\sigma_I^2 = \mathbb{E}[I^2]$.

Similar results can be derived for p_{MD} (see [27, Chapter 4] for further details).

3.7.2.2 Energy sensing

If the structure of the primary signal is unknown, its presence in (3.92) may be simply detected by measuring the received energy and comparing its value against a suitable threshold which depends on the noise floor. The decision metric Y is built again out of a sequence of N received signal samples,

$$Y \triangleq \frac{1}{N} \sum_{n=0}^{N-1} |y_n|^2, \tag{3.98}$$

and compared against threshold θ. In this situation, the probability of a false alarm depends on the probability distributions of interference I and noise Z,

$$\mathsf{p}_{\mathrm{FA}} = \mathbb{E}_{\mathrm{I,Z}}\mathbb{P}\left(\frac{1}{N}\sum_{n=0}^{N-1}|i_n + z_n|^2 > \theta\right)$$

$$= \mathbb{E}_{\mathrm{I,Z}}\mathbb{P}\left(\frac{2}{\sigma_z^2}\sum_{n=0}^{N-1}|i_n + z_n|^2 > \frac{2N}{\sigma_z^2}\theta\right).$$

The random variable appearing on the left of the inequality sign in last equation has a *noncentral chi-square* distribution with $2N$ degrees of freedom and noncentrality parameter

$$\lambda_0(I) \triangleq \frac{2}{\sigma_z^2} \sum_{n=0}^{N-1} |i_n|^2. \tag{3.99}$$

Therefore,

$$p_{FA} = \mathbb{E}_I Q_N \left(\sqrt{\lambda_0(I)}, \sqrt{2N\theta/\sigma_z^2} \right), \tag{3.100}$$

where Q_N denotes the generalized Marcum Q-function [238, §4.2.2]

$$Q_N(s, t) \triangleq \frac{1}{s^{N-1}} \int_t^\infty x^N \exp\left(-\frac{x^2 + s^2}{2} \right) I_{N-1}(sx)\,dx, \tag{3.101}$$

with I_ν the νth-order modified Bessel function of the first kind.

The conditional probability of missed detection can be derived in a similar way, yielding

$$p_{MD} = 1 - \mathbb{E}_I Q_N \left(\sqrt{\lambda_1(I)}, \sqrt{2N\theta/\sigma_z^2} \right), \tag{3.102}$$

where now

$$\lambda_1(I) \triangleq \frac{2}{\sigma_z^2} \sum_{n=0}^{N-1} |x_n + i_n|^2. \tag{3.103}$$

Sharp upper and lower bounds on p_{FA} and p_{MD} can be obtained under the assumption that $\mathbb{E}[i_n] = 0$, which allows us to obtain the moments

$$\mathbb{E}[\lambda_0(I)] = 2N \frac{\sigma_I^2}{\sigma_z^2} \quad \text{and} \quad \mathbb{E}[\lambda_1(I)] = 2N \frac{P + \sigma_I^2}{\sigma_z^2}, \tag{3.104}$$

where P denotes the average primary-signal power. The moment space relevant to our problem is obtained as the convex hulls of the curves

$$\left(x, Q_N \left(\sqrt{2Nx/\sigma_z^2}, \sqrt{2N\theta/\sigma_z^2} \right) \right) \quad \text{(for } p_{FA} \text{)}$$

and

$$\left(x, 1 - Q_N \left(\sqrt{2N(P+x)/\sigma_z^2}, \sqrt{2N\theta/\sigma_z^2} \right) \right) \quad \text{(for } p_{MD} \text{).}$$

See [24], [27, Chapter 4] for further details and extensions.

Remark 3.12. We have examined sensing under two extreme assumptions about the primary signal: this is either perfectly known (coherent sensing) or totally unknown (energy sensing). The intermediate situation of an incompletely known primary signal may be dealt with by using a detector which has the coherent sensor and the energy sensor as special cases: this is the *linear–quadratic* (LQ) detector studied in [30]. Its performance approaches that of the linear detector when the uncertainty on the primary signal is small, and that of the energy detector in the opposite case.

3.8 Quadrature-rule approximations and bounds

In this section we describe how the quantity $\mathbb{E}_{\mathbb{P}}[\varphi]$, with φ a known function, can be approximated, and eventually upper- and lower-bounded, when only the support set $\mathfrak{X}$ and some moments of $\mathbb{P}$ are available. We start assuming a compact $\mathfrak{X} = [a, b]$, while generalizations will be examined later. The approach here is based on quadrature rules. The idea is to apply numerical integration to the calculation of

$$\mathbb{E}_{\mathbb{P}}[\varphi(X)] = \int_{\mathfrak{X}} \varphi(x) f(x) \, dx, \tag{3.105}$$

where f is the pdf associated with probability measure $\mathbb{P}$.

3.8.1 The algebraic moment problem

To describe the quadrature-rule approach, we first examine the *algebraic moment problem*, which consists of finding a discrete probability measure $\mathbb{P}_d$ with support set $\mathfrak{X} = \{x_1, \dots, x_n\}$, values $\mathbb{P}[x_i] = w_i$, and representing a given power-moment sequence $\boldsymbol{\mu} = (\mu_0, \mu_1, \dots, \mu_N)$. This is tantamount to saying that the following system of "moment-matching" equations holds:

$$\sum_{i=1}^{n} w_i x_i^k = \mu_k, \qquad k = 0, \dots, N. \tag{3.106}$$

Once a $\mathbb{P}_d$ is found which satisfies (3.106), one may think of approximating the integral in (3.105) using a quadrature rule with "weights" w_i and "abscissas" x_i, that is,

$$\mathbb{E}_{\mathbb{P}}[\varphi] \approx \mathbb{E}_{\mathbb{P}_d}[\varphi]$$
$$= \sum_{i=1}^{n} w_i \varphi(x_i). \tag{3.107}$$

Since we want a tight approximation, it makes sense to choose quadrature rules yielding a remainder

$$R_n[\varphi] \triangleq \mathbb{E}_{\mathbb{P}}[\varphi] - \mathbb{E}_{\mathbb{P}_d}[\varphi] \tag{3.108}$$

as small as possible, in some sense to be defined. In addition, if the sign of $R_n[\varphi]$ can be determined, approximation (3.107) yields an upper or a lower bound.

The theory of quadrature rules defines the *degree of exactness* of (3.107) as the integer δ such that [93, p. 20]

$$R_n[\varphi] = 0 \quad \text{for } \varphi \text{ any polynomial of degree} \leqslant \delta. \tag{3.109}$$

It is easy to see that, if we use a quadrature rule with $\delta \geqslant N$, condition (3.106) is satisfied, and hence $\mathbb{P}_d$ satisfies the moment constraints, i.e., represents $\boldsymbol{\mu}$. As we shall see, quadrature rules with N abscissas and weights can be found having $\delta = 2N - 1$, which is the maximum achievable degree of exactness.

These "optimal" quadrature rules, having N abscissas and weights and degree of exactness $2N - 1$, are called *Gauss Quadrature Rules* (GQRs) with respect to $\mathbb{P}$. We shall see that their abscissas are the zeros of suitable "orthogonal" polynomials of degree N which can be constructed explicitly. A property of practical importance in our context is that, in addition to yielding a satisfactory approximation to $\mathbb{E}_{\mathbb{P}}[\varphi]$, GQRs can be efficiently computed from the knowledge of power moments alone.

Theorem 3.5 ([99]). *GQRs' abscissas are the zeros of polynomials $P_k(x)$ that are orthogonal with respect to the pdf f associated with the probability measure $\mathbb{P}$, i.e., satisfy the orthogonality condition*

$$(P_k, P_\ell)_{\mathbb{P}} = \delta_{k,\ell}, \tag{3.110}$$

where we have defined the scalar product of two real functions $h(\cdot)$, $k(\cdot)$ as

$$(h, k)_{\mathbb{P}} \triangleq \mathbb{E}_{\mathbb{P}}[h(X)k(X)] = \int_{\mathcal{X}} h(x)k(x)f(x)\,dx, \tag{3.111}$$

and the "Kronecker delta" function $\delta_{k,\ell}$ takes value 1 if $k = \ell$, and 0 otherwise. If the support of f is a continuous set (a condition which can be relaxed), polynomials satisfying (3.110) exist, and are uniquely determined by the property that, for every k, $P_k(x)$ has degree k, and the coefficient of x^k is positive.

Remark 3.13. Here we are interested only in the zeros of these polynomials, and hence they can be rescaled by any nonzero quantity. We see that condition (3.110) requires the polynomials be orthonormal. Another choice could be of polynomials that are monic, i.e., whose leading coefficient is equal to 1.

Key properties of orthogonal polynomials are summarized by the following theorem [93, p. 7], [99, pp. 221–222], [271, Chapter 2]:

Theorem 3.6.   (i)   *For any n, all roots of $P_n(x) = k_n \prod_{i=1}^{n}(x - x_i)$, $k_n > 0$, are real, simple, and located in the interior of $\mathcal{X}$. Moreover, the $n + 1$ roots of $P_{n+1}(x)$ interleave those of $P_n(x)$, that is, there is exactly one root of $P_n(x)$ between any two adjacent roots of $P_{n+1}(x)$.*

(ii)   *If φ has in $[a, b]$ continuous derivative of order $2N$, denoted $\varphi^{(2N)}$, then the GQR yields*

$$\int_{\mathcal{X}} \varphi(x)f(x)\,dx = \sum_{j=1}^{N} w_j \varphi(x_j) + \frac{\varphi^{(2N)}(\xi)}{(2N)!k_N^2}, \qquad a < \xi < b, \tag{3.112}$$

where, indicating the first derivative with a prime,

$$w_j = -\frac{k_{N+1}}{k_N} \frac{1}{P_{N+1}(x_j)P_N'(x_j)}, \tag{3.113}$$

and hence is exact for all polynomials of degree $\leqslant 2N - 1$.

(iii) *Any set of orthogonal polynomials satisfies a three-term recurrence of the form*

$$P_j(x) = (a_j x + b_j) P_{j-1}(x) - c_j P_{j-2}(x), \qquad j = 1, 2, \ldots, \qquad (3.114)$$

with $P_{-1}(x) \equiv 0$, $P_0(x) \equiv 1$, *and* $a_i > 0$, $c_i > 0$.

Before proceeding further, we describe some methods for the calculation of orthogonal polynomials.

3.8.2 Čebyšev form of orthogonal polynomials

A closed form of orthogonal polynomials can be obtained as follows [56, §12.6], [120, pp. 13–14], [72, p. 7].

Define the determinant of the Hankel moment matrix

$$\Delta_n \triangleq \det \begin{bmatrix} \mu_0 & \mu_1 & \cdots & \mu_n \\ \mu_1 & \mu_2 & \cdots & \mu_{n+1} \\ \vdots & \vdots & \ddots & \vdots \\ \mu_{n-1} & \mu_n & \cdots & \mu_{2n-1} \\ \mu_n & \mu_{n+1} & \cdots & \mu_{2n} \end{bmatrix}. \qquad (3.115)$$

To prove that Δ_n is positive, consider the polynomial $u(x) \triangleq u_0 + u_1 x + \cdots + u_n x^n$, and the quadratic form in $n + 1$ variables

$$\mathbb{E}[u^2(X)] = \sum_{i,k=0}^{n} \mu_{i+k} u_i u_k. \qquad (3.116)$$

If the CDF $F_X(x)$ has at least $n + 1$ points of increase (which is trivially satisfied if $F_X(x)$ is continuous), at least one of these must be different from all the n zeros of $u(x)$, so the expectation is positive unless $u(x)$ is identically zero. It follows that the quadratic form (3.116) must be positive definite, and hence $\Delta_n > 0$ [56, §11.10]. If $F_X(x)$ has only n_0 points of increase, then the determinants are still positive for $n < n_0$, but $\Delta_n = 0$ for $n \geqslant n_0$ [56, p. 132].

The polynomials

$$P_n(x) \triangleq \frac{1}{\Delta_{n-1}} \det \begin{bmatrix} \mu_0 & \mu_1 & \cdots & \mu_n \\ \mu_1 & \mu_2 & \cdots & \mu_{n+1} \\ \vdots & \vdots & \ddots & \vdots \\ \mu_{n-1} & \mu_n & \cdots & \mu_{2n-1} \\ 1 & x & \cdots & x^n \end{bmatrix} \qquad (3.117)$$

have degree n, as we can prove by expanding the determinant along its last row. Moreover, they are orthogonal. In fact, consider $(P_n, P_i)_{\mathbb{P}}$, $i \leqslant n$. Since $P_i(x)$ is a linear combination of powers of x with exponent $\leqslant i$, we have $(P_n, P_i)_{\mathbb{P}} = 0$ if $(P_n, x^i)_{\mathbb{P}} = 0$

with $i < n$. Now, $(P_n, x^i)_{\mathbb{P}}$ is the determinant of the matrix in the definition of P_n whose last row is replaced by $(\mu_i, \mu_{i+1}, \ldots, \mu_{i+n})$. But this matrix has two equal rows, and hence its determinant is zero. Finally, $(P_n, P_n)_{\mathbb{P}} = (P_n, x^n)_{\mathbb{P}} = \Delta_n / \Delta_{n-1}$, which is nonzero because $\Delta_n \neq 0$.

Remark 3.14. This procedure generates a uniquely determined sequence of orthogonal polynomials corresponding to any distribution with continuous support set. Notice that orthogonal polynomials obtained this way may be neither monic nor orthonormal.

Example 3.17. With $\mu_0 = 1$ and zero odd-order moments, we obtain

$$
\begin{aligned}
&P_0(x) = 1, \\
&P_1(x) = x, \\
&P_2(x) = \mu_2 x^2 - \mu_2^2, \\
&P_3(x) = (\mu_2 \mu_4 - \mu_2^3) x^3 - (\mu_4^2 - \mu_2^2 \mu_4) x.
\end{aligned}
\tag{3.118}
$$

These polynomial are orthogonal under the condition that μ_2 and μ_4 are consistent moments. For example, for a uniform distribution on $\mathcal{X} = [-1, 1]$, we have $\mu_2 = 1/3$ and $\mu_4 = 1/5$, which yields $P_2(x) \propto x^2 - 1/3$ and $P_3(x) \propto x^3 - (3/5)x$, the orthogonal "Legendre" polynomials.   ◇

To find the roots of $P_n(x)$, one may proceed as follows [120]: transform the matrix in (3.117) by subtracting column $i - 1$ multiplied by x from column i, $i = 2, \ldots, n$. We obtain the matrix

$$
\mathbf{L} = \begin{bmatrix}
\mu_0 & \mu_1 - \mu_0 x & \cdots & \mu_n - \mu_{n-1} x \\
\mu_1 & \mu_2 - \mu_1 x & \cdots & \mu_{n+1} - \mu_n x \\
\vdots & \vdots & \ddots & \vdots \\
\mu_{n-1} & \mu_n - \mu_{n-1} x & \cdots & \mu_{2n-1} - \mu_{2n-2} x \\
1 & 0 & \cdots & 0
\end{bmatrix}
\tag{3.119}
$$

whose determinant is $\Delta_{n-1} P_n(x)$. We can write

$$
\det \mathbf{L} = \det(\mathbf{L}_1 - x\mathbf{L}_0),
\tag{3.120}
$$

where

$$
\mathbf{L}_0 \triangleq \begin{bmatrix}
\mu_0 & \mu_1 & \cdots & \mu_{n-1} \\
\mu_1 & \mu_2 & \cdots & \mu_n \\
\vdots & \vdots & \ddots & \vdots \\
\mu_{n-1} & \mu_n & \cdots & \mu_{2n-2}
\end{bmatrix}, \quad
\mathbf{L}_1 \triangleq \begin{bmatrix}
\mu_1 & \mu_2 & \cdots & \mu_n \\
\mu_2 & \mu_3 & \cdots & \mu_{n+1} \\
\vdots & \vdots & \ddots & \vdots \\
\mu_n & \mu_{n+1} & \cdots & \mu_{2n-1}
\end{bmatrix}
\tag{3.121}
$$

are symmetric matrices. The roots of $P_n(x)$ are obtained by solving for x the equation $\det(\mathbf{L}_1 - x\mathbf{L}_0) = 0$. This is a symmetric generalized eigenvalue problem [154, §12.4], whose solution yields the roots $x_1, \ldots, x_n$ as the eigenvalues of the symmetric matrix $\mathbf{L}_0^{-1/2}\mathbf{L}_1\mathbf{L}_0^{-1/2}$. To prove this, decompose $\mathbf{L}_0$ as $\mathbf{L}_0 = \mathbf{L}_0^{1/2}\mathbf{L}_0^{1/2}$, and observe that the eigenvalue equation $\mathbf{L}_1\mathbf{u} = x\mathbf{L}_0\mathbf{u}$ is equivalent to $\mathbf{L}_0^{-1/2}\mathbf{L}_1\mathbf{L}_0^{-1/2}\mathbf{L}_0^{1/2}\mathbf{u} = x\mathbf{L}_0^{1/2}\mathbf{u}$. Defining $\mathbf{w} \triangleq \mathbf{L}_0^{1/2}\mathbf{u}$, we see that the roots of $P_n(x)$ are the solutions of the "classical" eigenvalue equation

$$\mathbf{L}_0^{-1/2}\mathbf{L}_1\mathbf{L}_0^{-1/2}\mathbf{w} = x\mathbf{w}. \tag{3.122}$$

Since $\mathbf{L}_0^{-1/2}\mathbf{L}_1\mathbf{L}_0^{-1/2}$ is symmetric, this equation has n real solutions.

If the abscissas x_i are obtained through the procedure just described, the explicit evaluation of orthogonal polynomials becomes unnecessary. In fact, the weights w_i can be obtained by solving the "moment-matching" equations (3.106), which can be rewritten in matrix form as

$$\mathbf{X}\mathbf{w} = \boldsymbol{\mu}, \tag{3.123}$$

where $\mathbf{X}$ is the $(n + 1) \times n$ matrix

$$\mathbf{X} \triangleq \begin{bmatrix} x_1^0 & x_2^0 & \cdots & x_n^0 \\ x_1^1 & x_2^1 & \cdots & x_n^1 \\ \vdots & \vdots & \ddots & \vdots \\ x_1^n & x_2^n & \cdots & x_n^n \end{bmatrix}. \tag{3.124}$$

Example 3.18. With zero odd-order moments, and $\mu_2 = 1/3$, $\mu_4 = 1/5$ as in Example 3.17, for $n = 4$ we obtain the three roots of the third-order Legendre polynomial $P_3(x)$, namely $x_1 = -\sqrt{3/5}$, $x_2 = 0$, and $x_3 = \sqrt{3/5}$. The corresponding weights are $5/18$, $4/9$, and $5/18$.   ◇

Example 3.19. Using the standard location–scale transformation of an RV $X \to (X - \mu_1)/\sigma$ (σ the standard deviation of X), we can deal with an RV whose first three moments are $\mu_1 = 0$, $\mu_2 = 1$, and μ_3 (the "skewness"). We have

$$\mathbf{L}_0 = \begin{bmatrix} 1 & 0 \\ 0 & 1 \end{bmatrix}, \qquad \mathbf{L}_1 = \begin{bmatrix} 0 & 1 \\ 1 & \mu_3 \end{bmatrix}, \tag{3.125}$$

which yields

$$\mathbf{L}_0^{-1/2}\mathbf{L}_1\mathbf{L}_0^{-1/2} = \mathbf{L}_1 \tag{3.126}$$

whose eigenvalues are $x_{1,2} = \frac{1}{2}\left(\mu_3 \mp \sqrt{4 - \mu_3^2}\right)$. The solution of the system

$$\begin{aligned} w_1 + w_2 &= 1, \\ w_1 x_1 + w_2 x_2 &= 0 \end{aligned} \tag{3.127}$$

yields

$$w_1 = \frac{x_2}{x_2 - x_1}, \qquad w_2 = -\frac{x_1}{x_2 - x_1}. \qquad (3.128)$$

3.8.3 Recursive generation of orthogonal polynomials

The polynomials $P_n(x)$ can be generated recursively by orthogonalizing powers $1, x, x^2, \ldots$ This method works provided that these powers are linearly independent in the function space where the scalar product of two functions $h(x), k(x)$ is defined by (3.111). Now, observe that, if $p(x)$ is an arbitrary real polynomial, then, with a continuous support set $\mathfrak{X}$,

$$(p, p)_{\mathbb{P}} = \int_{\mathfrak{X}} p^2(x) f(x)\, dx = 0, \qquad (3.129)$$

only if p is identically zero [247, p. 26].

We proceed using recursively the three-term relationship (3.114), rescaled in order to have *monic* polynomials [198, §4.6],

$$P_{j+1}(x) = (x - a_j) P_j(x) - b_j P_{j-1}(x), \qquad j = 0, 1, \ldots, \qquad (3.130)$$

with initial conditions $P_{-1}(x) \equiv 0$ and $P_0(x) \equiv 1$. Taking the scalar product of (3.130) by $P_j(x)$, and using the orthogonality property (3.110), we obtain

$$a_j = \frac{(x P_j, P_j)_{\mathbb{P}}}{(P_j, P_j)_{\mathbb{P}}}, \qquad j = 0, 1, \ldots \qquad (3.131)$$

Similarly, taking the scalar product by $P_{j-1}(x)$ yields

$$b_j = \frac{(x P_j, P_{j-1})_{\mathbb{P}}}{(P_{j-1}, P_{j-1})_{\mathbb{P}}} = \frac{(P_j, x P_{j-1})_{\mathbb{P}}}{(P_{j-1}, P_{j-1})_{\mathbb{P}}}, \qquad j = 1, 2, \ldots \qquad (3.132)$$

Moreover, observe that we can write $x P_{j-1}(x) = P_j(x) + u(x)$, where $u(x)$ is a polynomial with degree $< j$, so that $(u, P_j)_{\mathbb{P}} = 0$. Thus, we rewrite b_j in the form

$$b_j = \frac{(P_j, P_j)_{\mathbb{P}}}{(P_{j-1}, P_{j-1})_{\mathbb{P}}}, \qquad j = 1, 2, \ldots, \qquad (3.133)$$

and the coefficient b_0 may take value zero. The calculations above involve only power moments of $\mathbb{P}$, which shows again how the generation of polynomials orthogonal with respect to $\mathbb{P}$ needs no more than the power moments of $\mathbb{P}$. Polynomials generated this way are monic, while dividing each polynomial P_j by $\sqrt{(P_j, P_j)_{\mathbb{P}}}$ we can make them orthonormal.

This procedure, albeit simple, may not be that useful with large-order polynomials, as it may become rather unstable numerically. Another way of deriving the weights when the (monic) orthogonal polynomials are explicitly available consists of using equality [198, p. 182]

$$w_i = \frac{(P_{n-1}, P_{n-1})_{\mathbb{P}}}{P'_n(x_i) P_{n-1}(x_i)}, \tag{3.134}$$

with a prime indicating derivative.

3.8.4 Numerical generation of orthogonal polynomials

Following [99], [271, §2.5], we write the recurrence relationship (3.114) in the equivalent form

$$x P_{j-1}(x) = \frac{c_j}{a_j} P_{j-2}(x) - \frac{b_j}{a_j} P_{j-1}(x) + \frac{1}{a_j} P_j(x), \qquad j = 1, 2, \dots, \tag{3.135}$$

which leads to the following matrix equation:

$$x\mathbf{p}(x) = \mathbf{T}\mathbf{p}(x) + (1/a_n) P_n(x) \mathbf{e}_n, \tag{3.136}$$

where

$$\mathbf{p}(x) \triangleq \begin{bmatrix} P_0(x) \\ P_1(x) \\ \vdots \\ P_{n-1}(x) \end{bmatrix}. \tag{3.137}$$

The above $\mathbf{T}$ is the tridiagonal matrix

$$\mathbf{T} \triangleq \begin{bmatrix} -b_1/a_1 & 1/a_1 & 0 & & \\ c_2/a_2 & -b_2/a_2 & 1/a_2 & & \\ 0 & \ddots & \ddots & \ddots & \\ & & & & 1/a_{n-1} \\ & & & c_n/a_n & -b_n/a_n \end{bmatrix} \tag{3.138}$$

and

$$\mathbf{e}_n \triangleq \begin{bmatrix} 0 \\ 0 \\ \vdots \\ 0 \\ 1 \end{bmatrix}. \tag{3.139}$$

Thus, $P_n(x_j) = 0$ if and only if $x_j \mathbf{p}(x_j) = \mathbf{T}\mathbf{p}(x_j)$, i.e., x_j is an eigenvalue of $\mathbf{T}$. The matrix $\mathbf{T}$ is symmetric if the polynomials are orthogonal. Otherwise, a diagonal similarity transformation yields a symmetric tridiagonal "Jacobi" matrix $\mathbf{J}_n$ with the same eigenvalues of $\mathbf{T}$,

$$\mathbf{J}_n \triangleq \mathbf{DTD}^{-1} = \begin{bmatrix} \alpha_1 & \beta_1 & 0 & & \\ \beta_1 & \alpha_2 & \beta_2 & & \\ & & \ddots & \ddots & \\ & & \beta_{n-2} & \alpha_{n-1} & \beta_{n-1} \\ & & 0 & \beta_{n-1} & \alpha_n \end{bmatrix}, \tag{3.140}$$

where explicitly

$$\alpha_i \triangleq -\frac{b_i}{a_i}, \qquad \beta_i \triangleq \sqrt{\frac{c_{i+1}}{a_i a_{i+1}}}. \tag{3.141}$$

As for the calculation of the weights w_j, the following equality can be proved [271, §2.5]:

$$w_i \mathbf{p}^\mathsf{T}(x_i)\mathbf{p}(x_i) = 1, \qquad i = 1, \ldots, n, \tag{3.142}$$

where $\mathbf{p}(x_i)$ denotes the eigenvector of $\mathbf{J}_n$ associated with eigenvalue x_i. If the eigenvectors of $\mathbf{T}$ are normalized, that is,

$$\mathbf{J}_n \mathbf{q}_i = x_i \mathbf{q}_i, \qquad i = 1, \ldots, n, \qquad \mathbf{q}_i^\mathsf{T} \mathbf{q}_i = 1, \tag{3.143}$$

then, by writing

$$\mathbf{q}_i^\mathsf{T} = (q_{1,i}, q_{2,i}, \ldots, q_{n,i}), \tag{3.144}$$

it follows that (3.142) yields $q_{1,i}^2 = w_i P_0^2(x_i)$, and hence, with k_0 defined as in Theorem 3.6(i) *supra*,

$$w_i = \frac{q_{1,i}^2}{P_0^2(x_i)} = \frac{q_{1,i}^2}{k_0^2} = q_{1,i}^2 \int_X f(x)\,dx = q_{1,i}^2. \tag{3.145}$$

As a consequence, the GQR is fully determined by the eigenvalues of $\mathbf{T}$ and the first component of its orthonormal eigenvectors. This computation is efficiently performed by using the QR decomposition of $\mathbf{J}_n$ as described, for example, in [99].

To compute the entries of $\mathbf{J}_n$, an efficient procedure is based on the Cholesky decomposition of the Hankel matrix (3.53),

$$\mathbf{H}_n(\boldsymbol{\mu}) = \boldsymbol{\Gamma}^\mathsf{T}\boldsymbol{\Gamma}, \tag{3.146}$$

where $\boldsymbol{\Gamma}$ is a lower triangular matrix with positive diagonal entries [98, p. 88]. Let

$$\boldsymbol{\Gamma} \triangleq \begin{bmatrix} \gamma_{11} & \gamma_{12} & \cdots & \gamma_{1,n+1} \\ & \gamma_{22} & \cdots & \gamma_{2,n+1} \\ & & \ddots & \vdots \\ & & & \gamma_{n+1,n+1} \end{bmatrix}. \tag{3.147}$$

The entries of $\mathbf{J}_n$ are obtained using

$$\alpha_i = \frac{\gamma_{i,i+1}}{\gamma_{i,i}} - \frac{\gamma_{i-1,i}}{\gamma_{i-1,i-1}}, \qquad i = 1, \dots, n,$$

$$\beta_i = \frac{\gamma_{i+1,i+1}}{\gamma_{i,i}}, \qquad i = 1, \dots, n-1, \tag{3.148}$$

with $\gamma_{0,0} = 1$ and $\gamma_{0,1} = 0$.

Remark 3.15. Notice how $\mathbf{H}_n(\boldsymbol{\mu})$ includes also μ_{2n} among its entries. However, the actual value of μ_{2n} is irrelevant, provided that $\mathbf{H}_n(\boldsymbol{\mu})$ is positive definite. In fact, the role of μ_{2n} is just that of normalizing the polynomial $P_n(x)$, and its value affects neither the abscissas nor the weights of the GQR [92, p. 256].

In summary, the computational procedure for the derivation of an n-point GQR runs as follows:

(i)   Given n and the moment vector $\boldsymbol{\mu} = (\mu_0, \mu_1, \dots, \mu_{2n})$, form the $(n+1) \times (n+1)$ moment matrix $\mathbf{H}_n(\boldsymbol{\mu})$ as defined in (3.53).

(ii) Do Cholesky decomposition $\mathbf{H}_n(\boldsymbol{\mu}) = \boldsymbol{\Gamma}^{\mathsf{T}}\boldsymbol{\Gamma}$, with $\boldsymbol{\Gamma}$ in (3.147).

(iii) Using (3.148), compute $\alpha_i, i = 1, \dots, n$, and $\beta_i, i = 1, \dots, n-1$, and use these to generate the $n \times n$ symmetric tridiagonal matrix $\mathbf{J}_n$ in (3.140).

(iv) The eigenvalues of $\mathbf{J}_n$ are the abscissas, and the first components of the normalized eigenvectors of $\mathbf{J}_n$ are the weights.

Remark 3.16. Perusal of (3.112) may suggest that, if a sufficiently large number of moments is available, a GQR can be made as accurate as desired by increasing the number N of its abscissas and weights. In fact, as $N \to \infty$, the RHS of (3.112) converges to the value of its LHS "for almost any conceivable function which one meets in practice" [245, p. 13]. However, when the moments are not known with unlimited accuracy, it may happen that the moment matrix loses its positive definiteness due to roundoff errors, and Cholesky decomposition fails.

Example 3.20. Choose $\mathcal{X} = [-1, 1]$ and the moments $\mu_0 = 1$, $\mu_1 = \mu_3 = \mu_5 = 0$, $\mu_2 = 1/3$, $\mu_4 = 1/5$, and $\mu_6 = 1/7$ as generated by the uniform pdf. We have

$$\boldsymbol{\Gamma} = \begin{bmatrix} 1 & 0 & 1/3 & 0 \\ 0 & 1/\sqrt{3} & 0 & \sqrt{3}/5 \\ 0 & 0 & 2/3\sqrt{5} & 0 \\ 0 & 0 & 0 & 2/5\sqrt{7} \end{bmatrix} \tag{3.149}$$

and

$$\mathbf{J}_3 = \begin{bmatrix} 0 & 1/\sqrt{3} & 0 \\ 1/\sqrt{3} & 0 & 2/\sqrt{15} \\ 0 & 2/\sqrt{15} & 0 \end{bmatrix} \tag{3.150}$$

whose eigenvalues are $-\sqrt{3/5}, 0, \sqrt{3/5}$ and normalized eigenvalues have first components $5/18, 4/9, 5/18$ as in Example 3.18. These abscissas and weights are consistent with the quadrature rule tabulated in [2, Table 25.4]. $\diamond$

3.9 Čebyšev systems and principal representations

Wrapping up the previous analyses and calculations, we are now ready to describe how Gaussian and related quadrature rules may be applied to the solution of moment-bound problems. The bridge linking quadrature rules and moment problems is provided by the theory of *Čebyšev systems* and *principal representations*.

We examine moment problems without restriction to *power* moments, and hence we start considering the moment space $\mathcal{M}_{N+1}$ containing the vectors $\boldsymbol{\mu}$ of moments of the continuous functions $h_0(x), h_1(x), \ldots, h_N(x)$, $x \in \mathcal{X} = [a, b]$, computed with respect to any possible probability measure. We have previously seen that $\mathcal{M}_{N+1}$ can be viewed as the convex hull of the parametric curve generated by functions $h_0(x), h_1(x), \ldots, h_N(x)$, as x runs in $[a, b]$.

We recall that a representation of $\boldsymbol{\mu} = (\mu_0, \mu_1, \ldots, \mu_N)$ is a probability measure $\mathbb{P}_d$ of a discrete RV X taking values in the support set $\{x_1, \ldots, x_p\}$ with probabilities $\{w_1, \ldots, w_p\}$, so that we can write

$$\mu_i = \mathbb{E}_{\mathbb{P}_d}[h_i(X)], \qquad i = 0, 1, \ldots, N, \tag{3.151}$$

or, equivalently,

$$\mu_i = \sum_{j=1}^{p} w_j h_i(x_j), \qquad w_j \geqslant 0, \quad a \leqslant x_j \leqslant b. \tag{3.152}$$

We have also seen how quadrature rules can be used to approximate expectations of a known function of an RV whose moments are known. We are now ready to show how quadrature rules also play a central role in the derivation of upper and lower bounds on those expectations. For this reason we shall refer to the elements x_j of the support set of $\mathbb{P}_d$ as to "abscissas," and to the probabilities w_j as to "weights."

Among the probability measures $\mathbb{P}_d$ representing $\boldsymbol{\mu}$, we are interested in those with the smallest cardinality p of the support set. Carathéodory Theorem 3.4 states that p can be chosen $\leqslant N + 2$. As we shall see, if the functions $h_0(x), h_1(x), \ldots, h_N(x)$ are chosen to form a *Čebyšev-system* (abbreviated as *T-system*), each $\boldsymbol{\mu}$ in the interior of $\mathcal{M}_{N+1}$ admits a representation involving approximately $(N + 2)/2$ terms.

Definition 3.1. The continuous functions $\{h_0, h_1, \ldots, h_N\}$ are said to form a *T-system* over $[a, b]$ if the determinants

$$D(x_0, x_1, \ldots, x_N) \triangleq \det \begin{bmatrix} h_0(x_0) & h_0(x_1) & \ldots & h_0(x_N) \\ h_1(x_0) & h_1(x_1) & \ldots & h_1(x_N) \\ \vdots & \vdots & \ddots & \vdots \\ h_N(x_0) & h_N(x_1) & \cdots & h_N(x_N) \end{bmatrix} \tag{3.153}$$

are positive for every choice of $a \leqslant x_0 < x_1 < \cdots < x_N \leqslant b$.

Remark 3.17. Another characterization of *T*-systems is the property that every real "polynomial" $\sum_{i=1}^{N} \alpha_i h_i(x)$ with $\sum_{i=0}^{N} \alpha_i^2 > 0$ has at most N distinct zeros in $[a, b]$ [136, p. 21], [129, p. 91].

Remark 3.18. In [149], a slightly weaker definition of a T-system is given, as one in which the determinant (3.153) has constant sign for any pairwise distinct $a \leqslant x_0 < x_1 < \cdots < x_N \leqslant b$. Systems satisfying Definition 3.1 are then called T_+-*systems*.

Remark 3.19. The normalization $\mu_0 = 1$ need not be imposed here. Without it, $\mathcal{M}_{N+1}$ turns out to be a closed convex cone in $\mathbb{R}^{N+1}$, i.e., $\boldsymbol{\mu} \in \mathcal{M}_{N+1}$ implies $\lambda\boldsymbol{\mu} \in \mathcal{M}_{N+1}$ for all $\lambda \geqslant 0$ [136, p. 38].

Important T-systems on any interval $[a, b]$ are listed in [149, pp. 37–38] (see also [136, Chapter I, §3] and [129, p. 105]):

(i) $h_i(x) = x^i$ (for these functions, (3.153) is a Vandermonde determinant).
(ii) $h_i(x) = x^{\alpha_i}$ with $0 \leqslant \alpha_0 < \alpha_1 < \cdots < \alpha_N$, with $a > 0$.
(iii) $h_i(x) = e^{\alpha_i x}$ with $0 \leqslant \alpha_0 < \alpha_1 < \cdots < \alpha_N$.
(iv) $h_i(x) = 1, x, x^2, \ldots, x^{N-1}, \varphi(x)$, if the derivative $\varphi^{(N)}(x)$ of $\varphi(x)$ is continuous and positive in $[a, b]$.

T-system (i) is the most widely used in applications, along with its extended version (iv). Other T-systems can be generated by observing that the multiplication of each function $h_i(x)$ in a T-system by a continuous function $v(x) > 0$ yields the new T-system $\{v(x)h_i(x)\}_{i=0}^{N}$. If $\chi(y)$ is a continuous strictly increasing function on $[\alpha, \beta]$ with $\chi(\alpha) = a$ and $\chi(\beta) = b$, then the substitution $x = \chi(y)$ transforms the T-system $\{h_i(x)\}_{i=0}^{N}$ on $[a, b]$ into the T-system $\{h(\chi(y))\}_{i=0}^{N}$ on $[\alpha, \beta]$ [149, p. 33].

Now, consider the discrete representations $\mathbb{P}_d$ of points in the interior of the moment space $\mathcal{M}_{N+1}$ generated by a T-system. We classify them according to their *indices*, defined as follows:

$$\Omega[\mathbb{P}_d] \triangleq \sum_{i=1}^{p} \omega(x_i). \tag{3.154}$$

where

$$\omega(x_i) \triangleq \begin{cases} 1/2 & \text{if } x_i = a \text{ or } x_i = b, \\ 1 & \text{if } a < x_i < b, \end{cases} \tag{3.155}$$

i.e., $\omega(x_i) = 1$ if x_i is internal to $[a, b]$, and $\omega(x_i) = 1/2$ if x_i is at one extreme of the interval. In words, this is the number of abscissas in the representation, with a and b counting as "half abscissas." As we shall see, $\Omega[\mathbb{P}_d]$ is the minimum number of abscissas needed to represent a point in the moment space generated by a T-system.

We call $\mathbb{P}_d$ *canonical* if its index does not exceed $(N + 2)/2$, and *principal* if its index is exactly $(N + 1)/2$. In addition, a canonical or principal representation is called *upper* if it includes the endpoint b, and *lower* if it does not include b. It is shown in [136, pp. 44–46] that any $\boldsymbol{\mu}$ in the interior of $\mathcal{M}_{N+1}$ admits a unique canonical representation involving a preassigned abscissa $x^\star$ in $[a, b]$. Special choices of $x^\star$ yield principal representations, one lower and one upper. Prescribing $x^\star = a$ or $x^\star = b$, we have the following result [136, p. 47]:

Theorem 3.7. *If μ is an interior point of $\mathcal{M}_{N+1}$, there exist exactly two principal representations. If N is odd, which we write $N = 2v - 1$, the principal representations (PRs) of μ can only take one of these two forms: either*

$$a < \underline{x}_1 < \underline{x}_2 < \cdots < \underline{x}_v < b \tag{3.156}$$

or

$$a = \overline{x}_1 < \overline{x}_2 < \cdots < \overline{x}_{v+1} = b. \tag{3.157}$$

These are lower and upper PRs, respectively. If N is even, and $N = 2v$, the two PRs are

$$a = \underline{x}_1 < \underline{x}_2 < \cdots < \underline{x}_{v+1} < b \tag{3.158}$$

(the lower PR) and

$$a < \overline{x}_1 < \overline{x}_2 < \cdots < \overline{x}_{v+1} = b \tag{3.159}$$

(the upper PR).

The situation is summarized in Table 3.1.

Table 3.1 Number and locations of abscissas of principal representations (PRs) with $\mathcal{X} = [a, b]$.

Case	Upper PR	Lower PR
N odd	$(N - 1)/2$ abscissas in (a, b), one at a, one at b	$(N + 1)/2$ abscissas in (a, b)
N even	$N/2$ abscissas in (a, b), one at b	$N/2$ abscissas in (a, b), one at a

To derive a principal representation, observe that there are $N + 1$ abscissas and weights to be determined, and these satisfy the $N + 1$ constraints expressed by the nonlinear equations

$$\mu_i = \sum_{j=0}^{N} w_j h_i(x_j), \qquad i = 0, 1, \ldots, N. \tag{3.160}$$

3.9.0.1 Semiinfinite intervals

Much of the above can be extended to semiinfinite intervals, in particular those involving nonnegative RVs with no specific upper bound. However, the consideration of $\mathcal{X} = [0, \infty)$ requires some additional conditions to hold [75, p. 137], [105, p. 342], [136, Chapter V]. Specifically, it is required that the both $\{h_0, \ldots, h_{N-1}\}$ and

$\{h_0, \ldots, h_{N-1}, h_N\}$ be T-systems which satisfy the following conditions:

$$\lim_{x \to \infty} \frac{h_i(x)}{h_{N-1}(x)} = 0, \qquad i < N-1,$$

$$\lim_{x \to \infty} \frac{h_N(x)}{h_{N-1}(x)} < \infty. \tag{3.161}$$

The principal representations become

Table 3.2 Number and locations of abscissas of principal representations (PRs) with $\mathcal{X} = [0, \infty)$

Case	Upper PR	Lower PR
N odd	$(N-1)/2$ abscissas in $(0, \infty)$, one at 0	$(N+1)/2$ abscissas in $(0, \infty)$
N even	$N/2$ abscissas in $(0, \infty)$	$N/2$ abscissas in $(0, \infty)$, one at 0

Remark 3.20. Comparing Tables 3.1 and 3.2, we can observe that the lower PRs on $[0, \infty)$ are identical to the lower PRs on $[0, b]$. Moreover, as $b \to \infty$, the upper PRs converge weakly to the upper PRs on $[0, \infty)$ [129, p. 103].

Remark 3.21. To determine the PRs of Table 3.2, where for the upper PR there is one less unknown than in Table 3.1, one less moment is used. The condition pertaining to the Nth moment is satisfied by placing an infinitesimal weight at infinity [136, §§V.4–V.5].

Example 3.21. With $\mathcal{X} = [a, b]$, if only the mean value μ_1 is known, which for consistency must take a value in $[a, b]$, we have $\underline{x}_1 = \mu_1$ and $\overline{x}_1 = a, \overline{x}_2 = b$. $\quad\diamond$

Example 3.22. With $N = 2$, $\mathcal{X} = [a, b]$, and known mean value μ_1 and variance σ^2 (which for consistency must satisfy $\sigma^2 < (b - \mu_1)(\mu_1 - a)$—see Example 3.11 *supra*), we obtain

$$\underline{x}_1 = a, \qquad \underline{x}_2 = \mu_1 + \frac{\sigma^2}{\mu_1 - a}, \tag{3.162}$$

and

$$\overline{x}_1 = \mu_1 - \frac{\sigma^2}{b - \mu_1}, \qquad \overline{x}_2 = b. \tag{3.163}$$

$\diamond$

Example 3.23. With $N = 2$, $\mathcal{X} = [0, b]$, and known power moments μ_1 and μ_2, the lower PR is the two-point rule with abscissas and weights

$$\begin{aligned} \underline{x}_1 &= 0, & \underline{w}_1 &= (\mu_2 - \mu_1)^2/\mu_2, \\ \underline{x}_2 &= \mu_2/\mu_1, & \underline{w}_2 &= \mu_1^2/\mu_2. \end{aligned} \tag{3.164}$$

The upper PR is the two-point rule with abscissas and weights

$$\overline{x}_1 = \mu_1 - \frac{\mu_2 - \mu_1^2}{b - \mu_1}, \qquad \overline{w}_1 = \frac{b - \mu_1^2}{\mu_2 - \mu_1^2 + (b - \mu_1)^2},$$
$$\overline{x}_2 = b, \qquad \overline{w}_2 = 1 - \overline{w}_1. \tag{3.165}$$

With $\mathcal{X} = [0, \infty)$, the lower PR is the same as for $\mathcal{X} = [0, b]$. The upper PR, which depends only on μ_1, has $\overline{x}_1 = \mu_1$ and $\overline{w}_1 = 1$. ◇

Example 3.24. With $N = 3$, $\mathcal{X} = [0, b]$, known power moments μ_1, μ_2, μ_3, and the definitions

$$\alpha \triangleq \frac{\mu_3 - \mu_1 \mu_2}{\mu_2 - \mu_1^2},$$
$$\beta \triangleq \frac{\mu_2^2 - \mu_1 \mu_3}{\mu_2 - \mu_1^2}, \tag{3.166}$$
$$r \triangleq \sqrt{\alpha^2 + 4\beta},$$

the lower PR has the following abscissas and weights:

$$\underline{x}_1 = \tfrac{1}{2}(\alpha - r), \quad \underline{w}_1 = \frac{\alpha + r - 2\mu_1}{2r},$$
$$\underline{x}_2 = \tfrac{1}{2}(\alpha + r), \quad \underline{w}_2 = 1 - \underline{w}_1, \tag{3.167}$$

and the upper PR has

$$\overline{x}_1 = 0, \qquad \overline{w}_1 = 1 - \overline{w}_2 - \overline{w}_3,$$
$$\overline{x}_2 = \frac{\mu_2 - \overline{w}_3 b^2}{\mu_1 - \overline{w}_3 b}, \qquad \overline{w}_2 = \frac{(\mu_1 - \overline{w}_3 b)^2}{\mu_2 - \overline{w}_3 b},$$
$$\overline{x}_3 = b, \qquad \overline{w}_3 = \frac{\mu_1 \mu_3 - \mu_2^2}{\mu_1 b^3 - 2\mu_2 b^2 + \mu_3 b}. \tag{3.168}$$

With $\mathcal{X} = [0, \infty)$, the lower PR is the same as with $\mathcal{X} = [0, b]$. The upper PR is the same as the lower PR for the 2-moment case [129, p. 104]. ◇

3.9.1 Principal and canonical representations as quadrature rules

The actual computation of principal representations is connected to the construction of weights and abscissas of quadrature rules (QRs) with maximum degree of exactness. For simplicity, we restrict our consideration to functions $h_i(x) = x^i$, and hence to power moments. Recall that a QR with abscissas x_i and weights w_i, $i = 1, \ldots, p$, has degree of exactness δ if it yields an exact result for integrands being powers x^j, $j = 0, 1, \ldots, \delta$, but not necessarily for $x^{\delta+1}$. Thus, the four principal representations defined above correspond to Gauss quadrature rules with a minimal number of points, approximately $(N + 1)/2$. The addition of one of the endpoints of $[a, b]$ to the representation yields a canonical representation of the same moment vector which can be thought of as a GQR with one or two points fixed a priori, i.e., "Gauss–Radau" or "Gauss–Lobatto" quadrature rules. In fact, observe that the computation of a canonical representation is tantamount to the solution of the system of nonlinear equations

$$\sum_{i=1}^{n} w_i x_i^k = \mu_k, \qquad k = 0, 1, \dots, N, \tag{3.169}$$

for x_i and w_i, $i = 1, \dots, n$, under the constraint that the index of the representation should not exceed $(N + 2)/2$, which involves the choice of one of the endpoints as an abscissa. Thus, (3.169) can be reformulated as

$$\sum_{i=1}^{p} w_i x_i^k + \sum_{j=1}^{s} v_j y_j^k = \mu_k, \qquad k = 0, 1, \dots, N, \tag{3.170}$$

where y_j, $j = 1, \dots, s$, denote the predetermined abscissas. The unknowns in (3.170) are the "free" abscissas x_i, the weights w_i, $i = 1, \dots, n$, and the weights v_j, $j = 1, \dots, s$, corresponding to the predetermined abscissas. The relation among N, n, and s reflects the fact that there are as many equations as unknowns,

$$N + 1 = 2n + s. \tag{3.171}$$

With this formulation, solving (3.170) is equivalent to finding a quadrature rule with s predetermined abscissas and maximum degree of exactness.

3.9.1.1 Gauss–Radau quadrature rules

As with GQRs, we first determine a polynomial whose roots are the abscissas. To do this, we compute the orthogonal polynomial $P_{n+1}(x)$ including either a or b among its roots. Consider first the case $P_{n+1}(a) = 0$. Assuming orthonormality, we may write the recurrence relationship among polynomials in the matrix form [97, §7]

$$x\mathbf{p}(x) = \mathbf{J}_n \mathbf{p}(x) + \beta_n P_n(x)\mathbf{e}_n, \tag{3.172}$$

where $\mathbf{p}$, $\mathbf{J}_n$, and $\mathbf{e}_n$ are defined in (3.137), (3.140), and (3.139), respectively. This corresponds to the scalar form

$$\beta_j P_j(x) = (x - \alpha_j)P_{j-1}(x) - \beta_{j-1}P_{j-2}(x), \qquad j = 1, 2, \dots, \tag{3.173}$$

with $P_{-1}(x) \equiv 0$ and $P_0(x) \equiv 1$. The condition $P_{n+1}(a) = 0$ implies that, from (3.173),

$$(a - \alpha_{n+1})P_n(a) - \beta_n P_{n-1}(a) = 0 \tag{3.174}$$

and hence

$$\alpha_{n+1} = a - \beta_n \frac{P_{n-1}(a)}{P_n(a)}. \tag{3.175}$$

Equation (3.172) yields

$$(\mathbf{J}_n - a\mathbf{I}_n)\mathbf{p}(a) = -\beta_n P_n(a)\mathbf{e}_n, \tag{3.176}$$

which is equivalent to

$$(\mathbf{J}_n - a\mathbf{I}_n)\boldsymbol{\delta}(a) = \beta_n^2 \mathbf{e}_n, \tag{3.177}$$

where $\boldsymbol{\delta}(a)$ has components

$$\delta_j(a) = -\beta_n \frac{P_{n-1}(a)}{P_n(a)}. \tag{3.178}$$

Thus, from (3.175) we obtain

$$\alpha_{n+1} = a + \delta_n(a). \tag{3.179}$$

In conclusion, the procedure for the generation of the Gauss–Radau quadrature rule is:

- **(i)** Generate the matrix $\mathbf{J}_n$ and the element $\beta_n \triangleq \gamma_{n+1,n+1}/\gamma_{n,n}$.
- **(ii)** Solve the system of equations (3.177) for $\delta_n(a)$.
- **(iii)** Compute α_{n+1} by (3.179) and use it to replace entry $(n+1, n+1)$ of $\mathbf{J}_{n+1}$.
- **(iv)** Compute the eigenvalues and the first element of the eigenvectors of the tridiagonal matrix

$$\tilde{\mathbf{J}}_{n+1} \triangleq \left[\begin{array}{c|c} \mathbf{J}_n & \beta_n \mathbf{e}_n \\ \hline \beta_n \mathbf{e}_n^{\mathsf{T}} & \alpha_{n+1} \end{array} \right] \tag{3.180}$$

(one of the eigenvalues of $\mathbf{J}_{n+1}$ must be equal to a).

The procedure for the case $P_{n+1}(b) = 0$ is the same, except for a being replaced by b in (3.177)–(3.179).

Example 3.25. With the same data as in Example 3.20, and $y_1 = a = -1$, we have $\beta_3 = 3/\sqrt{35}$, $\delta_3(-1) = 3/7$, and $\alpha_4 = -4/7$. Thus,

$$\tilde{\mathbf{J}}_4 = \begin{bmatrix} 0 & 1/\sqrt{3} & 0 & 0 \\ 1/\sqrt{3} & 0 & 2/\sqrt{15} & 0 \\ 0 & 2/\sqrt{15} & 0 & 3/\sqrt{35} \\ 0 & 0 & 3/\sqrt{35} & -4/7 \end{bmatrix} \tag{3.181}$$

whose eigenvalues yield the abscissas $-1, -0.575319, 0.181066, 0.822824$ and normalized eigenvalues first components yield the weights

$$0.0625, 0.328844, 0.388193, 0.220462. \qquad \diamond$$

3.9.1.2 Gauss–Lobatto quadrature rules

Here we first compute a matrix $\tilde{\mathbf{J}}_{n+1}$ such that its minimum and maximum eigenvalues satisfy

$$\lambda_{\min}(\tilde{\mathbf{J}}_{n+1}) = a \quad \text{and} \quad \lambda_{\max}(\tilde{\mathbf{J}}_{n+1}) = b. \tag{3.182}$$

Next, we determine $P_{n+1}(x)$ such that

$$P_{n+1}(a) = P_{n+1}(b) = 0. \tag{3.183}$$

From (3.114) we have

$$\beta_{n+1} P_{n+1}(x) = (x - \alpha_{n+1}) P_n(x) - \beta_n P_{n-1}(x), \tag{3.184}$$

so that from (3.183) we obtain

$$\begin{aligned}\alpha_{n+1} P_n(a) + \beta_n P_{n+1}(a) &= a P_n(a), \\ \alpha_{n+1} P_n(b) + \beta_n P_{n+1}(b) &= a P_n(b).\end{aligned} \tag{3.185}$$

Using (3.136), if

$$(\mathbf{J}_n - a\mathbf{I})\boldsymbol{\gamma} = \mathbf{e}_n \quad \text{and} \quad (\mathbf{J}_n - b\mathbf{I})\boldsymbol{\zeta} = \mathbf{e}_n \tag{3.186}$$

then the components of vectors $\boldsymbol{\gamma}$ and $\boldsymbol{\zeta}$ are

$$\begin{aligned}\gamma_j &= -\frac{1}{\beta_n}\frac{P_{j-1}(a)}{P_n(a)}, \\ \zeta_j &= -\frac{1}{\beta_n}\frac{P_{j-1}(b)}{P_n(b)}.\end{aligned} \tag{3.187}$$

Thus, (3.185) is equivalent to the system of equations

$$\begin{aligned}\alpha_{n+1} - \gamma_n \beta_n^2 &= a, \\ \alpha_{n+1} - \zeta_n \beta_n^2 &= b.\end{aligned} \tag{3.188}$$

Hence the procedure is:

(i)   Generate the matrix $\mathbf{J}_n$.
(ii) Solve the system of equations (3.186) for γ_n and ζ_n.
(iii) Solve (3.188) for α_{n+1} and β_n^2.
(iv) Compute the eigenvalues and first element of the eigenvector of the tridiagonal matrix

$$\tilde{\mathbf{J}}_{n+1} \triangleq \left[\begin{array}{c|c} \mathbf{J}_n & \beta_n \mathbf{e}_n \\ \hline \beta_n \mathbf{e}_n^{\mathsf{T}} & \alpha_{n+1} \end{array}\right]. \tag{3.189}$$

Example 3.26. With the same data as in Example 3.20, we obtain $\gamma_3 = 5/3$, $\zeta_3 = -5/3$, $\alpha_4 = 0$, and $\beta_3 = \sqrt{3/5}$. Thus,

$$
\tilde{\mathbf{J}} = \begin{bmatrix} 0 & 1/\sqrt{3} & 0 & 0 \\ 1/\sqrt{3} & 0 & 2/\sqrt{15} & 0 \\ 0 & 2/\sqrt{15} & 0 & \sqrt{3/5} \\ 0 & 0 & \sqrt{3/5} & 0 \end{bmatrix} \tag{3.190}
$$

whose eigenvalues yield the abscissas $-1, -0.447214, 0.447214, 1$ and normalized eigenvectors first components yield the weights

$$0.0833333, \ 0.416667, \ 0.416667, \ 0.0833333.$$

These abscissas and weights are consistent with the quadrature rule tabulated in [2, Table 25.6]. $\Diamond$

3.9.2 Moment bounds and quadrature rules

We are now ready to describe the solution of problems (3.18). Let μ be contained in the interior of the moment space $\mathcal{M}_{N+1}$ induced by functions $h_i(x)$, $0 \leqslant i \leqslant N$. We look for the extremal values

$$
\max_{\mathcal{F}(\mu)} \int_a^b \varphi(x) f(x) \, dx, \tag{3.191a}
$$

$$
\min_{\mathcal{F}(\mu)} \int_a^b \varphi(x) f(x) \, dx, \tag{3.191b}
$$

where $\mathcal{F}(\mu)$ is the set of PDFs f whose moments are equal to the components of μ. The key result is expressed by the following theorem [136, p. 80]:

Theorem 3.8. *If the set of functions* $\{h_0, h_1, \ldots, h_N\}$ *and the augmented set of functions* $\{h_0, h_1, \ldots, h_N, \varphi\}$ *form T-systems in* $[a, b]$, *then the pdfs corresponding to the upper and lower principal representations of* μ *yield* (3.191a) *and* (3.191b), *respectively.*

Remark 3.22. Notice how the maximizing and minimizing representations do not depend on $\varphi(x)$, as the only constraint is that both $\{h_0, h_1, \ldots, h_N\}$ and the augmented set $\{h_0, h_1, \ldots, h_N, \varphi\}$ form T-systems.

Example 3.27. With $\mathcal{X} = [a, b]$, and three known moments $\mu \triangleq \mathbb{E}[X]$, variance σ^2, and $\kappa \triangleq \mathbb{E}[(X - \mu)^3]$, and assuming that $\varphi(x)$ is four-time differentiable with $\varphi^{(4)} \geqslant 0$ in $\mathcal{X}$, we have [40]

$$
\varphi(x_1)q + \varphi(x_2)(1-q) \leqslant \mathbb{E}[\varphi(X)] \leqslant \varphi(a)p_1 + \varphi(x_3)p_2 + \varphi(b)(1 - p_1 - p_2), \tag{3.192}
$$

where

$$x_1 = \mu + \frac{\kappa - \sqrt{\kappa^2 + 4\sigma^6}}{2\sigma^2},$$

$$x_2 = \mu + \frac{\kappa + \sqrt{\kappa^2 + 4\sigma^6}}{2\sigma^2},$$

$$x_3 = \mu + \frac{\kappa - (a + b - 2\mu)\sigma^2}{(a - \mu)(b - \mu) + \sigma^2},$$

$$q = \frac{1}{2} + \frac{\kappa}{2\sqrt{\kappa^2 + 4\sigma^6}},$$ \hfill (3.193)

$$p_1 = \frac{\sigma^2 + (x_3 - \mu)(b - \mu)}{(b - a)(x_3 - a)},$$

$$p_2 = \frac{\sigma^2 + (a - \mu)(b - \mu)}{(x_3 - b)(x_3 - a)}.$$ \hfill $\diamond$

Example 3.28. Consider the calculation of upper and lower bounds on error probabilities affected by random disturbances whose statistics are known only through a limited number of moments (see, for example, [276] and the references therein). Assume that the error probability is given the form

$$\mathsf{p} = \mathbb{E}_X Q\left(\frac{X + T}{\sigma}\right),$$ \hfill (3.194)

with $Q(\cdot)$ the Gaussian tail function defined in (2.17), and the moments $\mu_i = \mathbb{E}_X[X^i]$ known for $i = 0, 1, \ldots, N$. For $\{x^0, x^1, \ldots, x^N, g(x)\}$ to form a T-system, with

$$g(x) = Q\left(\frac{x + T}{\sigma}\right),$$ \hfill (3.195)

we must verify that $Q^{(N+1)}(x)$, the $(N + 1)$th derivative of $Q(x)$, is positive in $[a, b]$. Observe first that [2, §7.1.9]

$$\frac{d^m}{dx^m} Q\left(\frac{x}{\sigma}\right) = (-1)^m \frac{2^{-m/2}}{\sqrt{\pi}} \exp(-x^2/2\sigma^2) H_{m-1}\left(\frac{x^2}{\sqrt{2}\sigma}\right), \qquad m = 1, 2 \ldots,$$ \hfill (3.196)

where $H_{m-1}(x)$ denotes Hermite polynomial [2, §22]. The sign of this derivative is the same as that of $(-1)^m H_{m-1}(x^2/\sqrt{2}\sigma)$, which will not change if $[a, b]$ does not contain a root of $H_{m-1}(x^2/\sqrt{2}\sigma)$ as an internal point. A simple sufficient condition for this to hold is that the largest root of $H_{m-1}(x^2/\sqrt{2}\sigma)$ be smaller than a. Using Theorem 3.8, we can obtain the bounds sought. Reference [276] provides numerical examples, comparisons with other bounds, and computational details. \hfill $\diamond$

3.9.3 Bounds on CDFs and quadrature rules

The results above can be extended to obtain moment bounds on cumulative distribution functions. For example, if the task is to evaluate the probability $\mathbb{P}[X > T]$ that an RV X with measure $\mathbb{P}$ exceeds a given threshold T, or the probability $\mathbb{P}[T_1 < X < T_2]$ that X takes on values in interval (T_1, T_2), then its solution may be formulated as the search for sharp upper and lower bounds on $\mathbb{E}_\mathbb{P}[\varphi(X)]$, where φ is the indicator function of intervals (T, ∞) and (T_1, T_2), respectively. Direct application of the theory to a function $\varphi(x)$ taking value 1 for $x \leqslant \xi$ and 0 elsewhere would yield $\mathbb{E}_\mathbb{P}[\varphi(X)] = F(\xi)$, but this procedure is not possible as the set $\{x^0, x^1, \ldots, x_N, \varphi\}$ cannot be a T-system since this φ is not a continuous function. However, a suitable application of the theory of canonical representations yields bounds on the integral of a function over a subinterval $[a, \xi]$ of $[a, b]$ [129, p. 97], [136, p. 82], [149, Chapter 4], [193].

The following results hold. Consider the canonical representations of μ with respect to a T-system $\{h_0, \ldots, h_N\}$ on $[a, b]$. Given any $\xi \in [a, b]$, a unique canonical representation exists having ξ as an abscissa and whose extremal properties are summarized in the following theorem.

Theorem 3.9. *If $\{h_0, \ldots, h_k\}$ and $\{h_0, \ldots, h_k, \varphi\}$, with $\varphi(x) > 0$ for $x \in [a, b]$, are T-systems for $k = 0, \ldots, N$, then, for any $\xi \in [a, b]$ and any CDF of an RV with moments μ, we have*

$$\sum_{j:x_j < \xi} w_j \leqslant F(\xi) \leqslant \sum_{j:x_j \leqslant \xi} w_j, \tag{3.197}$$

where $\{x_j, w_j\}$ correspond to the unique canonical representation of μ having ξ among its abscissas. The bounds are sharp.

Example 3.29. With X being an RV with support set $[a, b]$ and mean value μ_1, where for consistency $\mu_1 \in [a, b]$, we have $\overline{x}_1 = a$, $\underline{x}_1 = \mu_1$, and $\overline{x}_2 = b$. Consequently,

$$0 \leqslant F_X(x) \leqslant \frac{b - \mu_1}{b - x}, \qquad a \leqslant x \leqslant \mu_1,$$
$$\frac{x - \mu_1}{x - a} \leqslant F_X(x) \leqslant 1, \qquad \mu_1 < x \leqslant b. \tag{3.198}$$

$\diamond$

Example 3.30. Assume $N = 2$, and that the mean μ_1 and the variance $\sigma_X^2 \triangleq \mathbb{E}[X - \mu_1]^2$ are known for an RV X taking values in the finite interval $[a, b]$ (we must have $\sigma_X^2 \leqslant (b - \mu_1)(\mu_1 - a)$ for consistency—see Example 3.11 *supra*). With $\underline{x}_1, \underline{x}_2, \overline{x}_1$, and $\overline{x}_2$ as in (3.162)–(3.163), we have, for $x \leqslant \overline{x}_1$,

$$0 \leqslant F_X(x) \leqslant \frac{\sigma_X^2}{(\mu_1 - x)^2 + \sigma_X^2}, \tag{3.199}$$

while for $\overline{x}_1 \leqslant x \leqslant \underline{x}_2$,

$$F_X(x) \geqslant \frac{(x - \mu_1)(b - \mu_1) + \sigma_X^2}{(x - a)(b - a)} \tag{3.200}$$

and

$$F_X(x) \leqslant 1 - \frac{(\mu_1 - x)(\mu_1 - a) + \sigma_X^2}{(b - x)(x - a)}. \tag{3.201}$$

Finally, for $\underline{x}_2 \leqslant x \leqslant b$,

$$\frac{(x - \mu_1)^2}{(x - \mu_1)^2 + \sigma_X^2} \leqslant F_X(x) \leqslant 1. \tag{3.202}$$

Fig. 3.4 shows upper and lower moment bounds on the CDF of an RV defined on $[0, 1]$ with expected value $\mu_1 = 0.2$ and variance $\sigma_X^2 = 0.125$.

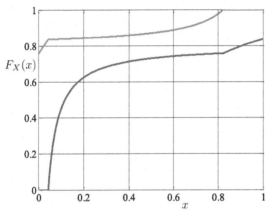

Figure 3.4 Upper and lower moment bounds on the cdf of a RV X, defined in $[0, 1]$, that is known only through its expected value $\mu_1 = 0.2$ and variance $\sigma_X^2 = 0.125$.

Closed-form equations for the case $N = 3$ can be found in [32]. For $N > 3$, numerical techniques should be used [32, p. 333], similar to those available in the context of principal representations [97, Section 7]. ◇

3.9.4 Bounds on Laplace–Stieltjes transform

Given an RV X whose support set is the finite interval $[0, b]$, the *Laplace–Stieltjes transform* of its probability density function is defined as

$$\phi(s) \triangleq \mathbb{E}\left[e^{-sX}\right], \quad s \in \mathbb{R}^+. \tag{3.203}$$

Upper and lower moment bounds on $\phi(s)$ can be derived under power moment constraints after observing that the set $\{1, x, x^2, \ldots, x^N, \pm e^{-sx}\}$ forms a T-system if the $(N + 1)$th derivative of $\pm e^{-sx}$ is positive on $[0, b]$. Observing that the mth derivative of e^{-sx} is $(-1)^m s^m e^{-sx}$, we see that $\{1, x, x^2, \ldots, x^N, e^{-sx}\}$ is a T-system for $s \neq 0$ if $N + 1$ is even, while, if $N + 1$ is odd, $\{1, x, x^2, \ldots, t^N, e^{-sx}\}$ is a T-system for

$s < 0$, and $\{1, x, x^2, \ldots, x^N, -e^{-sx}\}$ is a T-system for $s > 0$. The following bounds can be derived in closed form [75] using the principal representations of Table 3.1:

(i) If μ_1, σ^2, and b are known, then

$$
\begin{aligned}
\phi_1(s) \leqslant \phi(s) \leqslant \phi_2(s), & \quad s \leqslant 0, \\
\phi_2(s) \leqslant \phi(s) < \phi_1(s), & \quad s \geqslant 0,
\end{aligned}
\tag{3.204}
$$

where

$$
\begin{aligned}
\phi_1(s) &= \frac{\sigma^2}{\sigma^2 + \mu_1^2} + \frac{\mu_1^2}{\sigma^2 + \mu_1^2} e^{-(\mu_1 + \sigma^2/\mu_1)s}, \\
\phi_2(s) &= \frac{\sigma^2}{\sigma^2 + (b - \mu_1)^2} e^{-bs} + \frac{(b - \mu_1)^2}{\sigma^2 + (b - \mu_1)^2} e^{-[\mu_1 - \sigma^2/(b - \mu_1)]s}.
\end{aligned}
\tag{3.205}
$$

(ii) If μ_1, μ_2, μ_3, and b are known, then for all $s \neq 0$,

$$
pe^{-xs} + (1 - p)e^{-ys} \leqslant \phi(s) \leqslant 1 - p' - q' + p'e^{-zs} + q'e^{-bs} \tag{3.206}
$$

whose parameters x, y, z, p, p', q, and q' are computed from

$$
\begin{aligned}
p &= \frac{r + \alpha - 2\mu_1}{2r}, & x &= \frac{\alpha - r}{2}, & y &= \frac{\alpha + r}{2}, \\
r &= \sqrt{\alpha^2 + 4\beta}, & \alpha &= \frac{\mu_3 - \mu_1\mu_2}{\mu_2 - \mu_1^2}, & \beta &= \frac{\mu_2^2 - \mu_1\mu_3}{\mu_2 - \mu_1^2}, \\
q' &= \frac{\mu_1\mu_3 - \mu_2^2}{\mu_1 b^3 - 2\mu_2 b^2 + \mu_3 b}, & p' &= \frac{(\mu_1 - q'b)^2}{\mu_2 - q'b^2}, & z &= \frac{\mu_2 - q'b^2}{\mu_1 - q'b}.
\end{aligned}
\tag{3.207}
$$

The results above can be extended to the case of an RV X defined in $[0, \infty)$ and $s \geqslant 0$ [75]. Since $\{1, x, x^2, -e^{-sx}\}$ satisfies conditions (3.161) for $s > 0$, sharp bounds can be derived using the principal representations of Table 3.2:

(i) If μ_1 and σ^2 are known, then, for $s > 0$,

$$
e^{-\mu_1 s} \leqslant \phi(s) \leqslant \frac{\sigma^2}{\sigma^2 + \mu_1^2} + \frac{\mu_1^2}{\sigma^2 + \mu_1^2} e^{-(\mu_1 + \sigma^2/\mu_1)s}. \tag{3.208}
$$

(ii) Similarly, if μ_1, μ_2, μ_3, and b are known, then, for $s > 0$,

$$
pe^{-xs} + (1 - p)e^{-ys} \leqslant \phi(s) \leqslant \frac{\mu_2 - \mu_1^2}{\mu_2} + \frac{\mu_1^2}{\mu_2} e^{-(\mu_2/\mu_1)s}, \tag{3.209}
$$

with p, x, and y computed via (3.207).

Remark 3.23. As shown in [75], the results derived above can be applied to bound certain quantities in queueing and traffic theory: blocking probabilities, queue sizes, delays, relaxation times, and busy periods.

3.9.5 Extension to nondifferentiable functions: the method of contact polynomials

The requirement of differentiability of function $\varphi(\cdot)$ in moment-bound applications relying on Čebyšev systems can be relaxed in some cases by using the geometric approach advocated in [136, Chapter XII] and [141,142], and based on *contact polynomials*.

Consider again moments $\boldsymbol{\mu} = (\mu_0, \ldots, \mu_N)$ with $\mu_i = \mathbb{E}[h_i(X)]$, an RV X with support set $[a, b]$, and the expectation $\mathbb{E}_{\mathbb{P}}[\varphi(X)]$, where $\varphi(x)$ is a bounded continuous function and $\mathbb{P}$ is known only through $\boldsymbol{\mu}$. Here we do not assume that the functions h_i form a T-system, the only requirement being that they are linearly independent. We look for sharp upper and lower bounds on $\mathbb{E}_{\mathbb{P}}[\varphi(X)]$. We limit our description to the derivation of an upper bound, as a lower bound can be obtained by straightforward modifications of the ensuing arguments. Define the family $\mathcal{P}_+$ of generalized polynomials $P_N(x) = \sum_{i=0}^{N} p_i h_i(x)$ taking values greater than or equal to $\varphi(x)$ in $[a, b]$,

$$\mathcal{P}_+ \triangleq \{P_N(x) \mid P_N(x) \geqslant \varphi(x), x \in [a, b]\}, \tag{3.210}$$

and assume that $\mathcal{P}_+$ is nonvoid. If $P_N(x)$ is in $\mathcal{P}_+$, then

$$\mathbb{E}_{\mathbb{P}}[P_N(X)] = \sum_{i=0}^{N} p_i \mu_i \geqslant \mathbb{E}_{\mathbb{P}}[\varphi(X)], \tag{3.211}$$

and hence the sum $\sum_{i=0}^{N} p_i \mu_i$ yields an upper bound to $\mathbb{E}_{\mathbb{P}}[\varphi(X)]$. It seems natural to search for the tightest upper bound obtained by selecting the "best" $P_N(x)$. Due to the geometry of moment spaces, a polynomial $P_N(x) \in \mathcal{P}_+$ exists (called *upper contact polynomial*) which touches the graph of $\varphi(x)$ from above in an *upper contact set*

$$\mathcal{Z}_+ \triangleq \{x \mid P_N(x) = \varphi(x)\}. \tag{3.212}$$

Moreover, a pdf $f_+(x)$ exists which has support set $\mathcal{Z}_+$ and satisfies the moment constraints. For this pdf,

$$\begin{aligned}
\int_a^b \varphi(x) f_+(x) \, dx &= \int_{\mathcal{Z}_+} \varphi(x) f_+(x) \, dx \\
&= \int_{\mathcal{Z}_+} P_N(x) f_+(x) \, dx \\
&= \sum_{i=0}^{N} p_i \mu_i
\end{aligned} \tag{3.213}$$

so that the sought upper bound is given by

$$\sup \mathbb{E}_{\mathbb{P}}[\varphi(X)] = \int \varphi(x) f_+(x) \, dx. \tag{3.214}$$

Similar arguments lead to the definition of a *lower contact polynomial*, a *lower contact set* $\mathcal{Z}_-$, and to a lower bound. The bounding technique obtained from the above results consists of computing $\mathcal{Z}_+$ and $\mathcal{Z}_-$ along with two general pdfs with these supports, then solving the equations expressing the moment conditions for these pdfs to obtain $f_+(x)$ and $f_-(x)$. The bounds obtained are sharp.

Remark 3.24. If $\mathcal{Z}_+$ and $\mathcal{Z}_-$ contain only a finite number of points, it does not exceed $N + 2$ [136, p. 474].

Example 3.31. Consider first the convex function $\varphi(x)$ shown in Fig. 3.5(a), and assume that $\mu_1 = \mathbb{E}_\mathbb{P}[X]$ is known. Here $\mathcal{P}_+$ includes degree-1 polynomials $p_0 + p_1 x$. The upper contact polynomial, denoted $P_+(x)$, has 2 points of contact in $x = a$ and $x = b$. Thus, the maximizing $\mathbb{P}$ has support $\{a, b\}$, with these two values taken with probabilities q and $1 - q$. The moment constraint

$$\mu_1 = qa + (1 - q)b \tag{3.215}$$

yields $q = (b - \mu_1)/(b - a)$, and hence the upper bound

$$\mathbb{E}_\mathbb{P}[\varphi(X)] \leqslant q\varphi(a) + (1 - q)\varphi(b). \tag{3.216}$$

The lower contact polynomial $P_-(x)$, touches $\varphi(x)$ in a single point, denoted by t in Fig. 3.5(a). Thus, the minimizing $\mathbb{P}$ has support $\{t\}$, which yields $t = \mu_1$ due to the moment constraint, and

$$\mathbb{E}_\mathbb{P}[\varphi(X)] \geqslant \varphi(\mu_1). \tag{3.217}$$

Consider next the convex function $\varphi(x)$ of Fig. 3.5(b) (a function with similar features is for example the "stop-loss function" $\varphi(x) = \max(x - c, 0)$ [40,120]). The derivation of upper bound (3.216) is the same as above, while the polynomial $P_-(x)$ has a contact set which is now an interval. In this interval, $P_-(x)$ coincides with $\varphi(x)$, which in $\mathcal{Z}_-$

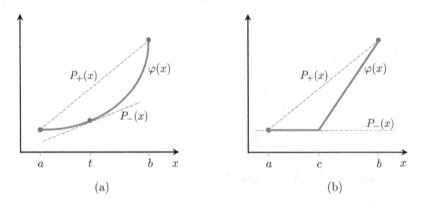

(a)   (b)

Figure 3.5 Moment bounds from upper and lower contact polynomials.

is an affine function, and hence of the form $\varphi(x) = p_0 + p_1 x$. Thus, the lower bound is

$$\int_{\mathcal{Z}_-} \varphi(x) f_-(x)\, dx = \int_{\mathcal{Z}_-} (p_0 + p_1 x) f_-(x)\, dx = p_0 + p_1 \mu_1 = \varphi(\mu_1), \quad (3.218)$$

i.e., the same as above. $\diamond$

Example 3.32. Quadratic contact polynomials can be used when the moment constraints involve μ_1 and μ_2. Reference [120, Chapter II] contains an extensive analysis of these polynomials as used with piecewise functions $\varphi(x)$. Two examples of application involve the stop-loss function $\varphi(x) = \max(x - c, 0)$ and the two-layer stop-loss function, defined for $x \in [a, b]$,

$$\varphi(x) = r \max(0, x - \alpha) + (1 - r) \max(x - \beta), \qquad 0 < r < 1, \ a < \alpha < \beta < b, \tag{3.219}$$

illustrated in Fig. 3.6.

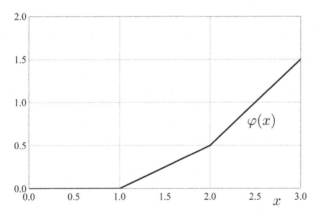

Figure 3.6 Two-layer stop-loss function with $a = 0$, $b = 3$, $\alpha = 1$, $\beta = 2$, and $r = 0.5$.

Quadratic contact polynomials are derived and tabulated in [170] for the derivation of upper and lower bounds on the error probability p of digital communication systems, where

$$\mathsf{p} = \mathbb{P}\big[|Y + Z| > K\big], \tag{3.220}$$

$Z \sim \mathcal{N}(0, \sigma_Z^2)$, and Y is an RV with support set $[-I, I]$ and first power moments $\mathbb{E}_Y[Y] = 0$ and $\mathbb{E}_Y[Y^2] = \sigma^2$. Under these assumptions, we can write

$$\mathsf{p} = \mathbb{E}_Y[\varphi(Y)], \tag{3.221}$$

where

$$\varphi(x) \triangleq Q\left(\frac{K-x}{\sigma_Z}\right) + Q\left(\frac{K+x}{\sigma_Z}\right). \qquad (3.222)$$

$$\diamond$$

Example 3.33 (Glasser inequality). In [136, pp. 481–482] a lower bound is derived on $\mathbb{P}[-T_1 < Y < T_2]$, where Y is an RV with support set $\mathbb{R}$ and $\mathbb{E}[Y] = 0$ and $\mathbb{E}[|Y|] = 1$. The problem is tantamount to the search for an upper bound on

$$\mathbb{P}[\{Y \in (-\infty, -T_1] \cup [T_2, \infty)\}] \qquad (3.223)$$

with the given moment constraints. With $\varphi(x)$ the indicator function of the set $\{(-\infty, -T_1] \cup [T_2, \infty)\}$, the candidates to be upper contact polynomials have the form

$$P_+(x) = a|x| + bx + c \geqslant \varphi(x) \qquad (3.224)$$

as shown in Fig. 3.7. We obtain the bound

$$\mathbb{P}[-T_1 < Y < T_2] \geqslant \max\left[1 - \left(T_1^{-1} + T_2^{-1}\right)/2, 0\right] \qquad (3.225)$$

achieved with equality by a discrete RV taking values in $\{-T_1, T_2\}$ with probabilities $1/2T_1$ and $1/2T_2$.

$$\diamond$$

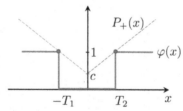

Figure 3.7 Finding an upper bound on $\mathbb{P}[X \in (-\infty, -T_1] \cup [T_2, \infty)]$ using contact polynomials.

3.10 Moment problems as semidefinite programs

Consider again the untruncated moment problem and Theorem 3.1. The condition that the components of a real vector $\boldsymbol{\mu}$ be moments of some probability measure on $\mathbb{R}$ (or the limit of the moments of some sequence of measures) can be expressed by assuming that $\mathbf{H} \geqslant \mathbf{0}$, with $\mathbf{H}$ defined in (3.53), together with the equality $\mu_0 = 1$ (here we use the notation $\mathbf{H} \geqslant \mathbf{0}$ to indicate that $\mathbf{H}$ is nonnegative definite). Using this fact, we can formulate some moment problems as Semidefinite Programs (SDP) [38, p. 168].

A *semidefinite program* (SDP) is an optimization problem cast in the form of the minimization (or maximization) of a linear function over a convex set of matrices. In our case, the problem is

minimize or maximize $(\mathbf{A}, \mathbf{X})$

subject to $\quad\quad\quad (\mathbf{B}_i, \mathbf{X}) = b_i, \quad i = 1, \ldots, m,$ $\quad\quad\quad$ (3.226)

$\quad\quad\quad\quad\quad\quad\quad \mathbf{X} \succcurlyeq \mathbf{0},$

where $\mathbf{A}$ and $\mathbf{B}_i$ are given square matrices, and the b_i are given real numbers. Here the maximization/minimization "variable" is the nonnegative-definite matrix $\mathbf{X}$, and the inner product of two matrices $\mathbf{C}$ and $\mathbf{D}$ is defined as

$$(\mathbf{C}, \mathbf{D}) = \mathrm{tr}\,(\mathbf{C}^{\mathsf{T}}\mathbf{D}). \tag{3.227}$$

The above is a *convex program*: the set of nonnegative-definite matrices is convex, and so is its intersection with the linear subspace defined by the equality constraints $(\mathbf{B}_i, \mathbf{X}) = b_i$. The upside of convex programs lies in the fact that they can be solved using computationally efficient algorithms and software packages. Given the above, the solution of a moment-bound problem using semidefinite programming may be reduced to finding the appropriate formulation of the constraints in the form of the definiteness of an appropriate matrix. Some examples will show how this can be done in a few relatively simple cases.

3.10.1 Moment problem in a parallelepiped

A situation in which SDPs can be used to solve a moment problem is offered by the calculation of bounds on the expected value of a polynomial in an RV when some moments of the RV are known to lie within a parallelepiped (we have examined this problem in Section 3.6 *supra*).

Let $P(x) = \sum_{i=0}^{N} p_i x^i$ be a real polynomial. The expected value of $P(X)$, with X an RV with support set $[a, b]$, is linear in the moments $\mu_i \triangleq \mathbb{E}_{\mathbb{P}}[X^i]$, i.e.,

$$\mathbb{E}_{\mathbb{P}}[P(X)] = \sum_{i=0}^{N} p_i \mu_i. \tag{3.228}$$

Upper and lower bounds on $\mathbb{E}_{\mathbb{P}}[P(X)]$ subject to parallelepiped bounds can be computed by solving an SDP. We first recall that, from Theorems 3.2 and 3.3, the conditions for $\boldsymbol{\mu}$ to be a moment sequence consist of the positive semidefiniteness of two suitable moment matrices. Let us call these $\mathbf{H}'$ and $\mathbf{H}''$ (the actual form of these matrices depends on the parity of the number of moments available). The SDP is then [48]

minimize or maximize $\quad p_1\mu_1 + \cdots + p_N\mu_N$
$\quad\quad\mathbb{P}\quad\quad\quad\quad\mathbb{P}$

subject to $\quad\quad\quad\underline{\mu}_k \leqslant \mu_k \leqslant \overline{\mu}_k, \quad k = 1, \ldots, N,$

$\quad\quad\quad\quad\quad\quad\quad \mathbf{H}' \geqslant \mathbf{0},$ $\quad\quad\quad\quad\quad\quad$ (3.229)

$\quad\quad\quad\quad\quad\quad\quad \mathbf{H}'' \geqslant \mathbf{0}.$

The solution of (3.229) yields bounds on $\mathbb{E}[P(X)]$ over all probability measures satisfying the moment constraints. The bounds are sharp in the usual sense.

Remark 3.25. With this approach, it is not necessary to verify before the computations that the diagonal vertices of the parallelepiped correspond to moment sequences, as in our treatment in Section 3.6.

Example 3.34. In [48], upper and lower moment bounds on the expected value of $Q((X + T)/\sigma)$ are obtained under the assumption that X is an RV taking values in $[-1, 1]$ and having first power moments known within a parallelepiped. The function $Q((X + T)/\sigma)$ is first bounded above and below using Čebyšev polynomials, and these bounds are used in (3.229).   ◇

3.10.2 Generalized Čebyšev bounds

We have introduced Čebyšev bounds in Example 3.3 *supra*. In their general form, these provide upper and lower bounds on the probability $\mathbb{P}[X \in S]$, where S is a subset of the support set X of the RV X, and some moments of X are known. Following [38, §7.4], we show how SDPs can be used to find the greatest lower bound to $\mathbb{P}[X \in S]$. We assume that the moments

$$\mathbb{E}_{\mathbb{P}}[h_i(X)] = \mu_i, \quad i = 1, \ldots, N, \tag{3.230}$$

are known. Consider the following linear combination of the functions h_i:

$$\varphi(x) \triangleq \sum_{i=0}^{N} p_i h_i(x), \tag{3.231}$$

with $h_0(x) = 1$, so that $\mathbb{E}_{\mathbb{P}}[h_0(X)] = \mu_0 = 1$. With obvious notations, we can write $\mathbb{E}_{\mathbb{P}}[\varphi(X)] = (\mathbf{p}, \boldsymbol{\mu})$. Now, if $\varphi(x)$ is chosen in such a way that $\varphi(x) \geqslant 1_S(x)$ for all $x \in S$, we have

$$\mathbb{E}_{\mathbb{P}}[\varphi(X)] = (\mathbf{p}, \boldsymbol{\mu}) \geqslant \mathbb{E}_{\mathbb{P}}[1_S(X)] = \mathbb{P}[X \in S]. \tag{3.232}$$

To find the tightest upper bound on $\mathbb{P}[X \in S]$, we solve the SDP

$$\text{minimize} \quad p_0 + p_1\mu_1 + \cdots + p_N\mu_N$$

$$\text{subject to} \quad \varphi(x) = \sum_{i=0}^{N} p_i h_i(x) \geqslant 1 \text{ for } x \in S,$$
$$\varphi(x) = \sum_{i=0}^{N} p_i h_i(x) \geqslant 0 \text{ for } x \in X, \ x \notin S, \tag{3.233}$$

with variables $p_0, \ldots, p_N$. This problem is convex, as the two constraints can be expressed by the requirement that the two convex functions $1 - \inf_{x \in S} \varphi(x)$ and $-\inf_{x \notin S} \varphi(x)$ be $\leqslant 0$.

Example 3.35 ([38, pp. 375–376]). If X has support set $\mathfrak{X} = \mathbb{R}^+$ and $\mathcal{S} = [1, \infty)$, with $h_0(x) = 1$, $h_1(x) = x$, and $\mathbb{E}[X] = \mu_1 \leqslant 1$, the constraint $\varphi(x) = p_0 + p_1 x \geqslant 0$ for $x \in \mathbb{R}^+$ reduces to $p_0 \geqslant 0$ and $p_1 \geqslant 0$. The constraint $\varphi(x) = p_0 + p_1 x \geqslant 1$ for $x \in \mathcal{S}$, i.e., for $x \geqslant 1$, becomes $p_0 + p_1 \geqslant 1$. Problem (3.233) becomes

$$
\begin{aligned}
\text{minimize} \quad & p_0 + \mu_1 p_1 \\
\text{subject to} \quad & p_0 \geqslant 0, \quad p_1 \geqslant 0, \\
& p_0 + p_1 \geqslant 1.
\end{aligned}
\tag{3.234}
$$

Since $0 \leqslant \mu_1 \leqslant 1$, the optimal point for this SDP has $p_0 = 0$ and hence $p_1 = 1$, which yields the Markov bound $\mathbb{P}[X \geqslant 1] \leqslant \mu_1$ as in (3.2). $\diamond$

3.11 Multidimensional moment bounds and approximations

The derivation of general moment bounds on the expected value $\mathbb{E}[\Phi(\mathbf{X})]$, where Φ is a known function of a vector RV $\mathbf{X}$ with d real components, is an extension of the scalar problem hindered by some technical difficulties that have not been fully overcome yet in the present state of the discipline. We proceed, as we did for the scalar case, with the derivation of conditions for a real vector to be a vector of joint moments of the components of $\mathbf{X}$. To do this, we define a moment matrix $\mathbf{M}_r(\boldsymbol{\mu})$, which generalizes the one with $d = 1$ and whose rows and columns are labeled by the components of the canonical basis

$$
(1, x_1, x_2, \dots, x_d, x_1^2, x_1 x_2, \dots, x_{d-1} x_d, x_d^2, \dots, x_1^r, \dots, x_d^r)
\tag{3.235}
$$

of the vector space of real polynomials with degree $\leqslant r$.

Example 3.36. With $d = r = 2$, the moment matrix is

$$
\mathbf{M}_2(\boldsymbol{\mu}) =
\begin{array}{c}
\begin{array}{cccccc}
1 & x_1 & x_2 & x_1^2 & x_1 x_2 & x_2^2
\end{array} \\
\left[
\begin{array}{cccccc}
\mu_{00} & \mu_{10} & \mu_{01} & \mu_{20} & \mu_{11} & \mu_{02} \\
\mu_{10} & \mu_{20} & \mu_{11} & \mu_{30} & \mu_{21} & \mu_{12} \\
\mu_{01} & \mu_{11} & \mu_{02} & \mu_{21} & \mu_{12} & \mu_{03} \\
\mu_{20} & \mu_{30} & \mu_{21} & \mu_{40} & \mu_{31} & \mu_{22} \\
\mu_{11} & \mu_{21} & \mu_{12} & \mu_{31} & \mu_{22} & \mu_{13} \\
\mu_{02} & \mu_{12} & \mu_{03} & \mu_{22} & \mu_{13} & \mu_{04}
\end{array}
\right]
\begin{array}{c}
1 \\ x_1 \\ x_2 \\ x_1^2 \\ x_1 x_2 \\ x_2^2
\end{array}
\end{array}.
$$

$\diamond$

Consistently with the scalar ($d = 1$) case, the (untruncated) moment problem asks for the conditions under which an infinite sequence $\boldsymbol{\mu}$ is a moment sequence. For example, with power moments a probability measure $\mathbb{P}$ should exist with support set in $\mathbb{R}^d$ such that

$$
\mathbb{E}_{\mathbb{P}}[\mathbf{X}^\nu] = \boldsymbol{\mu}, \qquad \nu \in \mathbb{N}^d,
\tag{3.236}
$$

where we denote by $\mathbf{x}^\nu$, with $\nu \triangleq (\nu_1, \dots, \nu_d) \in \mathbb{N}^d$, the monomial $\prod_{i=1}^d x^{\nu_i}$. If such $\mathbb{P}$ does exist, it is said to *represent* μ, and if it is unique (i.e., there is no other probability measure yielding the same moments) this is said to *determine* μ. One can similarly define the *truncated* moment problem when the sequence μ is finite.

In contrast with the scalar case, where the moment matrix is the Hankel matrix defined in (3.53), not every sequence μ whose matrix $\mathbf{M}_r(\mu)$ is nonnegative definite for every r has a representing probability measure on $\mathbb{R}^d$. Actually, the general moment problem is fairly complex, the main reason being that for $d > 1$ a nonnegative polynomial may not be a sum of squares as in (3.55). For comprehensive treatments of the problem and recent results, see monographs [4,153,232], and papers [15,203]. Necessary and sufficient conditions for the determinacy of a 2-dimensional power moment sequence are derived in [187], but these are not simple to verify. Some simple results are available regarding representing measures in some special cases. Among these, it can be proved that every probability measure with compact support set is determined by its moments [153, p. 59]. In [191] it is shown that a probability measure is determined if each of the one-dimensional marginal probability measures (i.e., the coordinate projections of the measure) are determined. See also [15, Theorem 5] for a related sufficient condition for determinacy.

3.11.1   Multidimensional moment bounds

We examine the following problem:

$$
\begin{aligned}
\underset{\mathbb{P}}{\text{maximize}} \quad & \mathbb{E}_{\mathbb{P}}[\Phi(\mathbf{X})] \\
\text{subject to} \quad & \mathbb{E}_{\mathbb{P}}[h_j(\mathbf{X})] = \mu_j, \ j \in \Gamma,
\end{aligned}
\tag{3.237}
$$

where $\Gamma \in \mathbb{N}^d$ is a d-dimensional index set, and $\mathbb{P}$ is a probability measure with support set $\mathcal{X} \subseteq \mathbb{R}^d$. Common assumptions here are that the support set $\mathcal{X}$ of $\mathbf{X}$ is compact, Φ is bounded and upper semicontinuous (for example, the indicator function of a closed set is upper semicontinuous, while the indicator function of an open set is lower semicontinuous), the functions h_j are continuous, and there exists an index $k \in \Gamma$ such that $h_k > 0$ on $\mathcal{X}$ [153, Theorem 1.3].

Important special cases occur when the constraints involve *polynomial* functions h_j, which leads to "traditional" power moments. Another special case occurs for $\Phi(\mathbf{x}) = \mathbb{1}_{\mathcal{S}}(\mathbf{x})$, the indicator function of set $\mathcal{S}$. We have that, if (3.237) has a feasible solution, then the optimal $\mathbb{P}$ has a discrete and finite support set.

In the balance of this section we shall first examine the derivation of Čebyšev-like bounds. Next, we shall describe a geometric bounding technique based on contact polynomials, and finally approximations obtained from cubature rules.

3.11.1.1   Multidimensional Čebyšev inequality

The following general technique is useful for obtaining inequalities on the probability that a vector RV lie in a set $\mathcal{S} \in \mathbb{R}^d$. Consider a nonnegative function $\Psi(\mathbf{x})$ such that

$\Psi(\mathbf{x}) \geqslant 1$ for all $\mathbf{x}$ in a set $\mathcal{S} \subset \mathbb{R}^d$. Then, if $f_{\mathbf{X}}(\mathbf{x})$ denotes the pdf of the d-dimensional random vector $\mathbf{X}$, we have the inequality chain

$$
\begin{aligned}
\mathbb{E}[\Psi(\mathbf{X})] &= \int_{\mathbf{x} \in \mathcal{S}} \Psi(\mathbf{x}) f_{\mathbf{X}}(\mathbf{x}) \, d\mathbf{x} + \int_{\mathbf{x} \notin \mathcal{S}} \Psi(\mathbf{x}) f_{\mathbf{X}}(\mathbf{x}) \, d\mathbf{x} \\
&\geqslant \int_{\mathbf{x} \in \mathcal{S}} \Psi(\mathbf{x}) f_{\mathbf{X}}(\mathbf{x}) \, d\mathbf{x} \\
&\geqslant \mathbb{P}[\mathbf{X} \in \mathcal{S}],
\end{aligned}
\tag{3.238}
$$

which yields an upper bound to $\mathbb{P}[\mathbf{X} \in \mathcal{S}]$ depending on the expected value of $\Psi(\mathbf{X})$. The choice of the function $\Psi(\mathbf{x})$ is based on the set of constraints imposed on $\mathbf{X}$: for example, if the constraints are in terms of first- and second-order moments, then $\Psi(\mathbf{x})$ should be quadratic, as in the case considered in [270, Theorem 1]. As with $d = 1$, a bound is said to be *sharp* if a probability measure $\mathbb{P}$ satisfying the given constraints exists for which the bound is satisfied with equality.

In [178, Theorem 3.1], it is proved that, for a random d-vector $\mathbf{X}$ whose component satisfy $\mathbb{E}[X_i^2] = \sigma_i^2$, $1 \leqslant i \leqslant d$, and a nonnegative homogeneous concave $\cap$ function Φ defined on the positive orthant of $\mathbb{R}^d$, the following holds for $T > 0$:

$$
\mathbb{P}\left[\Phi\left(X_1^2, X_2^2, \ldots, X_d^2\right) \geqslant T \right] \leqslant \Phi\left(\sigma_1^2, \sigma_2^2, \ldots, \sigma_d^2\right)/T,
\tag{3.239}
$$

and this inequality can be sharp.

References [50,254] provide the greatest lower bound on $\mathbb{P}[\mathbf{X} \in \mathcal{S}]$, where $\mathbf{X}$ is a d-vector RV with first and second moments $\mathbb{E}_{\mathbb{P}}[\mathbf{X}]$ and $\mathbb{E}_{\mathbb{P}}[\mathbf{X}\mathbf{X}^{\mathsf{T}}]$, and $\mathcal{S} \subseteq \mathbb{R}^d$ is a set defined by strict quadratic inequalities. The bound and the probability measure that achieves it are obtained as the solution to a convex optimization problem. As an application, it is shown that in the scalar case, with $\mathcal{S} = (-1, 1) = \{x \mid x^2 < 1\}$ and constraints on $\mu_1 \geqslant 0$ and μ_2,

$$
\mathbb{P}[|X| < 1] \geqslant
\begin{cases}
0, & 1 \leqslant \mu_2, \\
1 - \mu_2, & |\mu_1| \leqslant \mu_2 < 1, \\
\dfrac{(1 - |\mu_1|)^2}{(\mu_2 - 2|\mu_1| + 1)}, & \mu_2 < |\mu_1|.
\end{cases}
\tag{3.240}
$$

This is called *Selberg inequality*, and generalizes Čebyšev inequality (3.7).

Example 3.37. Given the d-dimensional random vector $\mathbf{X} = (X_1, \ldots, X_d)$ with known power moments $\mathbb{E}[X_i] = 0$ and $\sigma_{ij} \triangleq \mathbb{E}[X_i X_j]$, define

$$
\mathsf{p} \triangleq 1 - \mathbb{P}\left[|X_i| \leqslant T_i, 1 \leqslant i \leqslant d\right].
\tag{3.241}
$$

This is the probability that the vector value taken by $\mathbf{X}$ does not lie in the parallelepiped

$$
\Pi \triangleq \{\mathbf{x} \mid |x_i| \leqslant T_i, i = 1, \ldots, d\}.
\tag{3.242}
$$

The generalization of (3.7) to d dimensions yields

$$\mathsf{p} \leqslant \sum_{i=1}^{d} \frac{\sigma_{ii}}{T_i^2}. \tag{3.243}$$

This bound is the tightest possible if the RVs X_i are uncorrelated ($\sigma_{ij} = 0$ for $i \neq j$), but may be rather loose if they are highly correlated [270]. Notice that, if the RVs X_i were known to be independent, it would be immediate to derive from (3.7) the tighter inequality

$$\mathsf{p} \leqslant 1 - \prod_{i=1}^{d} \left[1 - \frac{\sigma_{ii}}{T_i^2} \right] \tag{3.244}$$

whose dependence on d is exponential, rather than linear as in (3.243). By choosing as $\mathcal{S}$ an ellipsoid lying entirely inside Π, and using (3.238), the following bound can be derived for the general case of $\mathbf{X}$ having mean zero and covariance matrix $\mathbf{C}$ [270, Theorem 1]:

$$\mathsf{p} \leqslant \mathrm{tr}\left[\mathbf{C}\mathbf{T}^{-1} \right], \tag{3.245}$$

where $\mathbf{T}$ is any positive-definite symmetric matrix with diagonal elements $\mathbf{T}|_{ii} = T_i^2$. In some simple cases it is possible to find explicitly the matrix $\mathbf{T}$ which minimizes the RHS of (3.245). For example, with $d = 2$, the matrix $\mathbf{T}$ depends on a single parameter σ_{12}. Minimizing with respect to this parameter, we obtain [270, p. 237]

$$\begin{aligned} \mathsf{p} &\leqslant \frac{\sigma_{11}}{T_1^2} + \frac{\sigma_{22}}{T_2^2} && \text{if } \sigma_{12} = 0, \\ \mathsf{p} &\leqslant \max\left[\frac{\sigma_{11}}{T_1^2}, \frac{\sigma_{22}}{T_2^2} \right] && \text{if } \sigma_{12}^2 = \sigma_{11}\sigma_{22}. \end{aligned} \tag{3.246}$$

$\diamond$

Example 3.38. A family of Čebyšev-like inequalities involving $\mathbb{P}[Y \geqslant 1]$, with $Y = \max_{1 \leqslant i \leqslant d} |X_i|$, $Y = \max_{1 \leqslant i \leqslant d} X_i$, $Y = \min_{1 \leqslant i \leqslant d} |X_i|$, $Y = \min_{1 \leqslant i \leqslant d} X_i$, $Y = \prod_{1 \leqslant i \leqslant d} |X_i|$, and $Y = \prod_{1 \leqslant i \leqslant d} X_i$, is derived in [168] and the references therein. $\diamond$

Example 3.39. One in a set of M possible signals $\mathcal{M} = \{\mathbf{s}_1, \ldots, \mathbf{s}_M\}$, $\mathbf{s}_i \in \mathbb{R}^d$, is transmitted over a noisy channel. The received signal is $\mathbf{Y} = \mathbf{s}_k + \mathbf{Z}$, where $\mathbf{Z} \sim \mathcal{N}(\mathbf{0}, \sigma^2 \mathbf{I})$ is a vector RV modeling the noise, and $\mathbf{s}_k$ is chosen at random from $\mathcal{M}$ with probability $1/M$. The receiver observes a realization $\mathbf{y}$ of $\mathbf{Y}$ and uses it for demodulation. The *minimum distance* demodulator picks the signal $\hat{\mathbf{s}} \in \mathcal{M}$ with the minimum Euclidean distance from $\mathbf{y}$,

$$\hat{\mathbf{s}} = \arg \min_{\mathbf{s} \in \mathcal{M}} |\mathbf{y} - \mathbf{s}|. \tag{3.247}$$

Denoting by $\mathcal{V}_k$ the *Voronoi region* of $\mathbf{s}_k$, i.e., the set of vectors in $\mathbb{R}^d$ whose distance from $\mathbf{s}_k$ is smaller than from any other $\mathbf{s} \in \mathcal{M}$, correct maximum-likelihood demodulation requires that, when $\mathbf{s}_k$ is transmitted, $\mathbf{y}$ lie in $\mathcal{V}_k$. Thus, the probability of a demodulation error when $\mathbf{s}_k$ is transmitted is

$$\mathbb{P}[\hat{\mathbf{s}} \neq \mathbf{s}_k \mid \mathbf{s}_k \text{ transmitted}] = 1 - \mathbb{P}[\mathbf{Y} \in \mathcal{V}_k] \tag{3.248}$$

and can be bounded in the form of a Čebyšev bound [38, §7.4.3], [50].   ◇

3.11.1.2 Method of contact polynomials

Consider the moment space spanned by the functions

$$h_0(\mathbf{x}) = 1, \quad h_i(\mathbf{x}) = x_i, \quad h_{ij}(\mathbf{x}) = x_i x_j, \qquad i = 1, \ldots, d, \ 1 \leqslant i \leqslant j \leqslant d. \tag{3.249}$$

We examine probability measures constrained by power moments of order up to 2. More specifically, we consider the following constraints:

$$\mathbb{E}_{\mathbb{P}}[X_i] = 0, \qquad \mathbb{E}_{\mathbb{P}}[X_i X_j] = \sigma_{ij}, \qquad i, j = 1, \ldots, d, \tag{3.250}$$

and such that the matrix $\mathbf{C}$ with entries σ_{ij} is positive definite. This yields an interior point of the moment space [136, p. 505]. Now, let $\Phi(\mathbf{x})$ be a nonnegative function, bounded on compact domains, such that $\Phi(\mathbf{x}) = \Phi(-\mathbf{x})$, and suppose that

$$\lim_{|\mathbf{x}|^2 \to \infty} \frac{\Phi(\mathbf{x})}{(\mathbf{A}\mathbf{x}, \mathbf{x})} = 0 \tag{3.251}$$

for any positive definite matrix $\mathbf{A}$. To determine bounds on $\mathbb{E}[\Phi(\mathbf{X})]$, consider polynomials $P(\mathbf{x})$ satisfying, for any $\mathbf{x} \in \mathbb{R}^d$, the inequality

$$p_0 + \sum_{i=1}^{d} p_i x_i + \sum_{i=1}^{d} \sum_{j=1}^{d} p_{ij} x_i x_j \geqslant \Phi(\mathbf{x}). \tag{3.252}$$

Due to the symmetry of $\Phi(\mathbf{x})$, we also have

$$p_0 - \sum_{i=1}^{d} p_i x_i + \sum_{i=1}^{d} \sum_{j=1}^{d} p_{ij} x_i x_j \geqslant \Phi(\mathbf{x}). \tag{3.253}$$

Adding (3.252) and (3.253) we obtain, for any $\mathbf{x} \in \mathbb{R}^d$,

$$p_0 + \sum_{i=1}^{d} \sum_{j=1}^{d} p_{ij} x_i x_j \geqslant \Phi(\mathbf{x}). \tag{3.254}$$

Since the expected values of the polynomials in (3.252)–(3.254) are equal, we can restrict our attention to polynomial (3.254). The following result holds [136, p. 506]:

$$\sup_{\mathbb{P}} \mathbb{E}_{\mathbb{P}}[\Phi(\mathbf{X})] = \inf [p_0 + \text{tr}(\mathbf{A}\mathbf{C})], \tag{3.255}$$

where $\mathbb{P}$ satisfies the moment constraints and the infimum is evaluated with respect to the set of all polynomials satisfying (3.254). Simplifications are possible when we assume that $\Phi(\mathbf{x})$ is the indicator function of a symmetric set $\mathcal{S}$, and hence takes only values 0 and 1. Other special choices of $\Phi(\mathbf{x})$, examined in [136, Chapter XIII], lead to Čebyšev-type inequalities for $\mathbb{P}[\max_{1 \leqslant i \leqslant d} |X_i| \geqslant 1]$ and $\mathbb{P}[\max_{1 \leqslant i \leqslant d} X_i \geqslant 1]$.

3.11.1.3 Method of contact polynomials: the bivariate case

To study the special case $d = 2$, we denote by $F_{XY}(x, y) \triangleq \mathbb{P}[X \leqslant x, Y \leqslant y]$ the joint CDF of the pair of RVs (X, Y) defined in $\mathbb{R}^2$ with marginal CDFs $F_X(x)$, $F_Y(y)$. If no moment constraints on $F_{XY}(x, y)$ are assigned, then the Fréchet–Hoeffding inequalities, to be illustrated in Chapter 6, provide sharp upper and lower bounds on $F_{XY}(x, y)$ in the form

$$\max[F_X(x) + F_Y(y) - 1, 0] \leqslant F_{XY}(x, y) \leqslant \min[F_X(x), F_Y(y)]. \qquad (3.256)$$

Assume that only the following quantities are known: mean values μ_X, μ_Y, variances σ_X^2, σ_Y^2, and linear correlation coefficient

$$\rho_{X,Y} \triangleq \frac{\text{cov}(X, Y)}{\sqrt{\text{var}(X)\text{var}(Y)}}, \qquad (3.257)$$

where $\text{cov}(X, Y) \triangleq \mathbb{E}[XY] - \mathbb{E}[X]\mathbb{E}[Y]$ denotes the covariance of X and Y, and the RVs X and Y are assumed to have finite variances $\text{var}(X)$ and $\text{var}(Y)$. Then, the method of contact polynomials can be used to bound $\mathbb{E}[\Phi(x, y)]$. We start from the construction of a bivariate polynomial

$$P(x, y) = p_{20}x^2 + p_{02}y^2 + p_{11}xy + p_{10}x + p_{01}y + p_{00} \qquad (3.258)$$

such that $P(x, y) \geqslant \Phi(x, y)$ (or $P(x, y) \leqslant \Phi(x, y)$). Reference [120, Theorem 2.1, Chapter V] derives the maximum value of $F_{XY}(x, y)$ with $(x, y) \in \mathbb{R}^2$ under the constraints above. This gives rise to the "Bivariate Čebyšev–Markov inequality."

Example 3.40. Consider the "standardized" pair of RVs X, Y with $\mu_X = \mu_Y = 0$, and $\sigma_X^2 = \sigma_Y^2 = 1$. The bivariate Čebyšev–Markov inequality for $F_{XY}(x, y)$ is derived in [120, p. 213] using quadratic contact polynomials. $\diamond$

Example 3.41. Given the pair X, Y of RVs with $\mu_X = \mu_Y = 0$, and the bivariate stop-loss function $\Phi(x, y) = \max(x + y - D, 0)$, $D \in \mathbb{R}$, define the following quadratic majorant of Φ:

$$P(x, y) = ax^2 + by^2 + cxy + dx + ey + f \qquad (3.259)$$

and the identities

$$\max(D - x - y, 0) = \int_0^D \mathbb{1}_{\{x, y \mid x \leqslant u, y \leqslant D - u\}} \, du, \qquad (3.260a)$$

$$\max(x + y - D, 0) = x + y - D + \max(D - y - y, 0). \tag{3.260b}$$

Inserting (3.260b) into (3.260a) and taking expectations, one obtains

$$\mathbb{E}\left[\max(X + Y - D, 0)\right] = \mathbb{E}[X] + \mathbb{E}[Y] - D + \int_{-\infty}^{D} F_{X,Y}(x, D - x)\,dx. \tag{3.261}$$

Upper and lower bounds on $\mathbb{E}[\Phi(X, Y)]$ can be obtained from (3.261) after a good deal of algebra [120, p. 217]. $\diamond$

3.11.1.4   The role of cubature rules

While quadrature rules, and especially GQRs, play a central role in the applications of one-dimensional moment-bound theory, the development of corresponding approximate integration rules in $d \geqslant 2$ dimensions, called *cubature rules*, is not sufficiently advanced to allow their general use to bound expectations of the form $\mathbb{E}[\Phi(\mathbf{X})]$ when $\mathbf{X}$ is only known through a number of its moments.

Cubature rules with n points and degree of exactness δ satisfy

$$\mathbb{E}_{\mathbb{P}}[P(\mathbf{X})] = \sum_{i=1}^{n} w_i P(\mathbf{x}_i) \tag{3.262}$$

for all polynomials P on $\mathbb{R}^d$ with degree $\leqslant \delta$. Cubature rules of degree of exactness δ are proved to exist for $\mathbb{P}$ where the $(\delta + 1)$th moments exist. In the case of compact support of $\mathbb{P}$, this result is known as *Tchakaloff theorem*. Reference [11] provides a simple proof of this theorem, while its extension to probability measures with non-compact support can be found in [202]. *Gauss cubature rules* can also be defined as those whose number n of abscissas is the smallest possible for a given δ, just like in the one-dimensional case.

For a given degree of exactness, a relevant problem is the search for a cubature rule with the minimum number of points. For $\delta = 2$, the minimum number of points is $d + 1$ [244, p. 79 ff.]. Third-degree rules with $2d$ points and fifth-degree rules with $d^2 + d + 2$ points are discussed in [244, §3.9] and [244, §3.10], respectively. An extensive list of cubature rules can be found in [244, Chapter 8]. One may also want to generalize the one-dimensional case and look for a simple way to construct cubature rules with maximum degree of exactness. However, this problem is unsolved in general, and only sparse results are available. The situation is further complicated by the fact that in d dimensions the region of integration may not be as simple as a product of one-dimensional intervals. The best lower bounds on the number of points in cubature rules of odd degree are listed in [51]. Rules attaining a known lower bound are available only for small values of d and low degrees of exactness. Moreover, the error made when a cubature rule of a given degree δ is applied to a function that is not a polynomial of degree smaller than or equal to δ may not be known, or hard to evaluate (see [51, pp. 11–12]).

Generally, one may settle for cubature rules whose degree of exactness is suboptimal. In principle, one can construct a cubature rule based on moments using a brute-force procedure, which consists of finding the solution to a system of nonlinear equations involving moments, abscissas, and weights.

Example 3.42. In [106], two-dimensional cubature rules are constructed with 12 points and $\delta = 7$, and 19 points and 18 point with $\delta = 9$ for planar regions and pdfs symmetric in each variable, i.e.,

$$f_{XY}(x, y) = f_{XY}(-x, y) = f_{XY}(x, -y). \tag{3.263}$$

$\diamond$

Example 3.43. In [179], a cubature rule with $\delta = 4$ and $n = 6$ is derived which is valid for pairs of RVs whose pdf is symmetric in each variable. With these assumptions, $\mu_{ik} = 0$ if i or k is an odd integer. The moments needed to derive the cubature rule are μ_{20}, μ_{02}, μ_{40}, μ_{04}, and μ_{22}. Under the same symmetry assumptions, in [192] a cubature rule with $\delta = 9$ and $n = 19$ is obtained using the moments μ_{20}, μ_{02}, μ_{40}, μ_{04}, μ_{22}, μ_{24}, μ_{42}, μ_{06}, μ_{60}, μ_{26}, μ_{62}, μ_{80}, μ_{08}, and μ_{44}. In [106], a method of constructing cubature rules with $\delta = 7$ and $n = 19$ and with $\delta = 9$ and $n = 18$ is given for planar regions and pdfs symmetric in each variable. $\diamond$

Simple cubature rules with an assigned degree of exactness δ can be obtained as products of one-dimensional GQRs. For example, if

$$\int \varphi(x)\,dx \approx \sum_{i=1}^{n} w_i \varphi(x_i) \tag{3.264}$$

is a quadrature rule with degree δ, then

$$\int \Phi(x, y)\,dx\,dy \approx \sum_{i=1}^{n} \sum_{j=1}^{n} w_i w_j \Phi(x_i, y_j) \tag{3.265}$$

is a cubature rule with the same degree of exactness δ. If X and Y are independent, then (3.265) can be determined from the knowledge of the moments of the individual RVs X and Y. The case of correlated RVs can be dealt with if the conditional moments of X given Y can be computed.

Orthogonal d-dimensional polynomials may be generated from the knowledge of moments, and used to construct cubature rules. The relevant result, summarized in [244, Theorem 3.7-2] and described here, for simplicity's sake, only in the case $d = 2$, shows that a cubature rule with weights $w_1, \ldots, w_n$ and abscissas $(x_1, y_1), \ldots, (x_n, y_n)$ exists such that

$$\mathbb{E}_{XY}[P(x, y)] = \sum_{i=1}^{n} w_i P(x_i, y_i) \tag{3.266}$$

for all polynomials $P(x, y)$ of degree $\leqslant \delta$. The abscissas (x_i, y_i) are the zeros of a set of orthogonal polynomials. In $\mathbb{R}^d$, there exist exactly $(d + \delta - 1)!/((d+1)!\delta!)$ linearly independent polynomials, $P_d(x_1, \ldots, x_d)$ each of which is orthogonal over $\mathbb{R}^d$ to all polynomials Q of degree $\leqslant \delta - 1$ [244, §3.5],

$$\mathbb{E}_{X_1 \cdots X_d}[Q(X_1, \ldots, X_d)P_d(X_1, \ldots, X_d)] = 0. \tag{3.267}$$

The conditions for the existence of such cubature rules are described in [244, p. 76], while an extensive treatment of the connection between orthogonal polynomials and cubature rules can be found in [53].

Example 3.44. The following four polynomials over $\mathbb{R}^2$, orthogonal to all bivariate polynomials of degree $\leqslant 2$:

$$
\begin{aligned}
p_{3,0}(x, y) &= a_{00} + a_{10}x + a_{01}y + a_{20}x^2 + a_{11}xy + a_{02}y^2 + x^3, \\
p_{2,1}(x, y) &= b_{00} + b_{10}x + b_{01}y + b_{20}x^2 + b_{11}xy + b_{02}y^2 + x^2y, \\
p_{1,2}(x, y) &= c_{00} + c_{10}x + c_{01}y + c_{20}x^2 + c_{11}xy + c_{02}y^2 + xy^2, \\
p_{0,3}(x, y) &= d_{00} + d_{10}x + d_{01}y + d_{20}x^2 + d_{11}xy + d_{02}y^2 + y^3,
\end{aligned} \tag{3.268}
$$

can be derived from the knowledge of a few power moments as follows. Take $p_{3,0}$. This is orthogonal to all bivariate polynomials with degree $\leqslant 2$ if and only if it is orthogonal to 1, x, y, x^2, xy, and y^2. This condition leads to six linear equations for the coefficients of $p_{3,0}(x)$:

$$
\begin{bmatrix}
\mu_{00} & \mu_{10} & \mu_{01} & \mu_{20} & \mu_{11} & \mu_{02} \\
\mu_{10} & \mu_{20} & \mu_{11} & \mu_{30} & \mu_{21} & \mu_{12} \\
\mu_{01} & \mu_{11} & \mu_{02} & \mu_{21} & \mu_{12} & \mu_{03} \\
\mu_{20} & \mu_{30} & \mu_{21} & \mu_{40} & \mu_{31} & \mu_{22} \\
\mu_{11} & \mu_{21} & \mu_{12} & \mu_{31} & \mu_{22} & \mu_{13} \\
\mu_{02} & \mu_{12} & \mu_{03} & \mu_{22} & \mu_{13} & \mu_{04}
\end{bmatrix}
\times
\begin{bmatrix}
a_{00} \\
a_{10} \\
a_{01} \\
a_{20} \\
a_{11} \\
a_{02}
\end{bmatrix}
=
\begin{bmatrix}
-\mu_{30} \\
-\mu_{40} \\
-\mu_{31} \\
-\mu_{50} \\
-\mu_{41} \\
-\mu_{32}
\end{bmatrix}. \tag{3.269}
$$

The 6×6 matrix in (3.269) is nonsingular [244, Theorem 3.5-1], so that the polynomial $p_{30}(x, y)$ exists and is unique. Similar calculations lead to the remaining polynomials in (3.268). Any third-degree polynomial orthogonal over $\mathbb{R}^2$ with respect to $f_{XY}(x, y)$ is a linear combination of those in (3.268) [244, Theorem 3.5-2], and for this reason they are called *basic third-degree orthogonal polynomials* for $\mathbb{R}^2$ and bivariate pdf $f_{XY}(x, y)$. $\diamond$

Example 3.45. The basic orthogonal polynomials in the d-cube $[-1, 1]^d$ are products of Legendre polynomials [244, p. 71]:

$$P_{\alpha_1, \alpha_2, \ldots, \alpha_d}(x_1, x_2, \ldots, x_d) = P_{\alpha_1}(x_1)P_{\alpha_2}(x_2) \cdots P_{\alpha_d}(x_d), \tag{3.270}$$

where $P_{\alpha_i}(x)$ is the monic Legendre polynomial of degree α_i for the interval $(-1, 1)$ [2, Chapter 22]. $\diamond$

Sources and parerga

1. The moment problem received its first systematic treatment in the works of P.L. Čebyšev, A.A. Markov, T.J. Stieltjes, and, later, H. Hamburger, R. Nevanlinna, M. Riesz, F. Hausdorff, T. Carleman, and M.H. Stone. To the author's knowledge, the first systematic treatment of the problem was published in 1943 [232]. This reference contains also a brief historical review of the problem. As the bibliography on moment bounds is tropically rich, in this chapter we limit ourselves to the citation of a relatively small sample of books and papers covering the subject.

2. The edited book [151] describes a number of theoretical and applied topics benefiting from results generated by the moment problem. Among these, there are applications to the prediction of stochastic processes, to the design of algorithms for signal-processing VLSI chips, to statistics, and to the spectral decomposition of operators.

3. The books by Karlin and Studden [136] and by Kreĭn and Nudel'man [149] are the standard references for Čebyšev systems. As these presentations are very general, the probabilistic interpretation and significance of the results may not be immediately apparent. Paper [129] contains a concise review of the results of Čebyšev-system theory relevant to probabilistic applications.

4. Several inequalities related to Markov and Čebyšev bounds are examined in [95]. Reference [220] lists a number of inequalities valid when additional restrictions on the distribution of X are assigned (e.g., unimodality).

5. In mathematical literature, authors often distinguish between the *classical moment problem* of deriving the existence of a probability measure with given specified moments and the *generalized moment problem* of finding bounds on the expected value of an RV whose moments are known.

6. We have chosen "Čebyšev" among the many transliterations of the name of the Russian mathematician Пафнутий Львович Чебышёв. As commented in Ralph P. Boas' "Spelling Lesson" [36],

> *Put letters in or leave them out,*
> *Garnish with accents round about,*
> *Finish the name with -eff or -off:*
> *There is no way to spell Чебышёв.*

7. Monograph [153] shows how, by restricting consideration to probability measures defined on special support sets called semialgebraic, the difficulties inherent in the solution of many moment problems in $\mathbb{R}^d$ can be alleviated, and how some of these may be solved by using a semidefinite-programming approach.

8. A generalized notion of unimodality may also be defined (see [40] and references therein). With this definition, a RV is said to be α-unimodal with α-mode m if X has the same distribution as $m + U^{1/\alpha}V$, where U and V are independent RVs and U is uniformly distributed on $[0, 1]$.

9. The SDP approach to moment problems is described, among other sources, in [18,196]. In [240], it is ascribed to [124] and other Isii papers. In [18], op-

timal bounds are computed for arbitrary distributions given any finite number of generalized moments.

10. For applications of linear programming to moment problems, see, for example, [38, p. 150].

11. A number of inequalities of Čebyšev type, subject to various constraints, are tabulated in [96] for $d = 1$, as well as for general d.

12. The development of cubature rules based on orthogonal polynomials was pioneered by Johann Radon in his 1948 influential paper [206] (although this is not the oldest paper).

13. In [51, pp. 2–3], an interesting story connecting Kepler, cubatures, and Austrian wine barrels is illustrated.

14. Book [241] contains a treatment of cubature rules based on functional-analysis techniques.

15. Reference [52] contains an extensive tabulation of known cubature rules, as well as the correction of some misprints in [244].

Interval analysis

4

Interval analysis (or interval arithmetic), which studies computations done with imprecise data, is an algebra of quantities whose exact values are only known to lie within given intervals. Interval analysis allows one to control the inaccuracies resulting from calculations affected by rounding errors or uncertain data. If a calculation is done using intervals rather than numbers, its result is an interval which is expected to include the exact value.

The key entities here are *interval numbers*, i.e., nonempty intervals representing the upper and lower values, denoted $\overline{x}$ and $\underline{x}$, respectively, of an imprecisely known real quantity x. An interval number $[\underline{x}, \overline{x}]$, which has the dual nature of a number pair and of a set, can be simply denoted by $[x]$. Formally,

$$[x] \triangleq \{x \in \mathbb{R} \mid \underline{x} \leqslant x \leqslant \overline{x}\}, \tag{4.1}$$

where $\underline{x} = -\infty$ and $\overline{x} = \infty$ are also allowed. In this context, a precise real number x is represented by the interval $[x, x]$. For interval analysis to be an extension of real analysis, its rules must be consistent with the latter whenever the interval reduces to a real number, i.e., when $\underline{x} = \overline{x} = x$.

Interval $[\underline{x}, \overline{x}]$ is also called an *enclosure* of x. Clearly, many enclosures of the same x can be defined, tighter enclosures being more satisfactory.

Example 4.1. Archimedes of Syracuse, in Proposition 3 of his treatise "Measurement of a Circle" [112], stated that the ratio of the circumference of any circle to its diameter is less than 22/7 but greater than 223/71. Using interval-analysis notations, this statement is expressed in the form $[\pi] = [223/71, 22/7]$. A tighter enclosure using an interval with rational endpoints is $[\pi] = [333/106, 355/113]$. Since π has no exact decimal representation, the use of interval number $[\pi]$ does not introduce uncontrolled inaccuracies in the final calculations. $\diamond$

4.1 Some definitions

The core of interval analysis is the development of a panoply of methods for performing operations on intervals. The accuracy achieved by representing number x as interval $[x]$ is described by the *width* of $[x]$, defined as $w([x]) \triangleq \overline{x} - \underline{x}$. If we are certain that the true value of x lies in $[x]$, then $w([x])$ quantifies the amount of error. The *absolute value* of $[x]$, denoted by $\|[x]\|$, is the maximum of the absolute values of its endpoints, $\|[x]\| \triangleq \max(|\underline{x}|, |\overline{x}|)$. The *midpoint* of $[x]$ is $m([x]) \triangleq (\underline{x} + \overline{x})/2$.

Remark 4.1. In interval analysis, the result of a computation with uncertain quantities lies within a set $\mathcal{S}$ which might not be an interval. If this is the case, to simplify

Dimensions of Uncertainty in Communication Engineering. https://doi.org/10.1016/B978-0-32-399275-6.00012-5

the computations, S may be eventually enclosed into an interval of $\mathbb{R}$, or, in a d-dimensional space, into an axis-aligned parallelepiped of $\mathbb{R}^d$ which forms an outer approximation of S. This choice may have consequences: one is pessimism, as the interval containing S may be larger than S, and the other is that properties holding true for set computations may no longer be true for computations with enclosures [125, p. 17]. This effect will be illustrated in Example 4.3 *infra*.

Equality of two interval numbers $[\underline{x}, \overline{x}] = [\underline{y}, \overline{y}]$ occurs when $\underline{x} = \underline{y}$ and $\overline{x} = \overline{y}$. A *partial order* in $[-\infty, \infty]$ may be defined by

$$[\underline{x}, \overline{x}] < [\underline{y}, \overline{y}] \quad \text{if and only if} \quad \overline{x} < \underline{y}. \tag{4.2}$$

A *distance* between interval numbers $[x]$ and $[y]$ may be defined as

$$d([x], [y]) \triangleq \max(|\underline{x} - \underline{y}|, |\overline{x} - \overline{y}|) \tag{4.3}$$

(so that, for example, $d([1, 3], [2, 4]) = \max(|1 - 2|, |3 - 4|) = 1$).

Remark 4.2. It can be proved that this definition of interval distance induces in the set of interval numbers a Hausdorff metric space, which allows the concepts of limit, continuity, and convergence to be extended to interval algebra [60, p. 15].

4.2 Set operations on intervals

Unary and binary algebraic operations on interval numbers can be defined to mimic corresponding set-theoretic definitions. Thus, the intersection $[x] \cap [y]$ of two intervals is empty if either $\overline{y} < \underline{x}$ or $\overline{x} < \underline{y}$, i.e., $[x]$ and $[y]$ have no points in common. Otherwise,

$$\begin{aligned} [x] \cap [y] &\triangleq [\{z \mid z \in [x] \text{ and } z \in [y]\}] \\ &= \left[\max\{\underline{x}, \underline{y}\}, \min\{\overline{x}, \overline{y}\} \right]. \end{aligned} \tag{4.4}$$

If $[x] \cap [y] \neq \emptyset$, then the union of two intervals is defined as

$$\begin{aligned} [x] \cup [y] &\triangleq \{z \mid z \in x \text{ or } z \in y\} \\ &= \left[\min\{\underline{x}, \underline{y}\}, \max\{\overline{x}, \overline{y}\} \right]. \end{aligned} \tag{4.5}$$

If $[x] \cap [y] = \emptyset$, the union above is not an interval, in which case we may define the *interval hull* of $[x]$ and $[y]$ as the enclosure

$$\begin{aligned} [x] \sqcup [y] &\triangleq [[x] \cup [y]] \\ &= [\min\{\underline{x}, \underline{y}\}, \max\{\overline{x}, \overline{y}\}], \end{aligned} \tag{4.6}$$

which is always an interval. Notice that $([x] \sqcup [y]) \supseteq ([x] \cup [y])$.

4.3 Algebraic operations on interval numbers

The negative of $[x]$ is defined as

$$-[x] \triangleq [-\overline{x}, -\underline{x}]. \tag{4.7}$$

The reciprocal of $[x]$ is defined, under the assumption that $0 \notin [x]$, as

$$1/[x] \triangleq \{x \mid 1/x \in [\underline{x}, \overline{x}]\} = [1/\overline{x}, 1/\underline{x}]. \tag{4.8}$$

For the multiplication by a real number α, we have

$$\alpha[x] \triangleq \begin{cases} [\alpha\underline{x}, \alpha\overline{x}], & \alpha \geqslant 0, \\ [\alpha\overline{x}, \alpha\underline{x}], & \alpha < 0. \end{cases} \tag{4.9}$$

For binary operations, generally denoted by $\circ$, that are monotone in both arguments, we may write

$$[x] \circ [y] \triangleq [\{x \circ y \mid x \in [x] \text{ and } y \in [y]\}]. \tag{4.10}$$

In particular, for the four binary operations in set $\{+, -, \cdot, \div\}$, with $x \circ y$ well defined for all $[\underline{x}, \overline{x}]$ and $[\underline{y}, \overline{y}]$:

$$[x] \circ [y] = [\min(\underline{x} \circ \underline{y}, \underline{x} \circ \overline{y}, \overline{x} \circ \underline{y}, \overline{x} \circ \overline{y}), \max(\underline{x} \circ \underline{y}, \underline{x} \circ \overline{y}, \overline{x} \circ \underline{y}, \overline{x} \circ \overline{y})]. \tag{4.11}$$

Eq. (4.11) specializes as follows:

(i)   Addition, $[x] + [y] = [\underline{x} + \underline{y}, \overline{x} + \overline{y}]$;
(ii) Subtraction, $[x] - [y] = [\underline{x} - \overline{y}, \overline{x} - \underline{y}]$;
(iii) Multiplication, $[x] \cdot [y] = [\min(\underline{x}\,\underline{y}, \underline{x}\,\overline{y}, \overline{x}\,\underline{y}, \overline{x}\,\overline{y}), \max(\underline{x}\,\underline{y}, \underline{x}\,\overline{y}, \overline{x}\,\underline{y}, \overline{x}\,\overline{y})]$;
(iv) Division, $[x] \div [y] = [x] \cdot 1/[y]$.

For powers, using the definition of interval multiplication (iii) above, we may define, for any integer n,

$$n = 0 \Rightarrow [x]^0 = [1, 1],$$
$$n > 0 \Rightarrow [x]^n = [x]^{n-1} \cdot [x], \tag{4.12}$$
$$n \geqslant 0 \text{ and } [x] \neq [0] \Rightarrow [x]^{-n} = 1/[x]^n.$$

The definition of interval power leads to an apparent inconsistency that may be resolved using the definition of power function to be provided in Example 4.6 *infra*. In fact, computing the square of $[-1, 1]$, we obtain

$$[-1, 1]^2 = [-1, 1] \cdot [-1, 1] = [-1, 1], \tag{4.13}$$

which is not consistent with the intuition that a square should always be nonnegative. Actually, using (iii), we have in general, if $0 \in [\underline{x}, \overline{x}]$,

$$
\begin{aligned}
[\underline{x}, \overline{x}]^2 &= [\underline{x}, \overline{x}] \cdot [\underline{x}, \overline{x}] \\
&= \left[\min(\underline{x}^2, \underline{x}\,\overline{x}, \overline{x}^2), \max(\underline{x}^2, \underline{x}\,\overline{x}, \overline{x}^2) \right] \\
&= \left[\underline{x}\,\overline{x}, \max(\underline{x}^2, \overline{x}^2) \right].
\end{aligned}
\tag{4.14}
$$

However, this is not a problem if we are willing to accept the outcome of an interval operation as correct whenever it contains all possible results, and hence the exact one [60, p. 12].

An important property of interval arithmetic is *inclusion monotonicity* [60, Theorem 2.13], which states that, if $[x_1] \subseteq [y_1]$ and $[x_2] \subseteq [y_2]$, then for $\circ \in \{+, \cdot\}$ we have

$$
[x_1] \circ [x_2] \subseteq [y_1] \circ [y_2].
\tag{4.15}
$$

To prove it, observe that, if (4.15) holds,

$$
\begin{aligned}
[x_1] \circ [x_2] &= \{x_1 \circ x_2 \mid x_1 \in [x_1], x_2 \in [x_2]\} \\
&\subseteq \{y_1 \circ y_2 \mid y_1 \in [y_1], y_2 \in [y_2]\} \\
&= [y_1] \circ [y_2].
\end{aligned}
\tag{4.16}
$$

4.3.1 Properties that may not be shared with ordinary real algebra

Interval addition and multiplication are commutative and associative:

$$
\begin{aligned}
[x] + [y] &= [y] + [x], & [x] + ([y] + [z]) &= ([x] + [y]) + [z], \\
[x] \cdot [y] &= [y] \cdot [x], & [x] \cdot ([y] \cdot [z]) &= ([x] \cdot [y]) \cdot [z],
\end{aligned}
\tag{4.17}
$$

and the interval numbers $[0, 0]$ and $[1, 1]$ are identity elements for addition and multiplication, respectively:

$$
\begin{aligned}
[0, 0] + [x] &= [x] + [0, 0] = [x], \\
[1, 1] \cdot [x] &= [x] \cdot [1, 1] = [x], \\
[0, 0] \cdot [x] &= [x] \cdot [0, 0] = [0, 0].
\end{aligned}
\tag{4.18}
$$

Interval number $-[x]$ is not the additive inverse of $[x]$. In fact,

$$
[x] + (-[x]) = [\underline{x}, \overline{x}] + [-\overline{x}, -\underline{x}] = [\underline{x} - \overline{x}, \overline{x} - \underline{x}],
\tag{4.19}
$$

which is equal to $[0, 0]$ only if $\underline{x} = \overline{x}$, i.e., $[x]$ has zero width. Similarly, interval number $1/[x]$ is not the multiplicative inverse of $[x]$, unless $[x]$ has zero width. In fact,

$$[x] \cdot 1/[x] = \begin{cases} [\underline{x}/\overline{x}, \overline{x}/\underline{x}], & \underline{x} > 0, \\ [\overline{x}/\underline{x}, \underline{x}/\overline{x}], & \overline{x} < 0. \end{cases} \tag{4.20}$$

Multiplication is not distributive with respect to addition. The *subdistributive property* holds:

$$[x] \cdot ([y] + [z]) \subseteq [x] \cdot [y] + [x] \cdot [z] \tag{4.21}$$

(this property is a consequence of the *interval dependence problem*, to be illustrated in Section 4.6 *infra*).

The cancellation law holds for addition [177, p. 33]:

$$[x] + [z] = [y] + [z] \Rightarrow [x] = [y], \tag{4.22}$$

but not for multiplication, as $[z] \cdot [x] = [z] \cdot [y]$ does not imply $[x] = [y]$.

4.4 Interval vectors and matrices

A d-dimensional real *interval vector* $[\mathbf{x}]$ is the Cartesian product of d closed intervals

$$[\mathbf{x}] \triangleq [x_1] \times [x_2] \times \cdots \times [x_d], \quad [x_i] = [\underline{x}_i, \overline{x}_i], \quad i = 1, \ldots, d. \tag{4.23}$$

Thus, nonempty interval vectors are d-dimensional parallelepipeds with sides aligned to the axes (see Fig. 4.1).

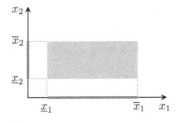

Figure 4.1 A 2-dimensional interval vector $[\mathbf{x}] = [x_1] \times [x_2]$.

Example 4.2. Fig. 4.2 shows the values taken by $[x + y, x - y]^\mathsf{T}$ with $x = [-1, 1]$ and $y = [-0.2, 0.2]$, as well as their two-dimensional square enclosure. ◇

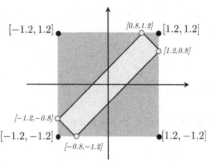

Figure 4.2 Two-dimensional range and square enclosure of the interval vector of Example 4.2.

Similarly, a real *interval m × n matrix* is the Cartesian product of *mn* closed intervals. We may write an interval matrix [**A**] in the equivalent forms

$$
\begin{aligned}
[\mathbf{A}] &= [a_{11}] \times [a_{12}] \times \cdots \times [a_{mn}] \\
&= \begin{bmatrix} [a_{11}] & \dots & [a_{1n}] \\ \vdots & \ddots & \vdots \\ [a_{m1}] & \dots & [a_{mn}] \end{bmatrix}.
\end{aligned}
\tag{4.24}
$$

Again, some of the properties of classical matrix calculus may not hold with interval matrices. For example, the product is no longer associative, so that the equalities

$$
([\mathbf{A}] \cdot [\mathbf{B}]) \cdot [\mathbf{C}] \overset{?}{=} [\mathbf{A}] \cdot ([\mathbf{B}] \cdot [\mathbf{C}])
\tag{4.25}
$$

and

$$
[\alpha] \cdot ([\mathbf{A}] \cdot [\mathbf{x}]) \overset{?}{=} [\mathbf{A}] \cdot ([\alpha] \cdot [\mathbf{x}])
\tag{4.26}
$$

do not hold in general. Part of this peculiar behavior is explained by the so-called *wrapping effect*, illustrated in the following example where $\mathbf{A}[\mathbf{x}] \supset \{\mathbf{Ax} \mid \mathbf{x} \in [\mathbf{x}]\}$.

Example 4.3. This example shows how we may have

$$
\mathbf{A}[\mathbf{x}] \supset \{\mathbf{Ax} \mid \mathbf{x} \in [\mathbf{x}]\}.
\tag{4.27}
$$

Taking

$$
\mathbf{A} = \begin{bmatrix} 1 & 1 \\ 0 & 1 \end{bmatrix} \quad \text{and} \quad [\mathbf{x}] = \begin{bmatrix} [-1, 0] \\ [1, 2] \end{bmatrix},
\tag{4.28}
$$

direct computation yields

$$A[\mathbf{x}] = \begin{bmatrix} 1 & 1 \\ 0 & 1 \end{bmatrix} \cdot \begin{bmatrix} [x_1] \\ [x_2] \end{bmatrix}$$

$$= \begin{bmatrix} [x_1] + [x_2] \\ [x_2] \end{bmatrix} \tag{4.29}$$

$$= \begin{bmatrix} [0, 2] \\ [1, 2] \end{bmatrix}.$$

This is represented, as $\mathbf{x} \in [\mathbf{x}]$, by the parallelepiped enclosure shown in Fig. 4.3. We see that, for example, $\mathbf{x} = (0, 2)^{\mathsf{T}}$ is in $A[\mathbf{x}]$, but not in $\{A\mathbf{x} \mid \mathbf{x} \in [\mathbf{x}]\}$. ◇

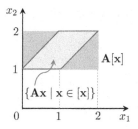

Figure 4.3 Illustration of wrapping effect: in Example 4.3, $A[\mathbf{x}] \supset \{A\mathbf{x} \mid \mathbf{x} \in [\mathbf{x}]\}$.

4.5 Interval functions

The problem of evaluating the interval image of a function of interval variables leads to the definition of *interval functions* (see Fig. 4.4). In some simple cases manipulation

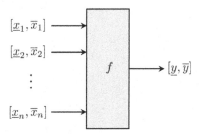

Figure 4.4 Interval function.

of interval functions is straightforward, as it occurs with functions including simple algebraic operations.

Example 4.4. The scalar function $f(x) = ax + b$, where x, a, and b are intervals, can be simply dealt with by computing its range: the interval $[\underline{a}, \overline{a}] \cdot [\underline{x}, \overline{x}] + [\underline{b}, \overline{b}]$

describes the range of $[f]$. We may also interpret f as a function of x with interval parameters $[a]$ and $[b]$, which allows the interval solutions of $f(x) = 0$ to be found in a straightforward way. From

$$[\underline{a}, \overline{a}]x + [\underline{b}, \overline{b}] = 0, \tag{4.30}$$

we obtain

$$[x] = \frac{[-\overline{b}, -\underline{b}]}{[\underline{a}, \overline{a}]}. \tag{4.31}$$

◇

More generally, consider a real-valued function f. We define

$$[f]([x]) \triangleq \{f(x) \mid x \in [x]\}. \tag{4.32}$$

If f is continuous in $[\underline{x}, \overline{x}]$, then

$$[f]([x]) = [\min_{x \in [x]} f(x), \max_{x \in [x]} f(x)]. \tag{4.33}$$

As a special case, for any nonempty $[x]$ and a monotone f, we have

$$[f]([x]) = \left[\min\big(f(\underline{x}), f(\overline{x})\big), \max\big(f(\underline{x}), f(\overline{x})\big)\right] \tag{4.34}$$

so that, for a nondecreasing f,

$$[f]([x]) = [f(\underline{x}), f(\overline{x})] \tag{4.35}$$

and, for a nonincreasing f,

$$[f]([x]) = [f(\overline{x}), f(\underline{x})] \tag{4.36}$$

(see Fig. 4.5(a)–(b)).

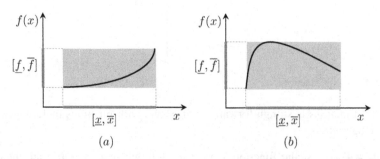

Figure 4.5 Interval function: (a) monotone f, (b) nonmonotone f.

Example 4.5. From Shannon channel-capacity equation $C = B \log_2(1 + \mathsf{snr})$, where B denotes bandwidth and snr signal-to-noise ratio, given $[B]$ and $[\mathsf{snr}]$, we may obtain

$$[C] = [B] \cdot [\log_2]([1 + \mathsf{snr}]) = \left[\underline{B} \log_2(1 + \underline{\mathsf{snr}}), \overline{B} \log_2(1 + \overline{\mathsf{snr}}) \right]. \qquad (4.37)$$

$$\diamond$$

In general, the result obtained through (4.33) can be extended, for functions that are piecewise monotone, if one examines the interval endpoints $\underline{x}$ and $\overline{x}$, along with the points where the slope of function f changes its sign.

Example 4.6. The interval values of function $f(x) = x^k$, with k a positive integer, are given by

$$[\underline{x}, \overline{x}]^k = \begin{cases} [\underline{x}^k, \overline{x}^k], & \underline{x} \geqslant 0, \\ [\overline{x}^k, \underline{x}^k], & \overline{x} \leqslant 0, \\ [0, \max(\underline{x}^k, \overline{x}^k)], & \underline{x} < 0 < \overline{x}. \end{cases} \qquad (4.38)$$

Notice how with this definition we obtain $[-1, 1]^2 = [0, 1]$, rather than $[-1, 1]$ as obtained from $[x] \cdot [x]$ in Eq. (4.13). $\qquad \diamond$

4.5.1 The vertex method

This conceptually simple method for the computation of $[f]([\mathbf{x}])$, suggested in [67], is based on (4.34). Given d intervals $[\underline{x}_i, \overline{x}_i]$, $1 \leqslant i \leqslant d$, form the interval parallelepiped $\bigotimes_{i=1}^{d}[\underline{x}_i, \overline{x}_i]$ as in Fig. 4.6, and compute the values of function f at all vertices. Then

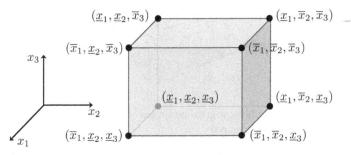

Figure 4.6 The 3-dimensional interval parallelepiped $\bigotimes_{i=1}^{3}[\underline{x}_i, \overline{x}_i]$.

we have

$$[f]([\mathbf{x}]) = \left[\min_j f(\boldsymbol{v}_j), \max_j f(\boldsymbol{v}_j) \right], \qquad (4.39)$$

where $\boldsymbol{v}_j$ includes the coordinates of the jth vertex, and min and max are computed for all 2^d vertices. This method yields the exact result provided that f is continuous and has no extreme point in the interior of the parallelepiped.

Example 4.7. Take $f(x_1, x_2, x_3) = x_1(x_2 - x_3)$, with $[x_1] = [1, 2]$, $x_2 = [2, 3]$, and $[x_3] = [1, 4]$. The values of f taken at the eight vertices are:

$f(1, 2, 1) = 1$,

$f(1, 2, 4) = -2$,

$f(1, 3, 1) = 2$,

$f(1, 3, 4) = -1$,

$f(2, 2, 1) = 2$,

$f(2, 2, 4) = -4$,

$f(2, 3, 1) = 4$,

$f(2, 3, 4) = -2$,

which yields

$$[f]([\mathbf{x}]) = [-4, 4].$$ ◇

Example 4.8. Fig. 4.7 illustrates an example in which the vertex method fails to provide the correct answer. Here we have $f(x_1, x_2) = (x_1^2 - x_1)(x_2^2 - x_2)$ with $\underline{x}_1 = \underline{x}_2 = 0$ and $\overline{x}_1 = \overline{x}_2 = 1$. We see that $f(0, 0) = f(0, 1) = f(1, 0) = f(1, 1) = 0$, while the maximum value achieved by f is $f(1/2, 1/2) = 1/16$. ◇

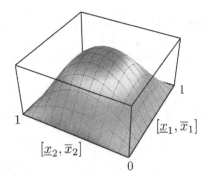

Figure 4.7 Vertex method fails because the maximum of function $f(x_1, x_2) = (x_1^2 - x_1)(x_2^2 - x_2)$ is within the interval parallelepiped.

Remark 4.3. In [67] it is described how the vertex method can be improved by adding, to the vertices $\boldsymbol{v}_j$ in (4.39), the extreme points of f.

4.6 The interval dependence problem

As mentioned before, the properties of interval algebra that differ from the corresponding properties of ordinary algebra create sometimes a major obstacle to the application of the theory. The main hindrance takes place when the same interval variable occurs more than once in a calculation, and each occurrence is taken independently. In this case, the interval resulting from the calculation may be erroneously widened. This gives rise to the so-called *interval dependence* effect, sometimes considered as "the main unsolved problem of the classical theory" of interval algebra [60, p. 34]. The following example serves as a motivation to introduce the interval dependence effect, and shows how this generates intervals whose width is wider than it should be.

Let us return to the calculation of $[-1, 1]^2$ using two different definitions as in (4.13) and Example 4.6 *supra*. When we form a product $[x] \cdot [y]$, it is assumed that the values in the first factor and the values in the second factor vary independently. This assumption remains implicitly valid when $[x] = [y]$. More generally, multiple occurrences of the variables in the same function yield pessimistic results if their dependence is not accounted for.

Example 4.9. The fact that $[x] - [x]$ is generally not equal to $[0, 0]$ can be viewed as an effect of interval dependence. In fact, the subtraction does not take into account the dependence of the two occurrences of $[x]$:

$$[x] - [x] = [\{x - y \mid x \in [x], y \in [y]\}]. \tag{4.40}$$

$\diamond$

Example 4.10. Consider $[x] = [-2, -1]$ and the function $f(x) = x^2 + x$, so that $[f]([x]) = [\underline{x}, \overline{x}]^2 + [\underline{x}, \overline{x}]$. Direct computation using (4.38) yields

$$\begin{aligned}
[f]([x]) &= [-2, -1]^2 + [-2, -1] \\
&= [1, 4] + [-2, -1] \\
&= [-1, 3],
\end{aligned} \tag{4.41}$$

while the range of $x^2 + x$ as $x \in [-1, 1]$ is interval $[0, 2]$, a subset of $[-1, 3]$. $\diamond$

Example 4.11. Assume that we want to solve the equation $[a] \cdot [x] = [b]$, where a and b are intervals. The solution yields x lying in $[b] \div [a]$. However, due to the dependence problem, we cannot write $[a] \cdot ([b] \div [a]) = [b]$. $\diamond$

Example 4.12. The calculation of $[f] = ([x] + [y]) - [x]$, done in two ways, yields different results:

$$\begin{aligned}
[f]_1 &\triangleq [\{x + y - x \mid x \in [x], y \in [y]\}] \\
&= [\{y \mid y \in [y]\}] \\
&= [y]
\end{aligned} \tag{4.42}$$

consistent with ordinary analysis, and

$$\begin{aligned}
[f]_2 &\triangleq [x] + [y] - [x] \\
&= \{x_1 + y - x_2 \mid x_1 \in [x], y \in [y], x_2 \in [x]\}.
\end{aligned} \tag{4.43}$$

It happens that, since the dependence between x_1 and x_2 (in fact, $x_1 = x_2$) has been overlooked in (4.43), the calculation of $[f]_2$ yields a pessimistic result: in fact, we see that $[f]_2 \supseteq [f]_1$. ◇

Example 4.13. The function $f(x) = x/(1+x)$ with interval data can be computed in the form $f(x) = 1/(1+1/x)$ to avoid interval dependence. With $x = [7,8]$, we obtain

$$\frac{[x]}{1+[x]} = [7/9, 1] \subset [0.77, 1],$$

$$\frac{1}{1+\dfrac{1}{[x]}} = [7/8, 8/9] \subset [0.875, 0.89]. \tag{4.44}$$

◇

4.7 Integrals

Assume a real-valued function $f(t)$ of the real variable t, and its integral $I \triangleq \int_a^b f(t)\,dt$. As an enclosure of $f(t)$, we take an interval-valued function $[f]([t])$ of $[t]$ with the property

$$f(t) \in [f]([t]) \qquad \forall t \in T; \tag{4.45}$$

$[I]$ is generated by finding a tight enclosure of $f(t)$ and integrating it.

The starting point here is the mean-value theorem for definite integrals, which states that if f is continuous in $[a,b]$ then

$$\int_a^b f(t)\,dt = (b-a)f(\xi) = w([a,b])f(\xi) \tag{4.46}$$

for some $\xi \in [a,b]$, so that

$$f(\xi) \in [f]([a,b]). \tag{4.47}$$

Consider next the interval enclosure of the continuous function f obtained from two continuous real-valued functions $\underline{f}(t)$, $\overline{f}(t)$ such that, for $t \in [a,b]$, $[f(t)] = [\underline{f}(t), \overline{f}(t)]$. We can write

$$[I] = \left[\int_a^b \underline{f}(t)\,dt, \int_a^b \overline{f}(t)\,dt \right]. \tag{4.48}$$

The simplest example of such an enclosure of $f(t)$ is the *constant interval enclosure* $[\min_{t\in[a,b]} f(t), \max_{t\in[a,b]} f(t)]$, which yields [60, Theorem 5.4]

$$\left[\int_a^b f(t)\,dt \right] \in \left[w([a,b]) \min_{t\in[a,b]} f(t), \, w([a,b]) \max_{t\in[a,b]} f(t) \right]. \tag{4.49}$$

In particular,

$$\left[\int_a^b f(t)\,dt\right] \in \begin{cases} [w([a,b])\,f(a),\,w([a,b])\,f(b)], & f \text{ nondecreasing,} \\ [w([a,b])\,f(b),\,w([a,b])\,f(a)], & f \text{ nonincreasing.} \end{cases} \tag{4.50}$$

For piecewise-monotone functions, (4.49) can be applied after dividing $[a,b]$ into subintervals in which f is either nondecreasing or nonincreasing, and using the additivity property of definite integrals. This technique may also be used with monotone functions, yielding narrower intervals [60, p. 60].

Example 4.14. Let $0 < a < b$ and $f(t) = [\alpha]t^2$, where $[\alpha] = [\underline{\alpha}, \overline{\alpha}]$. We have

$$\left[\int_a^b f(t)\,dt\right] = \left[\int_a^b \underline{\alpha}t^2\,dt,\,\int_a^b \overline{\alpha}t^2\,dt\right]$$

$$= \left[\underline{\alpha}\frac{b^3 - a^3}{3},\,\overline{\alpha}\frac{b^3 - a^3}{3}\right] \tag{4.51}$$

$$= [\alpha]\frac{b^3 - a^3}{3}. \qquad \qquad \diamond$$

Example 4.15. Fig. 4.8(a) shows a constant enclosure for function $f(x) = x^2$, $[x] = [0,2]$, to be integrated in $[0,2]$. The result is $I \in [2 \cdot 0, 2 \cdot 4] = [0,8]$. This result can be improved by considering a constant enclosure in $[0,1]$, and one in $[1,2]$, as in Fig. 4.8(b). In this case we obtain the narrower interval

$$I = \int_0^1 x^2\,dx + \int_1^2 x^2\,dx \in [0,1] + [1,4] = [1,5]. \tag{4.52}$$

$$\diamond$$

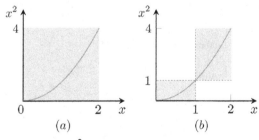

Figure 4.8 Integration of $f(x) = x^2$ in $[0,2]$: (a) constant interval enclosure in $[0,2]$; (b) constant interval enclosures in $[0,1]$ and $[1,2]$.

Another convenient enclosure of $f(t)$ is the *interval polynomial enclosure*, obtained as a polynomial in the variable t with interval coefficients. If $f(t)$ is an analytic function, assuming integration in $[0, 1]$, we may consider its McLaurin series expansion

$$f(t) = f(0) + \frac{f'(0)}{1!}t + \frac{f''(0)}{2!}t^2 + \cdots + \frac{f^{(n-1)}(0)}{(n-1)!}t^{n-1} + R_n(t), \qquad (4.53)$$

with $R_n(t)$ being the remainder whose Lagrange form is

$$R_n(t) = \frac{f^{(n)}(\xi)}{n!}t^n \qquad\qquad (4.54)$$

for ξ lying somewhere in $[0, 1]$. If we have $f^{(n)}(\xi) \in [\underline{\rho}_n, \overline{\rho}_n]$, then a polynomial enclosure of $f(t)$ is given by

$$f(t) \in f(0) + \frac{f'(0)}{1!}t + \frac{f''(0)}{2!}t^2 + \cdots + \frac{f^{(n-1)}(0)}{(n-1)!}t^{n-1} + \frac{1}{n!}[\underline{\rho}_n, \overline{\rho}_n]t^n. \quad (4.55)$$

Example 4.16. We have the polynomial enclosure of e^t given by

$$e^t \in 1 + t + \frac{1}{2}[1, e]t^2 \qquad \forall t \in [0, 1], \qquad\qquad (4.56)$$

shown in Fig. 4.9.

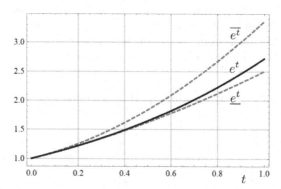

Figure 4.9 Interval polynomial enclosure of $f(t) = e^t$.

To find upper and lower bounds on

$$I = \int_0^1 e^t \, dt, \qquad\qquad (4.57)$$

use enclosure (4.56) to obtain [177, §9.3]

$$\begin{aligned} I &\in [1 + 1/2 + 1/6, \, 1 + 1/2 + e/6] \\ &= [1.\overline{6}, 1.5 + e/6] \subset [1.66, 1.96], \end{aligned} \qquad (4.58)$$

to be compared with the result, rounded to 2 decimal places, $I = 1.73$.   ◇

4.8   Choosing a representative in an interval

If the real number x is only known to lie in interval $[x]$, a natural way of choosing a representative of the interval, to be used as an approximation to x, if needed, is to pick a single real number $x^\star$ within $[x]$. A simple choice for $x^\star$ is the midpoint $m([x])$, which yields

$$|x - x^\star| \leqslant \left|w([x])\right|/2, \tag{4.59}$$

and hence provides an approximation to x along with an estimate of the error made with the approximation. A more general choice is

$$\begin{aligned} x^\star &= \varepsilon \underline{x} + (1 - \varepsilon)\overline{x} \\ &= \overline{x} - (\overline{x} - \underline{x})\varepsilon \end{aligned} \tag{4.60}$$

for a suitable choice of $\varepsilon \in [0, 1]$. Yager [274] advocates the choice of the value of ε based on its interpretation as the expected value of an RV Y with support set $[0, 1]$ and cumulative distribution function (CDF) $Q(y)$,

$$\varepsilon = \int_0^1 y \, dQ(y). \tag{4.61}$$

We denote by $x_Q^\star$ the resulting value of $x^\star$. Observe also that $Q_2 \geqslant Q_1$ yields $x_{Q_2}^\star \geqslant x_{Q_1}^\star$. In fact, from (4.60) and (4.61) we obtain

$$x_{Q_2}^\star - x_{Q_1}^\star = (\overline{x} - \underline{x}) \int_0^1 y \, d[Q_2(y) - Q_1(y)] \geqslant 0. \tag{4.62}$$

As suggested in [274] in the context of decision-making, the choice of function Q reflects the attitude of the person who chooses the representative of $[x]$.

Example 4.17.   Three simple choices for Q are

$$\begin{aligned} Q(y) = y &\implies x_Q^\star = \frac{\overline{x} + \underline{x}}{2} = m([x]), \\ Q(y) = u(y - q) &\implies x_Q^\star = q\underline{x} + (1 - q)\overline{x}, \\ Q(y) = y^r &\implies x_Q^\star = \frac{\overline{x} + r\underline{x}}{r + 1}, \end{aligned} \tag{4.63}$$

where $u(x)$ denotes the unit step function. Other possible choices are tabulated in [274]. $\diamond$

A choice of $x^\star$ more general that that in (4.60) can be obtained [274] by picking another CDF $G(z)$ defined in $[\underline{x}, \overline{x}]$, computing the inverse $H^{[-1]}$ of the "survival function"

$$H(x) \triangleq \int_x^{\overline{x}} dG(z), \qquad x \in [\underline{x}, \overline{x}], \tag{4.64}$$

and setting

$$x_{Q,G}^{\star} \triangleq \int_0^1 H^{[-1]}(y)\,dQ(y).$$

(4.65)

The choice of $G(z)$ models the weight one wants to associate with values in $[\underline{x}, \overline{x}]$, and hence is related to the relative importance assigned to those values.

Example 4.18. The simple choice $dG(z) = dz/(\overline{x} - \underline{x})$ yields

$$H(x) = \frac{1}{\overline{x} - \underline{x}} \int_x^{\overline{x}} dz = \frac{\overline{x} - x}{\overline{x} - \underline{x}},$$

(4.66)

whose inverse is $H^{[-1]}(y) = \overline{x} - (\overline{x} - \underline{x})y$, thus yielding

$$x_{Q,G}^{\star} = \int_0^1 [\overline{x} - (\overline{x} - \underline{x})y]\,dQ(y),$$

(4.67)

which is the same as in (4.60)–(4.61). $\Diamond$

Example 4.19. The choice

$$dG(z) = 2\frac{\overline{x} - z}{\overline{x} - \underline{x}}\,dz$$

(4.68)

yields

$$H(x) = \frac{(\overline{x} - x)^2}{(\overline{x} - \underline{x})^2},$$

(4.69)

whose inverse, under the constraint $x \in [\underline{x}, \overline{x}]$, is

$$H^{[-1]}(y) = \overline{x} - (\overline{x} - \underline{x})y^{1/2}$$

(4.70)

and hence

$$x_{Q,G}^{\star} = \overline{x} - (\overline{x} - \underline{x}) \int_0^1 y^{1/2}\,dQ(y).$$

(4.71)

If, in addition, we assume $dQ(y) = dy$, observing that $\int_0^1 y^{1/2}\,dy = 2/3$, we obtain

$$x_{Q,G}^{\star} = \frac{2\underline{x} + \overline{x}}{3},$$

(4.72)

which shows how more weight is assigned to lower values in $[\underline{x}, \overline{x}]$. $\Diamond$

Sources and parerga

1. Extensive treatments of interval analysis can be found in books [60,125,176,177].
   A succinct history of the discipline is in [60, pp. 4–5].
2. Some aspects of the theory are left out of this chapter (for example, complex
   interval numbers). For a deeper insight, the reader is referred to [118,176,177],
   with [125] containing a number of applications (e.g., to estimation, robust control, and robotics). Book [110] describes the application of interval analysis to
   optimization problems. Book [60], besides a self-contained exposition of the key
   concepts (including approaches to interval computations that are alternative to the
   one we followed in this chapter), has a study of the design of hardware operations
   for interval arithmetic.
3. In 1611, Johannes Kepler conjectured that no sphere-packing in $\mathbb{R}^3$ is denser than
   the "cannonball packing," or "face-centered-cubic lattice," which fills $\mathbb{R}^3$ with
   density $\pi/\sqrt{18} \approx 0.7405$. David Hilbert in 1900 included Kepler conjecture in his
   famous problem list. The conjecture was finally proved by Thomas C. Hales and
   Samuel P. Ferguson in 2005 [107,239] using a number of mathematical techniques
   which included interval analysis.
4. In [167, p. 13], a story is told illustrating the reluctance of decision makers to
   accept the presence of uncertainties causing data to appear in the interval form.
   The story is about "an economist's attempt to describe his uncertainty about a
   forecast to President Lyndon B. Johnson. The economist presented his forecast as
   a likely range of values for the quantity under discussion. Johnson is said to have
   replied, 'Ranges are for cattle. Give me a number'."

Probability boxes

5

In this chapter we elaborate on the concept of probability intervals, and see how it can be used in applications involving epistemic uncertainty, i.e., in situations in which more than one probability measure is compatible with the observations. When this is the case, if the probabilities associated with one or more events are described by intervals of values rather than by single numbers, epistemic uncertainty can be treated by using probability intervals, or *p-boxes*, which generalize the concept of interval from a pair of points to a pair of cumulative distribution functions (CDFs). For motivation purposes, we start with a simple example, borrowed from [143, p. 110 ff.], which shows how a probability interval can be generated when the amount of available information gathered about a probability distribution is incomplete.

Example 5.1. We have two random variables (RVs) X and Y taking values in $\{x_1, x_2\}$ and $\{y_1, y_2\}$, respectively. Let the information about the joint distribution of X and Y, viz., the values of $p_{ij} \triangleq \mathbb{P}(X = x_i, Y = y_j)$, $i, j = 1, 2$, be restricted to the knowledge of the four marginals

$$
\begin{aligned}
p_X(x_1) &= p_{11} + p_{12}, \\
p_X(x_2) &= 1 - p_X(x_1), \\
p_Y(y_1) &= p_{11} + p_{21}, \\
p_Y(y_2) &= 1 - p_Y(y_1).
\end{aligned}
\tag{5.1}
$$

To determine the joint distribution, we solve the following system of equations for the four unknowns p_{ij}:

$$
\begin{aligned}
p_{11} + p_{12} &= p_X(x_1), &\text{(5.2a)} \\
p_{21} + p_{22} &= 1 - p_X(x_1), &\text{(5.2b)} \\
p_{11} + p_{21} &= p_Y(y_1), &\text{(5.2c)} \\
p_{12} + p_{22} &= 1 - p_Y(y_1). &\text{(5.2d)}
\end{aligned}
$$

Eqs. (5.2a)–(5.2d) are not linearly independent. Indeed, we have

$$
(5.2d) = (5.2a) + (5.2b) - (5.2c).
$$

Removing (5.2d), we are left with 3 equations with 4 unknowns. Choosing, for example, p_{11} as a free variable, the joint distribution is

$$
\begin{aligned}
&p_{11}, \\
&p_{12} = p_X(x_1) - p_{11}, \\
&p_{21} = p_Y(y_1) - p_{11}, \\
&p_{22} = 1 - p_X(x_1) - p_Y(y_1) + p_{11},
\end{aligned}
\tag{5.3}
$$

Dimensions of Uncertainty in Communication Engineering. https://doi.org/10.1016/B978-0-32-399275-6.00013-7

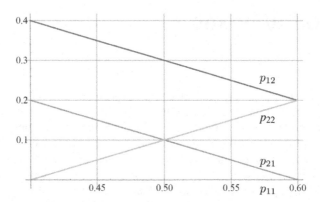

Figure 5.1 Range of possible values for p_{11}, p_{12}, p_{21}, and p_{22} in Example 5.1.

with p_{11} being nonnegative and constrained by the inequalities

$$\max[p_X(x_1) + p_Y(y_1) - 1, 0] \leqslant p_{11} \leqslant \min[p_X(x_1), p_Y(y_1)]. \tag{5.4}$$

For example, the assumption $p_X(x_1) = 0.8$ and $p_Y(y_1) = 0.6$ leads to the joint-distributions interval

$$\begin{aligned}
p_{11} &\in [0.4, 0.6], \\
p_{12} &= 0.8 - p_{11}, \\
p_{21} &= 0.6 - p_{11}, \\
p_{22} &= p_{11} - 0.4.
\end{aligned} \tag{5.5}$$

Fig. 5.1 illustrates how the uncertain knowledge of p_{11} is reflected into uncertain knowledge of all values of p_{ij}. $\diamond$

Example 5.1 illustrates how a probability may be confined to an interval, and how other probabilities, depending on the former one, become interval probabilities as well.

5.1 Interval probabilities

Give a space of events, it may happen that the choice of a precise probability measure on it is prevented by epistemic uncertainty. If this is the case, one may resort to the extension of classical probability provided by interval probability, which roughly consists of extending to probabilities the methods developed for interval analysis. An interval probability, used instead of a single real number $\mathbb{P}(A)$ to describe the probability of an event A, is an interval of the type $[\underline{\mathbb{P}}(A), \overline{\mathbb{P}}(A)]$. More precisely, we consider a family $\mathcal{P}$ of probability measures $\mathbb{P}$ such that, for all A, we have $\underline{\mathbb{P}}(A) \leqslant \mathbb{P}(A) \leqslant \overline{\mathbb{P}}(A)$. In this section we examine in particular intervals of probabilities of events $X \in (-\infty, x]$, where the probability measure of the RV X lies in $\mathcal{P}$. These intervals enclose CDFs, and are called *probability boxes*, or *p-boxes* for short.

Once upper and lower probabilities have been assigned to every event in the probability space, p-boxes are delimited by the two CDFs that circumscribe the imprecisely known $F_X(x)$ [207, p. 24]:

$$\overline{F}_X(x) \triangleq \overline{\mathbb{P}}(X \leqslant x) = 1 - \underline{\mathbb{P}}(X > x),$$
$$\underline{F}_X(x) \triangleq \underline{\mathbb{P}}(X \leqslant x) = 1 - \overline{\mathbb{P}}(X > x). \qquad (5.6)$$

(Unless otherwise specified, we restrict our attention to RVs defined on the nonnegative real axis $\mathbb{R}^+$.) In Chapter 7 we shall return on some subtleties involved in the definitions of $\overline{\mathbb{P}}$ and $\underline{\mathbb{P}}$. Here we limit ourselves to the consideration of $\overline{\mathbb{P}}$ and $\underline{\mathbb{P}}$ as obtained as sup and inf of probability measures in a set $\mathcal{P}$ defined by the constraints which express the knowledge about $\mathbb{P}$. Fig. 5.2 below shows an example of a p-box.

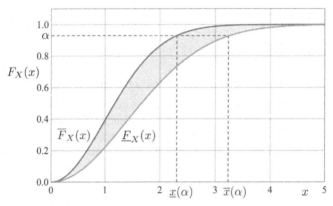

Figure 5.2 Upper and lower bound on the CDF $F_X(x)$; p-box $[\overline{F}_X, \underline{F}_X]$.

Remark 5.1. With p-boxes, we follow the notational convention of [81], which is derived from the concept of the *quantile function* $x(\alpha)$ of $F(x)$. This is defined as

$$x(\alpha) \triangleq \min \{x \mid F(x) \geqslant \alpha\}. \qquad (5.7)$$

If the CDF is continuous, $x(\alpha)$ is the value such that an RV whose CDF is $F(x)$ has probability α of being lower than $x(\alpha)$. If the CDF is continuous and strictly increasing, the quantile function is the inverse of $F(x)$; see Fig. 5.2 *supra*. In terms of p-boxes, $[\overline{F}_X, \underline{F}_X]$ is equivalent to $[\underline{x}(\alpha), \overline{x}(\alpha)]$, where $\underline{x}(\alpha)$ and $\overline{x}(\alpha)$ are the quantiles corresponding to $\overline{F}_X$ and $\underline{F}_X$, respectively. Thus, the left bound of interval $[\overline{F}_X, \underline{F}_X]$ is an upper bound on probabilities and a lower bound on quantiles, while the right bound is a lower bound on probabilities and an upper bound on quantiles. Other authors use a different notation; see, for example, [16].

Example 5.2. An interval $[a, b]$ may be identified by a p-box with $\overline{F}(x) = u(x - a)$ and $\underline{F}(x) = u(x - b)$, where $u(\cdot)$ denotes the unit-step function. A precise CDF is described by a p-box with $\underline{F} = \overline{F} = F$. $\diamond$

Example 5.3. Probability boxes may reflect the uncertainty coming from the measure of probabilities. For example, consider N observations x_i of values taken on by the RV X, and the empirical CDF $\langle F_X(x) \rangle$ obtained as the ratio of the number $n(x)$ of times the event $x_i \leqslant x$ is observed to the total number of observations,

$$\hat{\mathbb{P}}[X \leqslant x] \triangleq \langle F_X(x) \rangle = n(x)/N. \tag{5.8}$$

Uncertainty here is caused by the incomplete amount of statistical information about X. For this estimate of $F_X(x)$ to be reasonably accurate, N should be large enough. The inaccuracy may be measured as follows [143, p. 130]: assume $\overline{\hat{\mathbb{P}}}[X \leqslant x] = 0$ and $\hat{\mathbb{P}}[X \leqslant x] = 1$ for $N = 0$ (total epistemic uncertainty), and define, for $N > 0$, the estimators

$$\underline{\hat{\mathbb{P}}}[X \leqslant x] = \frac{n(x)}{N+c}, \tag{5.9a}$$

$$\overline{\hat{\mathbb{P}}}[X \leqslant x] = \frac{n(x)+c}{N+c}, \tag{5.9b}$$

with $c \geqslant 1$. The choice of c reflects how quickly the epistemic uncertainty decreases while N increases. The uncertainty may be measured by the difference $\overline{\hat{\mathbb{P}}}(x) - \underline{\hat{\mathbb{P}}}(x)$ between upper and lower estimates of $\mathbb{P}[X \leqslant x]$. Enclosing the empirical histogram within confidence bands, one can obtain a p-box of interval data, as shown in Fig. 5.3. We should observe that the p-box obtained this way is not certain, but rather statisti-

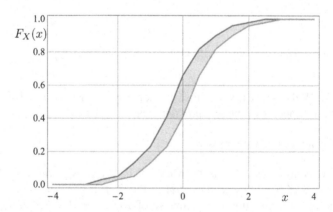

Figure 5.3 P-box in the form of cumulative distribution of interval data.

cal. For a more accurate analysis, one should extend the p-box using distribution-free bounds about an empirical CDF [82, pp. 53–55], as those provided by Kolmogorov–Smirnov tests [10], [66, p. 283 ff.], [81, pp. 53–55]. ◇

Probability boxes can handle several problems, like model uncertainties, poorly known or unknown dependencies of RVs, or imprecisely specified distributions [81, pp. 12–14]. The key point here is that the use of a p-box in lieu of an individual CDF results in performance analyses not relying on unwarranted distribution assumptions. With this approach, instead of selecting a single CDF from a class of distributions matching a set of constraints, as we did, for example, in Section 1.6, the entire class is used. In addition, instead of deriving upper and lower bounds on performance parameters only at the end of calculations (see, for example, [33]), all relevant calculations keep track of the uncertainties implicit in the entities involved, possibly using interval analysis as described in Chapter 4.

5.2   Generating probability boxes

Probability boxes can be generated whenever upper and lower bounds on the CDF of an RV can be obtained on the basis of the knowledge available about the RV itself. The width of a p-box yields a quantitative indication of the effect of the model uncertainty on the probability measure of the RV, and eventually on the performance parameters derived from it. In particular, the shape of the p-box reflects the aleatory uncertainty, while the area enclosed between the edges of the p-box describes the extent of what is ignored, and hence epistemic uncertainty. Hence, a wide p-box may not be caused by a weakness of the theory, but rather reflects the amount of model uncertainty. P-boxes can also show the robustness of a performance parameter against the choice of a model. Here we examine some examples of constraint sets generating p-boxes [81, §3.4.8]. Later on, we shall examine how to propagate the uncertainty intrinsic in a p-box to the calculation of the performance parameters of interest.

  (i)   The RV has a known pdf whose support set has boundaries with interval values.
 (ii)   The RV has a known CDF given in parametric form, and some parameters are known within an interval of values. In this case a p-box can be generated as the envelope of all CDFs whose parameters lie in that interval.
(iii)   The pdf of the RV has known support set and median value.
(iv)   The RV has known values of its first moments.
 (v)   The CDF of the RV is the marginal of a joint CDF of two RVs whose dependence is unknown.

The addition of qualitative information about the shape of the CDF (for example, convexity, or concavity) may tighten the p-box. In case (iv), the p-box can be generated using the moment bounds described in Section 3.9.3. Case (v) will be examined in the next chapter.

Example 5.4. If the RV X is uniformly distributed in a real interval whose boundaries are imperfectly known, the p-box corresponding to this situation has the shape illustrated in Fig. 5.4.   ◇

Example 5.5. The p-box of an RV with support set $[1, 5]$ and mean value $\mu_1 = 2$ is obtained using the result of Example 3.29 and shown in Fig. 5.5. It is seen how the mean value alone may not bring much information about the structure of the CDF.
                                                                                                        ◇

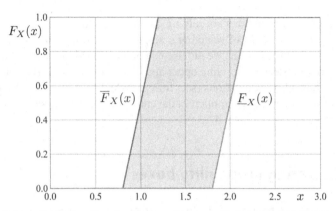

Figure 5.4 P-box of an RV uniformly distributed between [0.8, 1.2] and [1.8, 2.2].

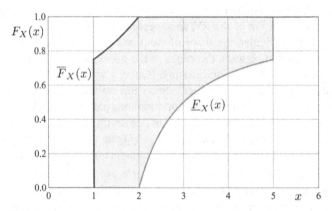

Figure 5.5 P-box of an RV with support set [1, 5] and mean value $\mu_1 = 2$.

Example 5.6. Let the RV X be known to be Gaussian with variance $\sigma^2 = 0.01$ and mean value in interval [0.5, 1.0]. We may represent this uncertainty by drawing the CDFs of two RVs distributed as $\mathcal{N}(0.5, 0.01)$ and $\mathcal{N}(1.0, 0.01)$, and considering all the Gaussian CDFs enclosed in the shaded area of Fig. 5.6. ◇

Example 5.7. Fig. 5.7 shows the p-box generated by Nakagami-m distributions (1.8) with parameters $m \in [0.5, 5.]$ and $\Omega \in [1., 2.]$. ◇

The p-box of Fig. 5.6, generated by a Gaussian CDF with interval parameters, includes the whole class of CDFs that can be drawn within the two bounding curves, and hence also nonsmooth, and possibly unrealistic, CDFs. The only limitation of the CDFs in the p-box is that $F(x) \in [\overline{F}(x), \underline{F}(x)]$, which does not reflect exactly the fact that we are only dealing with Gaussian CDFs. Now, in some cases it might be desirable to restrict consideration to CDFs sharing the same smoothness features of bounds $\underline{F}$ and $\overline{F}$. To do this, consider two CDFs F_1 and F_2, described by their quantile functions

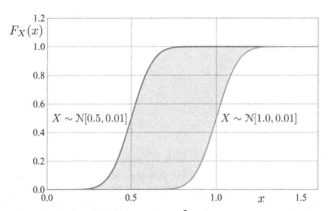

Figure 5.6 CDFs of Gaussian RVs with variance $\sigma^2 = 0.01$ and mean value in interval [0.5, 1.0] lie in the p-box of shaded area.

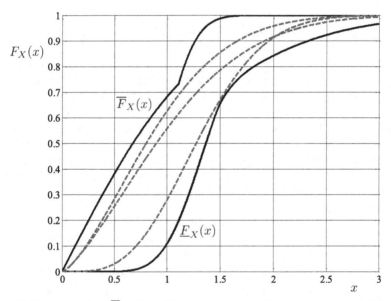

Figure 5.7 Probability box $[\overline{F}_X, \underline{F}_X]$ of Nakagami-m RVs with parameters $m \in [0.5, 5.]$, $\Omega \in [1., 2.]$. The dashed curves correspond to three Nakagami-m CDFs with randomly selected values of m and Ω within their intervals. Reproduced with permission from [25].

$x_1(\alpha)$ and $x_2(\alpha)$ defined in (5.7), and assume that the convex combination

$$x(\alpha) = \beta x_1(\alpha) + (1 - \beta) x_2(\alpha), \qquad \beta \in [0, 1], \tag{5.10}$$

corresponds to a possible CDF. If F_1 and F_2 are fixed, we obtain a family of CDFs, parametrized by β, which is a subset of all CDFs in $[\overline{F}(x), \underline{F}(x)]$. Returning to the Gaussian example above, let F_0 denote the CDF of the standardized Gaussian RV

$X \sim \mathcal{N}(0, 1)$ with quantile function $x_0(\alpha)$. A Gaussian RV $X \sim \mathcal{N}(\mu, \sigma^2)$ turns out to have the CDF

$$F(x) = F_0 \left(\frac{x - \mu}{\sigma} \right) \tag{5.11}$$

and hence quantile function

$$x(\alpha) = \mu + \sigma x_0(\alpha). \tag{5.12}$$

Now, if we have two Gaussian CDFs with $x_1(\alpha) = \mu_1 + \sigma_1 x_0(\alpha)$ and $x_2(\alpha) = \mu_2 + \sigma_2 x_0(\alpha)$, their convex combination (5.10) has the same form (5.12) with $\mu = \beta \mu_1 + (1-\beta)\mu_2$ and $\sigma = \beta \sigma_1 + (1-\beta)\sigma_2$. This family of CDFs, parametrized by β, contains only Gaussian CDFs. This procedure is generalized in [16] using concepts from "Info-Gap Decision Theory," as described in [14].

5.3 Aggregating probability boxes

Whenever possible, one should work directly with the p-box and carry on all the relevant calculations using interval analysis. Should this approach be computationally too intensive, one may want to choose a representative CDF within the p-box, and use it to compute the relevant performance metric (this common approach, called *aggregation*, was illustrated in Section 4.8 in the simple case of scalar intervals). For example, one may choose the *worst* CDF within the p-box, although this choice may be unduly pessimistic and hence detract from robustness (see, e.g., [257]). Another approach consists of choosing a small number of CDFs within the box, and running multiple separate performance analyses for each one. This approach does not allow an overall synthesis of the modeling effort.

Given a set of p-boxes referring to an individual modeling task, they can be aggregated in several ways as follows (see [81, p. 67 ff.] for examples and comments about the suitability and the limitations of each aggregation rule):

(i) **Intersection.** If it is assumed that all p-boxes contain the correct CDF, then one may generate their intersection. Given the n p-boxes $[\overline{F}_i(x), \underline{F}_i(x)]$, $1 \leqslant i \leqslant n$, the intersection p-box is $[\overline{F}^\star(x), \underline{F}^\star(x)]$, where

$$\begin{aligned} \underline{F}^\star(x) &= \max\left(\underline{F}_1(x), \ldots, \underline{F}_n(x) \right), \\ \overline{F}^\star(x) &= \min\left(\overline{F}_1(x), \ldots, \overline{F}_n(x) \right). \end{aligned} \tag{5.13}$$

This operation is defined when, for all x, $\underline{F}^\star(x) \leqslant \overline{F}^\star(x)$.

(ii) **Envelope.** If *at least one* of the p-boxes contains the correct model cdf, then the p-box $[\overline{F}^\star(x), \underline{F}^\star(x)]$ may be used, where

$$\begin{aligned} \underline{F}^\star(x) &= \min\left(\underline{F}_1(x), \ldots, \underline{F}_n(x) \right), \\ \overline{F}^\star(x) &= \max\left(\overline{F}_1(x), \ldots, \overline{F}_n(x) \right). \end{aligned} \tag{5.14}$$

(iii) **Mixture.** The p-box mixed from $[\overline{F}_1(x), \underline{F}_1(x)], \ldots, [\overline{F}_n(x), \underline{F}_n(x)]$ with respective nonnegative weights $w_1, \ldots, w_n$, $\sum_{i=1}^{n} w_i = 1$, is $[\overline{F}^{\star}(x), \underline{F}^{\star}(x)]$, where

$$\underline{F}^{\star}(x) = \sum_{i=1}^{n} w_i \underline{F}_i(x),$$

$$\overline{F}^{\star}(x) = \sum_{i=1}^{n} w_i \overline{F}_i(x).$$

(5.15)

As observed in [81, p. 87], the use of mixtures treats disagreements among estimates as a variability, and condenses these disagreements into a single p-box.

5.4 Combining probability boxes of random variables

An important application of p-boxes arises with an RV combines other RVs that are described through their p-boxes. In several instances, the resulting p-boxes can be computed by a simple application of the definitions.

Example 5.8. Consider the maximum and minimum of n independent RVs X_i, $i = 1, \ldots, n$, whose CDFs $F_i(x)$ are included in the p-boxes $[\overline{F}_i(x), \underline{F}_i(x)]$. The CDF of the maximum is, with obvious notations, $F_{\max}(x) = \prod_{i=1}^{n} F_i(x)$, and the CDF of the minimum is $F_{\min}(x) = 1 - \prod_{i=1}^{n}(1 - F_i(x))$. Simple interval arithmetic yields the p-boxes

$$F_{\max}(x) \in \left[\prod_{i=1}^{n} \overline{F}_i(x), \prod_{i=1}^{n} \underline{F}_i(x) \right],$$

$$F_{\min}(x) \in \left[1 - \prod_{i=1}^{n}(1 - \overline{F}_i(x)), 1 - \prod_{i=1}^{n}(1 - \underline{F}_i(x)) \right].$$

(5.16)

$\diamond$

General results can be obtained with binary operations that are monotone in both arguments, and in particular the four operations, denoted $\circ$, of the set $\{+, -, \times, \div\}$. Consider the case of two independent nonnegative RVs whose p-boxes are given. If X and Y have pdfs f_X and f_Y, and $Z = X \circ Y$, then the CDF of Z is given by

$$F_Z(z) = \mathbb{P}\big[(X, Y) \in L_z\big] = \iint_{L_z} f_X(u) f_Y(v) \, du \, dv,$$

(5.17)

where L_z is the set of the real pairs (u, v) such that $u \circ v \leqslant z$. For example, if $Z = X + Y$, (5.17) yields

$$F_{X+Y}(z) = \int_0^{\infty} f_Y(v) \, dv \int_0^{z-v} f_X(u) \, du$$

$$= \int_0^z f_Y(v) F_X(z - v) \, dv \tag{5.18}$$

$$= \int_0^z F_X(z - v) \, dF_Y(v).$$

Denote by $\oplus$, $\ominus$, $\otimes$, and $\oslash$, respectively, the "generalized convolutions" that, for independent nonnegative RVs, combine F_X and F_Y to generate $F_{X \circ Y}$ [243, Chapters 3–4]:

$$F_{X+Y}(z) = (F_X \oplus F_Y)(z) = \int_0^z F_X(z - v) \, dF_Y(v), \tag{5.19a}$$

$$F_{X-Y}(z) = (F_X \ominus F_Y)(z) = \int_{-z}^\infty F_X(z + v) \, dF_Y(v), \tag{5.19b}$$

$$F_{X \times Y}(z) = (F_X \otimes F_Y)(z) = \int_0^\infty F_X(z/v) \, dF_Y(v), \tag{5.19c}$$

$$F_{X \div Y}(z) = (F_X \oslash F_Y)(z) = \int_0^\infty F_X(zv) \, dF_Y(v). \tag{5.19d}$$

In these conditions the p-boxes for F_Z are given by [272, p. 100]

$$\underline{F}_{X+Y} = \underline{F}_X \oplus \underline{F}_Y, \qquad \overline{F}_{X+Y} = \overline{F}_X \oplus \overline{F}_Y, \tag{5.20a}$$

$$\underline{F}_{X-Y} = \underline{F}_X \ominus \overline{F}_Y, \qquad \overline{F}_{X-Y} = \overline{F}_X \ominus \underline{F}_Y, \tag{5.20b}$$

$$\underline{F}_{X \times Y} = \underline{F}_X \otimes \underline{F}_Y, \qquad \overline{F}_{X \times Y} = \overline{F}_X \otimes \overline{F}_Y, \tag{5.20c}$$

$$\underline{F}_{X \div Y} = \underline{F}_X \oslash \overline{F}_Y, \qquad \overline{F}_{X \div Y} = \overline{F}_X \oslash \underline{F}_Y. \tag{5.20d}$$

To prove this, consider for example (5.19a). Since $F_X \geqslant 0$, we have that, if F_Y is fixed,

$$\underline{F}_{X+Y}(x) \geqslant \int_0^z \underline{F}_X(z - v) \, dF_Y(v),$$

$$\overline{F}_{X+Y}(x) \leqslant \int_0^z \overline{F}_X(z - v) \, dF_Y(v). \tag{5.21}$$

By symmetry, considering F_X as fixed, we obtain

$$\underline{F}_{X+Y}(x) \geqslant \int_0^z F_X(z - v) \, d\underline{F}_Y(v),$$

$$\overline{F}_{X+Y}(x) \leqslant \int_0^z F_X(z - v) \, d\overline{F}_Y(v), \tag{5.22}$$

from which (5.20a) follows.

Example 5.9. Let X and Y be two independent RVs with common exponential distribution:

$$f(x) = ae^{-ax}, \qquad F(x) = 1 - e^{-ax}, \qquad x \geqslant 0, \tag{5.23}$$

and parameter a in $[1, 2]$. The sum $X + Y$ has CDF

$$F_{X+Y}(z) = 1 - e^{-az}(1 + az), \quad z \geqslant 0. \tag{5.24}$$

Thus, the p-box of $X + Y$ is delimited by the two curves obtained from (5.24) with $a = 1$ and $a = 2$. The difference $X - Y$, where X has parameter a_1 and Y has parameter a_2, has CDF

$$F_{X-Y}(z) = \frac{1}{a_1 + a_2} \left(a_1 e^{a_2 z} + (a_1 + a_2 - a_2 e^{-a_1 z} - a_1 e^{a_2 z}) u(z) \right), \tag{5.25}$$

and the p-box of $X - Y$ is delimited by the two curves obtained from (5.25) with $a_1 = 1$, $a_2 = 2$, and with $a_1 = 2$, $a_2 = 1$, respectively. Fig. 5.8 shows the p-boxes of X, $X + Y$, and $X - Y$. ◇

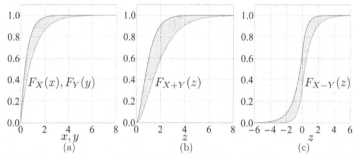

Figure 5.8 (a) P-box of exponentially distributed RVs X and Y with parameter $a \in [1, 2]$. (b) P-box of the sum $X + Y$, with X and Y independent RVs exponentially distributed with parameter $a \in [1, 2]$. (c) P-box of the difference $X - Y$, with X and Y independent RVs exponentially distributed with parameter $a \in [1, 2]$.

Example 5.10. Assume that two RVs X, Y have support sets $[0, 1]$ and $[0.5, 1.5]$, respectively. Nothing else is known. What can we say about the sum $X + Y$? The principle of insufficient reason, more about which will be told in Section 7.2.1 *infra*, suggests that our epistemic ignorance about X and Y can be expressed by choosing *independent and uniform* distributions for both RVs, $X \sim \mathcal{U}[0, 1]$ and $Y \sim \mathcal{U}[0.5, 1.5]$. The resulting pdf of $X + Y$ is triangular with support set $[0.5, 2.5]$ and mode 1.5. Or we may simply observe that the possible values taken by the sum are in $[0.5, 2.5]$, and assume $X + Y \sim \mathcal{U}[0.5, 2.5]$ (see Fig. 5.9). Otherwise, if one does not want to make any assumption that cannot be justified by available empirical data, the most neutral choice is simply to accept ignorance about the pdf, and do the ensuing calculations after accepting that we only know that the values taken by $X + Y$ are in $[0.5, 2.5]$, a more convincing characterization of the underlying uncertainty. Intermediate assumptions may accept that X and Y are uniformly distributed, and avoid an unwarranted independence assumption in favor of bounds on the distribution obtained from the Fréchet–Hoeffding bounds (see the next chapter). ◇

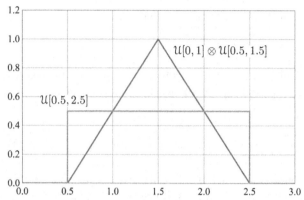

Figure 5.9 Possible probability density functions of the sum $X + Y$, when it is only known that the support set of X is [0, 1] and that of Y is [0.5, 1.5].

5.5   Using probability boxes in performance evaluation

Assume again, as we did throughout Chapter 2, that η, the performance parameter of interest, can be expressed as the expectation, with respect to the continuous RV X, of the known function $h(\cdot)$. If the CDF F_X is known to lie in the p-box $[\overline{F}_X, \underline{F}_X]$, how can we use this information to obtain the interval $[\underline{\eta}, \overline{\eta}]$?

This problem can be solved using the notion of *stochastic order* of distribution functions (see, e.g., [43]). Consider two RVs X and X' with CDFs F_X and $F_{X'}$, respectively. We say that X is *stochastically smaller* than X' if $F_X(x) \geqslant F_{X'}(x)$ for all x. Then, if $h(\cdot)$ is an a.s. strictly increasing function, we have [43], [256, Theorem 3] that $h(X)$ is also stochastically smaller than $h(X')$,

$$\mathbb{E}_X[h(X)] \leqslant \mathbb{E}_{X'}[h(X')]. \tag{5.26}$$

(The order is reversed if h is a decreasing function.) Using this result, and observing that in a p-box $\overline{F}_X \geqslant \underline{F}_X$, we have from (5.26)

$$\underline{\eta} = \int h(x) \, d\overline{F}(x) \quad \text{and} \quad \overline{\eta} = \int h(x) \, d\underline{F}(x). \tag{5.27}$$

Example 5.11. Consider the function

$$h(x) = \frac{1}{2}\text{erfc}\left(\sqrt{\text{snr}\, x}\right), \quad 0 \leqslant x < \infty, \tag{5.28}$$

which vanishes at infinity and whose derivative is

$$h'(x) = -\frac{\sqrt{\text{snr}}}{\sqrt{4\pi x}} \exp(-\text{snr}\, x) \leqslant 0, \tag{5.29}$$

so that, with X a nonnegative RV with CDF $F_X(x)$,

$$
\begin{aligned}
\eta &= \mathbb{E}_X[h(X)] \\
&= \int_0^\infty h(x)\, dF_X(x) \\
&= -\int_0^\infty h'(x) F_X(x)\, dx \\
&= \sqrt{\frac{\mathsf{snr}}{4\pi}} \int_0^\infty \exp(-\mathsf{snr}\, x) x^{-1/2} F_X(x)\, dx.
\end{aligned}
\tag{5.30}
$$

Replacing F_X with $\underline{F}_X$ and $\overline{F}_X$, from (5.27) we obtain $[\underline{\eta}, \overline{\eta}]$. ◇

Sources and parerga

1. Interval-type bounds on CDFs were introduced in [272,273], and called p-boxes in [81]. The 1989 PhD thesis report [272] refers to the calculations described in this chapter as "probabilistic arithmetic."
2. SIPTA, the *Society for Imprecise Probabilities*, has the website *sipta.org* which informs about events and provides basic information and resources.
3. Stochastic ordering, which quantifies the concept of an RV being bigger than another, was introduced in the context of dynamic inventory theory by Samuel Karlin [135].
4. A mathematical theory closely connected with imprecise probabilities is the *Theory of Random Sets*. This was originally conceived in the 1970s by Georges Matheron and David George Kendall. Its connection with belief functions of Dempster–Shafer theory is explored in [186]. Comprehensive coverages of this theory can be found in [101,163,169,175,186]. Monograph [28] describes some applications to communication engineering.

Dependence bounds

6

In this chapter we show how to deal with the epistemic uncertainty affecting the cumulative distribution function (CDF) $F_{\mathbf{X}}$ of the d-dimensional RV $\mathbf{X} = (X_1, \ldots, X_d)$ when only the marginal CDFs F_{X_i} are known. In particular, we examine the derivation of bounds on expectations of the form $\mathbb{E}[\Phi(\mathbf{X})]$, with Φ being a known function. Other topics involve finding bounds on the probability $\mathbb{P}[X \circ Y \leq z]$, where $\circ$ denotes a binary operation. The general problem of deriving these *dependence bounds* arises when the RVs combined are not independent, and their dependence is unknown or only partially known. It often happens that independence is assumed only for ease of treatment, in which case it is important to assess the possible performance penalty caused by an unjustified independence assumption. In fact, while for example maximum-entropy modeling, in the absence of additional epistemic information about the structure of joint pdfs, yields independent-variable models (Section 1.6), the error involved with this choice may be extremely relevant. To assess the cost of a wrong choice of the dependence model we derive (possibly sharp) upper and lower bounds on a final performance parameter under the assumption of an unknown dependence structure.

Most of what we are going to study in this chapter uses a probabilistic tool called *copula theory*, which studies the functions connecting the marginal CDFs of $\mathbf{X}$ to its joint CDF.

6.1 Copulas

Copulas allow the joint CDF $F_{\mathbf{X}}$ to be linked to the CDFs of the individual components X_i of $\mathbf{X}$. Thus, $F_{\mathbf{X}}$ is identified through a combination of two separate entities, one describing the marginal CDFs and the other describing the dependence structure. Copula theory achieves the separation of marginal-CDF modeling from the modeling of dependence structure, and hence allows a considerable simplification of the overall modeling task for multidimensional RVs. We start studying two-dimensional (or joint, or bivariate) copulas in some depth, next we extend the results to d-dimensional copulas, $d > 2$.

Let $\mathbb{I}$ denote the interval $[0, 1]$. Formally, a (bivariate) copula (see, e.g., [181]) is a function $\mathsf{C} : \mathbb{I}^2 \mapsto \mathbb{I}$ such that:

(i) For all $u \in \mathbb{I}$, the following boundary conditions hold:

$$\mathsf{C}(u, 0) = \mathsf{C}(0, u) = 0 \text{ and } \mathsf{C}(u, 1) = \mathsf{C}(1, u) = u. \tag{6.1}$$

(ii) For all $u_1, u_2, v_1, v_2 \in \mathbb{I}$ and $u_2 \geqslant u_1$, $v_2 \geqslant v_1$, the following monotonicity condition holds:

$$[\mathsf{C}(u_2, v_2) - \mathsf{C}(u_1, v_2)] \geqslant [\mathsf{C}(u_2, v_1) - \mathsf{C}(u_1, v_1)]. \tag{6.2}$$

Dimensions of Uncertainty in Communication Engineering. https://doi.org/10.1016/B978-0-32-399275-6.00014-9

Example 6.1. From the definition above, we have that the three functions

$$\mathsf{W}(u, v) \triangleq \max(u + v - 1, 0),\tag{6.3a}$$

$$\mathsf{M}(u, v) \triangleq \min(u, v),\tag{6.3b}$$

$$\mathsf{\Pi}(u, v) \triangleq uv\tag{6.3c}$$

are copulas. Their role, which is central in copula theory, will be illustrated soon. The graphs of these three functions are shown in Fig. 6.1.   ◇

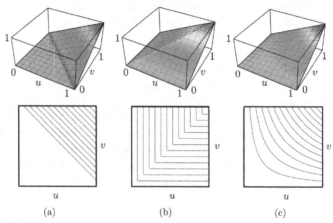

(a)   (b)   (c)

Figure 6.1 Illustration of the three basic two-dimensional copulas: (a) $\mathsf{W}(u, v) = \max(u + v - 1, 0)$, (b) $\mathsf{M}(u, v) = \min(u, v)$, and (c) $\mathsf{\Pi}(u, v) = uv$.

The following two theorems describe fundamental properties of copulas [5, p. 17]:

Theorem 6.1. *For any copula* C, *whenever* $u_2 \geqslant u_1$ *and* $v_2 \geqslant v_1$, *we have*

$$0 \leqslant \mathsf{C}(u_2, v_2) - \mathsf{C}(u_1, v_1) \leqslant u_2 - u_1 + v_2 - v_1,\tag{6.4}$$

that is, every copula in nondecreasing in each place and Lipschitz-continuous.

Proof. From (6.1)–(6.2), letting $u_1 = 0$, we obtain

$$\mathsf{C}(u_2, v_2) - \mathsf{C}(u_2, v_1) \geqslant 0.\tag{6.5}$$

Similarly, letting $v_1 = 0$ and renaming v_2 as v_1 gives

$$\mathsf{C}(u_2, v_1) - \mathsf{C}(u_1, v_1) \geqslant 0.\tag{6.6}$$

Summing the last two equations, we have, for $u_2 \geqslant u_1$ and $v_2 \geqslant v_1$,

$$\mathsf{C}(u_2, v_2) - \mathsf{C}(u_1, v_1) \geqslant 0,\tag{6.7}$$

which is the left-hand inequality in (6.4). Next, letting $u_2 = 1$ in (6.2), we have

$$v_2 - \mathsf{C}(u_1, v_2) - v_1 + \mathsf{C}(u_1, v_1) \geqslant 0 \tag{6.8}$$

and, after letting $v_2 = 1$ and renaming v_1 as v_2,

$$u_2 - u_1 - \mathsf{C}(u_2, v_2) + \mathsf{C}(u_1, v_2) \geqslant 0. \tag{6.9}$$

Summing the last two inequalities, we obtain the right-hand inequality in (6.4). $\quad\square$

Theorem 6.2. *Every copula* $\mathsf{C}(u, v)$ *with* $(u, v) \in \mathbb{I}^2$ *is bounded below and above by*

$$\mathsf{W}(u, v) \leqslant \mathsf{C}(u, v) \leqslant \mathsf{M}(u, v). \tag{6.10}$$

Proof. From (6.1)–(6.2), letting $u_2 = v_2 = 1$ and observing that $\mathsf{C} \geqslant 0$, we obtain the left-hand inequality in (6.10). Further, since $\mathsf{C}(u, v) \leqslant \mathsf{C}(u, 1) = u$ and $\mathsf{C}(u, v) \leqslant \mathsf{C}(1, v) = v$, we obtain the right-hand inequality in (6.10). $\quad\square$

6.1.1 Copulas and joint CDFs

The main reason why copulas are studied lies in *Sklar Theorem* [5, Theorem 1.4.3], [181, p. 18]:

Theorem 6.3. *Let* F_{XY} *be the joint CDF of two RVs* X *and* Y, *whose support sets are* $\mathcal{X}$ *and* $\mathcal{Y}$ *and whose marginal CDFs* F_X *and* F_Y *are continuous. Then there exists a unique copula* C_{XY} *such that*

$$F_{XY}(x, y) = \mathsf{C}_{XY}\big(F_X(x), F_Y(y)\big) \qquad \forall (x, y) \in \mathbb{R}^2. \tag{6.11}$$

Conversely, given two CDFs F_X, F_Y *and any copula* C, *the function* $\mathsf{C}(F_X, F_Y)$ *is a joint CDF with marginal CDFs* F_X *and* F_Y.

Remark 6.1. If F_X and F_Y are not continuous, then C is uniquely determined on the product $\mathsf{range}(F_X) \times \mathsf{range}(F_Y)$.

The following theorem, which we express for simplicity under the assumption of continuous CDFs, summarizes some fundamental facts about the use of copulas [5, p. 19]. In particular, it illustrates the role played by the three copulas W, M, and $\sqcap$:

Theorem 6.4. *Using the notations of Theorem 6.3, with* X, Y *RVs having support sets* $\mathcal{X}$, $\mathcal{Y}$, *respectively, we have:*

  (i) $\mathsf{C}_{XY} = \sqcap$ *if and only if* X *and* Y *are statistically independent.*
  (ii) $\mathsf{C}_{XY} = \mathsf{M}$ *if and only if* $Y = f(X)$ *a.s. where* f *is a.s. strictly increasing on* $\mathcal{X}$.
  (iii) $\mathsf{C}_{XY} = \mathsf{W}$ *if and only if* $Y = f(X)$ *a.s. where* f *is a.s. strictly decreasing on* $\mathcal{X}$.
  (iv)   *If* f *and* g *are a.s. strictly increasing on* $\mathcal{X}$ *and* $\mathcal{Y}$, *respectively, then*

$$\mathsf{C}_{f(X),g(Y)} = \mathsf{C}_{XY}. \tag{6.12}$$

(v) *If f and g are a.s. strictly decreasing on $\mathfrak{X}$ and $\mathfrak{Y}$, respectively, then*

$$\mathsf{C}_{f(X),Y}(x, y) = y - \mathsf{C}_{XY}(1 - x, y), \tag{6.13a}$$

$$\mathsf{C}_{X,g(Y)}(x, y) = x - \mathsf{C}_{XY}(x, 1 - y), \tag{6.13b}$$

$$\mathsf{C}_{f(X),g(Y)}(x, y) = x + y - 1 + \mathsf{C}_{XY}(1 - x, 1 - y). \tag{6.13c}$$

We see that (6.11) yields the bivariate CDF of two RVs in terms of their marginal CDFs and a copula. Inverting that, we derive an expression of the copula joining a bivariate CDF to its marginals. To do this properly, we need to define the inverse of a CDF $F(x)$ that is not strictly increasing. We define the *quasi-inverse* of F, and denote it by $F^{\leftarrow}$, as any function with domain $\mathbb{I}$ such that

(i) If $t \in \mathsf{range}(F)$, then $F^{\leftarrow}(t)$ is any $x \in \overline{\mathbb{R}}$ such that $F(x) = t$, and hence $F\big(F^{\leftarrow}(t)\big) = t$.

(ii) If $t \notin \mathsf{range}(F)$, then

$$F^{\leftarrow}(t) \triangleq \inf\{x \mid F(x) \geqslant t\} = \sup\{x \mid F(x) \leqslant t\}. \tag{6.14}$$

If F is strictly increasing, then $F^{\leftarrow}$ coincides with the ordinary inverse F^{-1}.

Example 6.2 ([181, p. 22]). If $F(x) = u(x - a)$, then

$$F^{\leftarrow}(t) = \begin{cases} a_0, & t = 0, \\ a, & t \in (0, 1), \\ a_1, & t = 1, \end{cases} \tag{6.15}$$

with a_0, a_1 any real numbers with $a_0 < a \leqslant a_1$. ◇

Using this definition, we obtain a relation which can be used as a tool to extract a copula from a bivariate pdf F_{XY} with continuous marginals F_X, F_Y:

$$\mathsf{C}(x, y) = F_{XY}\big[F_X^{\leftarrow}(x), F_Y^{\leftarrow}(y)\big]. \tag{6.16}$$

Example 6.3 (Bivariate exponential CDF [181, p. 23]). With

$$F_{XY}(x, y) = \begin{cases} 1 - e^{-x} + e^{-y} + e^{-(x+y+\theta xy)}, & x \geqslant 0, y \geqslant 0, \\ 0, & \text{otherwise}, \end{cases} \tag{6.17}$$

and $\theta \in \mathbb{I}$ a parameter, we have exponential marginal CDFs, and hence $F_X^{\leftarrow}(u) = -\log(1 - u)$ and $F_Y^{\leftarrow}(v) = -\log(1 - v)$ for $(u, v) \in \mathbb{I}^2$. The resulting copula is

$$\mathsf{C}(u, v) = u + v - 1 + (1 - u)(1 - v)e^{-\theta \log(1-u)\log(1-v)}. \tag{6.18}$$

◇

Since copulas may be viewed as CDFs, we can define their densities according to the usual definition of pdfs as derivatives of CDFs,

$$c(u, v) \triangleq \frac{\partial^2 \mathsf{C}(u, v)}{\partial u \partial v}. \tag{6.19}$$

Joint densities may not exist: for example, copulas $W(u, v)$ and $M(u, v)$, which are not absolutely continuous, have no densities.

Example 6.4. The *Student copula*, which depends on two parameters ρ and η, is defined by choosing marginal Student distributions, whose pdf is [231, p. 11]

$$f(x) = \frac{1}{\sqrt{\pi\eta}} \cdot \frac{\Gamma\left(\frac{\eta+1}{2}\right)}{\Gamma\left(\frac{\eta}{2}\right)} \cdot \left(1 + \frac{x^2}{\eta}\right)^{-(\eta+1)/2}, \tag{6.20}$$

where η is called the *tail parameter* ($\eta = 1$ yields the *Cauchy* pdf). This pdf exhibits tails that are heavier than those of the Gaussian distribution, corresponding to higher probabilities of extreme values. The Student copula, derived using (6.16), has density [231, p. 208]

$$c(u, v) \triangleq \frac{\partial^2}{\partial u \partial v} C(u, v) = \frac{\Gamma\left(\frac{\eta+2}{2}\right)\Gamma\left(\frac{\eta}{2}\right)}{\sqrt{1-\rho^2}\,\Gamma^2\left(\frac{\eta+1}{2}\right)} \cdot \frac{\left(\left(1 + \frac{u^2}{\eta}\right)\left(1 + \frac{v^2}{\eta}\right)\right)^{(\eta+1)/2}}{\left(1 + \frac{u^2 + v^2 - 2\rho u v}{\eta(1-\rho^2)}\right)^{(\eta+2)/2}}, \tag{6.21}$$

with $\rho \in (-1, 1)$ and $\eta \geqslant 2$. This copula is especially suitable in finance and risk management modeling, where pdfs of financial returns are usually heavy-tailed. ◇

6.1.2 Some explicit copula families

Some families of copulas are easy to construct, and share a number of interesting properties which make them useful for modeling joint distributions in two dimensions. In particular, certain copulas allow dependence among RVs to be modeled using a single parameter.

Example 6.5. The *Gumbel copula* is defined by [224]

$$C_\theta(u, v) = \exp\left[-\left((-\log u)^\theta + (-\log v)^\theta\right)^{1/\theta}\right], \tag{6.22}$$

where $\theta \geqslant 1$. Fig. 6.2 shows Gumbel copula for $\theta = 5$. The choice $\theta = 1$ yields copula Π of Fig. 6.1(c), while as $\theta \to \infty$ the Gumbel copula tends to the M copula of Fig. 6.1(b). This is a copula with tail dependence in one corner, as we shall discuss in Example 6.42 *infra*. ◇

Example 6.6. The *Clayton copula* is defined by [224]

$$C_\theta(u, v) = \left(\max\left(u^{-\theta} + v^{-\theta} - 1, 0\right)\right)^{-1/\theta}, \tag{6.23}$$

where $\theta \geqslant -1$ and $\theta \neq 0$ (Fig. 6.3). As $\theta \to 0$, we obtain the independence copula Π,

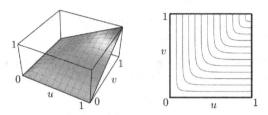

Figure 6.2 Illustration of Gumbel copula with parameter $\theta = 5$.

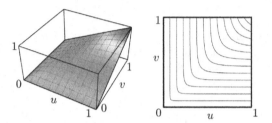

Figure 6.3 Illustration of Clayton copula with parameter $\theta = 5$.

while as $\theta \to \infty$ the Clayton copula tends to copula M. For $\theta = -1$, we obtain copula W. $\diamond$

The two examples above have in common a copula structure whose form is

$$\mathsf{C}(u, v) = \varphi^{-1}\big(\varphi(u) + \varphi(v)\big) \tag{6.24}$$

for a decreasing function $\varphi : \mathbb{I} \mapsto [0, \infty]$. In particular, Gumbel copula has $\varphi(u) = (-\log u)^{\theta}$, while Clayton copula has $\varphi(u) = (u^{-\theta} - 1)/\theta$. We have the following general result [181, Theorem 4.1.4], [224, Theorem 3.2]:

Theorem 6.5. *Any continuous, convex, and strictly decreasing function* $\varphi : \mathbb{I} \mapsto [0, \infty]$ *with* $\varphi(1) = 0$ *generates a copula with the form*

$$\mathsf{C}(u, v) = \begin{cases} \varphi^{-1}\big(\varphi(u) + \varphi(v)\big), & \text{if } \varphi(u) + \varphi(v) \leqslant \varphi(0), \\ 0, & \text{otherwise.} \end{cases} \tag{6.25}$$

Example 6.7. Choosing $\varphi(u) = \log(e^{-\theta} - 1) - \log(e^{-\theta u} - 1)$, we obtain the *Frank copula* [224]

$$\mathsf{C}_\theta(u, v) = -\frac{1}{\theta} \log\left(1 + \frac{(e^{-\theta u} - 1)(e^{-\theta v} - 1)}{e^{-\theta} - 1}\right) \tag{6.26}$$

for $\theta \in \mathbb{R}$ and $\theta \neq 0$. $\diamond$

Example 6.8. Let Z_1 and Z_2 be two independent RVs with common CDF F. Define $X \triangleq \min(Z_1, Z_2)$ and $Y \triangleq \max(Z_1, Z_2)$. The joint CDF of X and Y is given by [58, pp. 11–12]

$$F_{XY}(x, y) = 2F\big(\min(x, y)\big)F(y) - F^2\big(\min(x, y)\big). \tag{6.27}$$

The two marginal CDFs are

$$F_X(x) = 2F(x) - F^2(x),$$
$$F_Y(y) = F^2(y). \tag{6.28}$$

Thus, from (6.16) we obtain the copula

$$\mathsf{C}(u, v) = 2\min\big(1 - \sqrt{1 - u}, \sqrt{v}\big)\sqrt{v} - [\min\big(1 - \sqrt{1 - u}, \sqrt{v}\big)]^2. \tag{6.29}$$

$$\diamond$$

Remark 6.2. Using the definition (6.14) of the quasi-inverse of a CDF, we observe that, if $U \sim \mathcal{U}[0, 1]$, and F_X is a CDF, then

$$\mathbb{P}\big(F_X^{\leftarrow}(U) \leq x\big) = F_X(x). \tag{6.30}$$

Moreover, if F_X is a continuous CDF, then $F_X(X) \sim \mathcal{U}[0, 1]$. The "quantile transformation" operated by $F_X^{\leftarrow}$ in (6.30) is useful in Monte Carlo simulation of random events. This allows one to obtain a sequence of pseudorandom numbers with a given CDF F_X by first generating a sequence of pseudorandom numbers with uniform distribution, next applying to these a quantile transformation.

A part of Theorem 6.3 can be obtained by using the observation in Remark 6.2. Let F_X and F_Y be continuous marginal CDFs. We can express the joint CDF F_{XY} by noting that

$$\mathbb{P}\big(F_X(X) \leqslant u, F_Y(Y) \leqslant y\big) = \mathbb{P}\big(X \leqslant F_X^{\leftarrow}(u), Y \leqslant F_Y^{\leftarrow}(v)\big)$$
$$= F_{XY}\big(F_X^{\leftarrow}(u), F_Y^{\leftarrow}(v)\big). \tag{6.31}$$

Letting $u \triangleq F_X(x)$, $v \triangleq F_Y(y)$, we finally obtain

$$F_{XY}(x, y) = \mathsf{C}_{XY}\big(F_X(x), F_Y(y)\big), \qquad x, y \in \overline{\mathbb{R}}, \tag{6.32}$$

as well as

$$\mathsf{C}_{XY}(x, y) = F_{XY}\big(F_X^{\leftarrow}(x), F_Y^{\leftarrow}(y)\big). \tag{6.33}$$

This shows that every two-dimensional copula is the joint CDF of a pair of RVs marginally uniform in [0, 1]. Stated equivalently, we can define a copula as a joint CDF with uniform marginals, so that much can be obtained in the study of copulas by examining different dependences occurring among RVs that are uniformly distributed in [0, 1]. Thus, letting $U \sim \mathcal{U}(0, 1)$ and $V \sim \mathcal{U}(0, 1)$, *comonotone*, independent, and *countermonotone* continuous RVs can be defined as follows:

Comonotonicity. With $U = V$, we have

$$
\begin{aligned}
\mathsf{C}(u, v) &\triangleq \mathbb{P}(U \leqslant u, V \leqslant v) \\
&= \mathbb{P}(U \leqslant u, U \leqslant v) \\
&= \min(u, v) \\
&= \mathsf{M}(u, v).
\end{aligned}
\tag{6.34}
$$

As stated in Theorem 6.4, this is the copula corresponding to $V = T(U)$, where T is an a.s. strictly increasing transformation. RVs with this type of dependence are called *comonotone*.

Independence. Independence occurs with

$$
\mathsf{C}(u, v) = uv = \sqcap(u, v).
\tag{6.35}
$$

Countermonotonicity. This is obtained for

$$
U = 1 - V.
\tag{6.36}
$$

The related copula is, for $1 - v < u$,

$$
\begin{aligned}
\mathsf{C}(u, v) &\triangleq \mathbb{P}(U \leqslant u, V \leqslant v) \\
&= \mathbb{P}(1 - v \leqslant U \leqslant u).
\end{aligned}
\tag{6.37}
$$

Since $U \sim \mathcal{U}(0, 1)$, it follows that

$$
\mathsf{C}(u, v) = u + v - 1,
\tag{6.38}
$$

while $\mathsf{C}(u, v) = 0$ for $1 - v \geqslant u$. Thus, $\mathsf{C}(u, v) = \mathsf{W}(u, v)$. Countermonotonicity corresponds to perfect negative dependence, $V = T(U)$, with T a.s. strictly decreasing. RVs with this type of dependence are called *countermonotone*.

Example 6.9. A well-known result from probability theory states that, if two RVs are jointly Gaussian, they are also marginally Gaussian, but the converse is not true (see, e.g., [189, p. 128]). The second part of this result can be proved using copulas. In fact, if Ψ_ρ denotes the jointly Gaussian CDF with correlation coefficient ρ, and Φ denotes the Gaussian CDF

$$
\Phi(x) \triangleq \frac{1}{2} \operatorname{erfc}\left(-\frac{x}{\sqrt{2}}\right),
\tag{6.39}
$$

from (6.33) we obtain the Gaussian copula [173]

$$
\begin{aligned}
\mathsf{C}(u, v) &= \Psi_\rho\big(\Phi^{-1}(u), \Phi^{-1}(v)\big) \\
&= \frac{1}{2\pi\sqrt{1 - \rho^2}} \int_{-\infty}^{\Phi^{-1}(u)} \int_{-\infty}^{\Phi^{-1}(v)} \exp\left[-\frac{s^2 - 2\rho st + t^2}{2(1 - \rho^2)}\right] ds \, dt.
\end{aligned}
\tag{6.40}
$$

We can see in particular that with $\rho = 0$ we obtain $\mathsf{C}(u, v) = uv$, if $\rho = -1$ we obtain the countermonotonicity copula, and if $\rho = 1$ we obtain the comonotonicity copula [224, p. 8]. Thus, the bivariate Gaussian copula can be interpreted as describing a dependence that interpolates between perfect positive and negative dependence, with ρ describing the amount of dependence. Any copula other than (6.40) does not yield a jointly Gaussian CDF (see [147] for explicit examples of non-Gaussian joint CDFs whose marginals are Gaussian). $\diamond$

6.1.3 Fréchet–Hoeffding inequalities

By combining (6.34) and (6.37), the following general *Fréchet–Hoeffding inequalities*, bounding any copula from above and below, can be proved:

$$\mathsf{W}(x, y) \leqslant \mathsf{C}(x, y) \leqslant \mathsf{M}(x, y), \qquad \forall (x, y) \in \mathbb{I}^2, \tag{6.41}$$

with W and M defined in (6.3a) and (6.3b), respectively. Bounds (6.41) indicate that every two-dimensional copula has to lie inside of the pyramid bounded by the surfaces in Fig. 6.1, (a) and (b).

To prove the upper bound in (6.41), observe that, with $U \sim \mathcal{U}(0, 1)$ and $V \sim \mathcal{U}(0, 1)$,

$$\big\{\{U \leqslant u\} \cap \{V \leqslant v\}\big\} \subseteq \{U \leqslant u\} \text{ and } \big\{\{U \leqslant u\} \cap \{V \leqslant v\}\big\} \subseteq \{V \leqslant v\}, \tag{6.42}$$

while the lower bound follows from

$$\begin{aligned} \mathsf{C}(u, v) = \mathbb{P}(U \leqslant u, V \leqslant v) &= 1 - \mathbb{P}(\{U > u\} \cup \{V > v\}) \\ &\geqslant 1 - \mathbb{P}(\{U > u\}) - \mathbb{P}(\{V > v\}) \\ &= u + v - 1. \end{aligned} \tag{6.43}$$

Combining (6.41) with Theorem 6.3 *supra*, we obtain the Fréchet–Hoeffding bounds on a joint CDF in terms of its marginals [181, p. 30],

$$\max\left[F_X(x) + F_Y(y) - 1, 0\right] \leqslant F_{XY}(x, y) \leqslant \min\left[F_X(x), F_Y(y)\right]. \tag{6.44}$$

Recalling Theorem 6.4 *supra*, we have that the upper bound is achieved when Y is an a.s. strictly increasing function of X, while the lower bound is achieved when Y is an a.s. strictly decreasing function of X.

6.1.4 A version of Fréchet–Hoeffding bounds involving probabilities

To calculate probabilities that involve combinations of two events A and B, we need to know, in addition to $\mathbb{P}(A)$ and $\mathbb{P}(B)$, either $\mathbb{P}(A \cap B)$ or $\mathbb{P}(A \cup B)$. If these are unknown, we may bound them using the following *Fréchet inequalities* [214]:

$$\max\big(\mathbb{P}(A) + \mathbb{P}(B) - 1, 0\big) \leqslant \mathbb{P}(A \cap B) \leqslant \min\big(\mathbb{P}(A), \mathbb{P}(B)\big), \tag{6.45a}$$

$$\max\big(\mathbb{P}(A), \mathbb{P}(B)\big) \leqslant \mathbb{P}(A \cup B) \leqslant \min\big(\mathbb{P}(A) + \mathbb{P}(B), 1\big). \qquad (6.45b)$$

To prove the lower bound in (6.45a), observe that in general $\mathbb{P}(A \cup B) = \mathbb{P}(A) + \mathbb{P}(B) - \mathbb{P}(A \cap B)$. Since $\mathbb{P}(A \cup B) \leqslant 1$, we have $\mathbb{P}(A \cap B) \geqslant \mathbb{P}(A) + \mathbb{P}(B) - 1$, which in turn must be greater than or equal to 0. The upper bound in (6.45a) is obtained by observing that in general $\mathbb{P}(A \cap B) = \mathbb{P}(A \mid B)\mathbb{P}(B) = \mathbb{P}(B \mid A)\mathbb{P}(A)$. Since $\mathbb{P}(A \mid B) \leqslant 1$ and $\mathbb{P}(B \mid A) \leqslant 1$, it follows that $\mathbb{P}(A \cap B) \leqslant \mathbb{P}(A)$ and $\mathbb{P}(A \cap B) \leqslant \mathbb{P}(B)$. The lower bound in (6.45b) is proved by observing that $\{A \cup B\} \supseteq A$ and $\{A \cup B\} \supseteq B$. As for the upper bound in (6.45b), observe that $\mathbb{P}(A \cup B) = \mathbb{P}(A) + \mathbb{P}(B) - \mathbb{P}(A \cap B) \leqslant \mathbb{P}(A) + \mathbb{P}(B)$.

These bounds can be achieved by some dependence relations between events A and B, and hence are sharp [212].

Example 6.10. With two "rare" events A, B, each occurring with a small probability ε, (6.45a)–(6.45b) yield

$$0 \leqslant \mathbb{P}(A \cap B) \leqslant \varepsilon,$$
$$\varepsilon \leqslant \mathbb{P}(A \cup B) \leqslant 2\varepsilon.$$

Assuming independence, we would obtain $\mathbb{P}(A \cap B) = \varepsilon^2$ (a value close to the lower bound) and $\mathbb{P}(A \cup B) = 2\varepsilon - \varepsilon^2$ (a value close to the upper bound). $\qquad \diamond$

Remark 6.3. Fréchet inequalities can be generalized by induction to the multivariate case, yielding

$$\max\left(\sum_{i=1}^{d}\mathbb{P}(A_i) - (d-1), 0\right) \leqslant \mathbb{P}\left(\bigcap_{i=1}^{d}A_i\right) \leqslant \min\big(\mathbb{P}(A_1), \dots, \mathbb{P}(A_d)\big),$$
$$\max\big(\mathbb{P}(A_1), \dots, \mathbb{P}(A_d)\big) \leqslant \mathbb{P}\left(\bigcup_{i=1}^{d}A_i\right) \leqslant \min\left(\sum_{i=1}^{d}\mathbb{P}(A_i), 1\right). \qquad (6.46)$$

6.1.5 Dual of a copula and the survival copula

The *dual* C^∂ of copula C is defined as

$$\mathsf{C}^\partial(u, v) \triangleq u + v - \mathsf{C}(u, v). \qquad (6.47)$$

The motivation for this definition comes from the observation that [5, p. 21]

$$\mathsf{C}^\partial\big(F_X(x), F_Y(y)\big) = F_X(x) + F_Y(y) - \mathsf{C}\big(F_X(x), F_Y(y)\big)$$
$$= \mathbb{P}[\{X \leqslant x\} \cup \{Y \leqslant y\}], \qquad (6.48)$$

which yields a probabilistic interpretation of C^∂.

Example 6.11. Recalling the definition $W(u, v) \triangleq \max(u + v - 1, 0)$, we obtain

$$W^{\partial}(u, v) = u + v - \max(u + v - 1, 0)$$
$$= \min(u + v, 1). \tag{6.49}$$

$\diamond$

Example 6.12. The following inequalities hold [227, §6.4]:

$$M^{\partial} \leqslant C^{\partial} \leqslant W^{\partial}. \tag{6.50}$$

$\diamond$

The complement of the CDF F_X is called *survival function*, or *reliability function* [181, p. 32]. Formally,

$$\tilde{F}_X(x) \triangleq 1 - F_X(x) = \mathbb{P}(X > x). \tag{6.51}$$

For the pair (X, Y) of RVs whose joint CDF is F_{XY}, the joint survival function is defined as $\tilde{F}_{XY}(x, y) \triangleq \mathbb{P}(X > x, Y > y)$. The marginal survival CDFs are $\tilde{F}_{XY}(x, -\infty) = \tilde{F}_X(x)$ and $\tilde{F}_{XY}(-\infty, y) = \tilde{F}_Y(y)$. The version of Sklar Theorem involving survival functions is obtained by observing that

$$\begin{aligned}
\tilde{F}_{XY}(x, y) &\triangleq \mathbb{P}(\{X > x\} \cap \{Y > y\}) \\
&= 1 - \mathbb{P}(\{X \leqslant x\} \cup \{Y \leqslant y\}) \\
&= 1 - F_X(x) - F_Y(y) + F_{XY}(x, y) \\
&= \tilde{F}_X(x) + \tilde{F}_Y(y) - 1 + C(F_X(x), F_Y(y)) \\
&= \tilde{F}_X(x) + \tilde{F}_Y(y) - 1 + C(1 - \tilde{F}_X(x), 1 - \tilde{F}_Y(y))
\end{aligned} \tag{6.52}$$

and defining the function $\hat{C} : \mathbb{I}^2 \mapsto \mathbb{I}$ by

$$\hat{C}(u, v) \triangleq u + v - 1 + C(1 - u, 1 - v) \tag{6.53}$$

so that, finally,

$$\tilde{F}_{XY}(x, y) = \hat{C}(\tilde{F}_X(x), \tilde{F}_Y(y)). \tag{6.54}$$

It can be proved that $\hat{C}$ is itself a copula, called the *survival copula* of X and Y [181, p. 32].

Remark 6.4. The survival copula is not the joint survival function $\tilde{C}$ for two RVs with joint CDF C. Actually, using its definition (6.53), we have, with $U \sim \mathcal{U}[0, 1]$ and $V \sim \mathcal{U}[0, 1]$,

$$\begin{aligned}
\tilde{C}(u, v) &\triangleq \mathbb{P}(\{U > u\} \cap \{V > v\}) \\
&= 1 - u - v + C(u, v) \\
&= \hat{C}(1 - u, 1 - v).
\end{aligned} \tag{6.55}$$

Remark 6.5. A random vector $\mathbf{X} = (X_1, X_2)$ is called *radially symmetric* about the center of symmetry $\mathbf{c} = (c_1, c_2)$ if $\mathbf{X} - \mathbf{c}$ and $\mathbf{c} - \mathbf{X}$ have the same joint CDF. If $\mathbf{U}$ has components $\sim \mathcal{U}[0, 1]$, then $\mathbf{U}$ is radially symmetric with center of symmetry $(0.5, 0.5)$, and hence $\mathbf{U}$ has the same CDF as $\mathbf{1} - \mathbf{U}$. This implies that if a random vector is radially symmetric, then its copula is equal to its survival copula, $\hat{\mathsf{C}} = \mathsf{C}$.

6.1.6 d-Dimensional copulas

The main results concerning d-dimensional copulas generalize in a rather straightforward way those derived in the bivariate case. A d-dimensional copula $\mathsf{C} : \mathbb{I}^d \mapsto \mathbb{I}$ can be viewed as a CDF with uniform marginals. We have in particular, for $\mathsf{C}(\mathbf{u}) = \mathsf{C}(u_1, \ldots, u_d)$ [172, p. 185]:

  (i) $\mathsf{C}(u_1, \ldots, u_d)$ is nondecreasing in each argument u_1.
  (ii) The ith marginal CDF is obtained from

$$\mathsf{C}(1, 1, \ldots, 1, u_i, 1, \ldots, 1) = u_i. \tag{6.56}$$

  (iii) For all $(x_1, \ldots, x_d), (y_1, \ldots, y_d) \in \mathbb{I}^d$ with $x_i \leqslant y_i$,

$$\sum_{i_1=1}^{2} \cdots \sum_{i_d=1}^{2} (-1)^{i_1 + \cdots + i_d} \mathsf{C}(u_{1,i_1}, \ldots, u_{d,i_d}) \geqslant 0, \tag{6.57}$$

with $u_{j,1} = x_j$ and $u_{j,2} = y_j$ for all $j \in \{1, \ldots, d\}$.

Conversely, any function satisfying conditions (i)–(iii) is a d-dimensional copula. Property (i) follows from the definition of C as a multivariate CDF, while (ii) is required by uniform marginal CDFs. Property (iii), the "rectangle inequality," states that, if $U_i \sim \mathcal{U}[0, 1]$, $1 \leqslant i \leqslant d$, then $\mathbb{P}[U_i \in [x_i, y_i], \forall i] \geqslant 0$. Observe that the k-dimensional marginal CDFs of C are themselves copulas for $2 \leqslant k < d$.

The d-dimensional version of Sklar Theorem states the following [172, Theorem 5.3]:

Theorem 6.6. *To any d-dimensional CDF F with marginal CDFs $F_1, \ldots, F_d$ there corresponds a copula C such that, for all $(x_1, \ldots, x_d) \in \overline{\mathbb{R}}^d$,*

$$F(x_1, \ldots, x_d) = \mathsf{C}\big(F_1(x_1), \ldots, F_d(x_d)\big). \tag{6.58}$$

If all CDFs F_i are continuous, then C is unique. Otherwise, C is uniquely determined only on the product of the ranges of $F_1, \ldots, F_d$. Conversely, given a d-dimensional copula C and d CDFs $F_1, \ldots, F_d$, (6.58) defines a d-dimensional CDF whose marginals are $F_1, \ldots, F_d$. We also have, using definition (6.14),

$$\mathsf{C}(\mathbf{u}) = F\big(F_1^{\leftarrow}(u_1), \ldots, F_n^{\leftarrow}(u_d)\big). \tag{6.59}$$

The following result [172, Proposition 5.6], [224, Proposition 2.4] generalizes what was derived for $d = 2$: If the dependence structure among the RVs $X_1, \ldots, X_d$ is described by copula C, and T_i, $i = 1, \ldots, d$, are a.s. strictly increasing functions $T_i :$ $\mathbb{R} \mapsto \mathbb{R}$, then the dependence structure of the RVs $T_1(X_1), \ldots, T_d(X_d)$ is described by the same copula.

The 2-dimensional bounds (6.41) can be extended to general d by a straightforward generalization of the proof in (6.42)–(6.43) [172, Theorem 5.7]:

$$\max \left\{ \sum_{i=1}^{d} u_i + 1 - d, 0 \right\} \leqslant C(\mathbf{u}) \leqslant \min(u_1, \ldots, u_d), \tag{6.60}$$

while the d-dimensional Fréchet–Hoeffding bounds (6.44) become

$$\max \left\{ \sum_{i=1}^{d} F_i(x_i) + 1 - d, 0 \right\} \leqslant F_{X_1, \ldots, X_d}(x_1, \ldots, x_d) \leqslant \min \big(F_1(x_1), \ldots, F_d(x_d) \big).$$
$$\tag{6.61}$$

The upper bound in (6.60) is a copula, while the lower bound is not a copula for $d > 2$ [172, Example 5.21]. Thus, while the upper bound in (6.61) can be achieved for any d, the lower bound is achieved in general only for $d = 2$, and hence the bound is generally not sharp when $d > 2$.

The extensions of copulas W and M to d dimensions are given by

$$W_d(\mathbf{u}) \triangleq \max(u_1 + \cdots + u_d + 1 - d, 0), \tag{6.62a}$$

$$M_d(\mathbf{u}) \triangleq \min(u_1, \ldots, u_d), \tag{6.62b}$$

while the independence copula is given by

$$\sqcap_d(\mathbf{u}) = \prod_{i=1}^{d} u_i, \tag{6.63}$$

and we can say, using Sklar Theorem, that d random variables are independent if and only if their copula is the independence copula (6.63).

The dual of C is given by

$$C^{\partial}(u_1, \ldots, u_d) \triangleq P \big(\{U_1 \leqslant u_i\} \cup \{U_2 \leqslant u_2\} \cup \cdots \cup \{U_d \leqslant u_d\} \big). \tag{6.64}$$

Remark 6.6. If discrete RVs are dealt with, additional care should be exercised, because marginal CDFs may be joined by more than one copula to form the joint CDF (for an example, see [172, p. 188]).

Similarly to two-dimensional copulas, we say that the RVs $X_1, \ldots, X_d$ are *comonotone* if they are connected by the comonotonicity copula yielding the upper bound in (6.61). If their CDFs are continuous, then they are comonotone if and only if every pair X_i, X_j is related by an a.s. increasing transformation T_{ij}, i.e., $X_j = T_{ij}(X_i)$. For a proof, and a related result about comonotone RVs whose CDFs may not be continuous, see, e.g., [172, p. 199]. Notice also that the concept of countermonotonicity does not generalize to a number of dimensions $d > 2$. This is due to the fact, observed before, that the Fréchet–Hoeffding lower bound is not a copula.

Example 6.13. A d-dimensional Gaussian random vector $\mathbf{X}$ has a copula obtained as a straightforward generalization of the binary copula (6.40) [224]. Since copulas are invariant to changes of location or scale, we may take $\mathbf{X} \sim \mathcal{N}(\mathbf{0}, \mathbf{R})$, with $\mathbf{R}$ the matrix of correlation coefficients of the components of $\mathbf{X}$. The matrix $\mathbf{R}$ is specified by selecting its $d(d-1)/2$ off-diagonal entries, constrained by nonnegative definiteness. With $\mathbf{R} = \mathbf{I}$, we obtain the independence copula, while if $\mathbf{R}$ is the matrix all of whose entries are 1 we obtain the comonotonicity copula $\min(u_1, \ldots, u_d)$. The Gaussian copula with off-diagonal entries of matrix $\mathbf{R}$ taking the same value $\rho \in [0, 1]$ is called "one-factor Gaussian copula."   ◇

6.2 Dependence p-boxes from copula bounds

Let C be a copula, and $\circ$ a binary operation from $\mathbb{R}^+ \times \mathbb{R}^+$ onto $\mathbb{R}^+$ which is nondecreasing in each place and continuous on $\mathbb{R}^+ \times \mathbb{R}^+$ except possibly at points $(0, \infty)$ and $(\infty, 0)$. Further, denote by $\mathcal{D}$ the set of all CDFs that are left-continuous on $\mathbb{R}$, and by $\mathcal{D}_0$ the subset of $\mathcal{D}$ including the CDFs $F(x)$ such that $F(0) = 0$.

Let $F, G \in \mathcal{D}_0$. We define three functions with arguments F, G in $\mathcal{D}_0 \times \mathcal{D}_0$ and values

$$\tau_{\mathsf{C}, \circ}(F, G)(x) \triangleq \sup_{u \circ v = x} \mathsf{C}\big(F(u), G(v)\big), \tag{6.65a}$$

$$\rho_{\mathsf{C}, \circ}(F, G)(x) \triangleq \inf_{u \circ v = x} \mathsf{C}^\partial\big(F(u), G(v)\big), \tag{6.65b}$$

$$\sigma_{\mathsf{C}, \circ}(F, G)(x) \triangleq \iint_{u \circ v \leqslant x} d\mathsf{C}\big(F(u), G(v)\big), \tag{6.65c}$$

with $x \in (0, \infty)$. Notice that $\sigma_{\mathsf{C}, \circ}(F, G)(x)$ is the CDF of $X \circ Y$ where X, Y have joint CDF $F_{XY}(u, v) = \mathsf{C}\big(F(u), G(v)\big)$.

The following theorem (see [272, p. 76 ff.], [273, p. 99]) generalizes [89]:

Theorem 6.7. *Let X, Y be RVs with support sets in $\mathbb{R}^+$ having CDFs F_X, F_Y in $\mathcal{D}$. Further, let $F_{X \circ Y}$ denote the CDF of $X \circ Y$. The following lower and upper dependence bounds hold:*

$$\underline{F}_{X \circ Y}(z) \leqslant F_{X \circ Y}(z) \leqslant \overline{F}_{X \circ Y}(z), \tag{6.66}$$

where

$$
\begin{aligned}
\underline{F}_{X \circ Y}(z) &= \tau_{\mathsf{W}, \circ}(F_X, F_Y)(z) = \sup_{u \circ v = z} \mathsf{W}\big(F_X(u), F_Y(v)\big), \\
\overline{F}_{X \circ Y}(z) &= \rho_{\mathsf{W}, \circ}(F_X, F_Y)(z) = \inf_{u \circ v = z} \mathsf{W}^\partial\big(F_X(u), F_Y(v)\big).
\end{aligned}
\tag{6.67}
$$

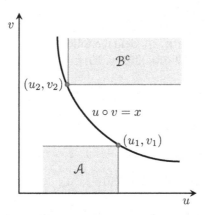

Figure 6.4 Illustration of the proof of Theorem 6.7.

Proof. For any given copula C and any pair of points (u_1, v_1) and (u_2, v_2) on $u \circ v = x$ (see Fig. 6.4), we have the following chain of inequalities: *

$$\mathsf{W}\big(F_X(u_1), F_Y(v_1)\big) \leqslant \mathsf{C}\big(F_X(u_1), F_Y(v_1)\big) \qquad \text{(from (6.41))}$$

$$= \iint_{\mathcal{A}} d\mathsf{C}\big(F_X(u), F_Y(v)\big) \qquad \text{(see Fig. 6.4)}$$

$$\leqslant \sigma_{\mathsf{C},\circ}(F_X, F_Y)(x) \qquad \text{(from definition (6.65c))}$$

$$\leqslant \iint_{\mathcal{B}} d\mathsf{C}\big(F_X(u), F_Y(v)\big) \qquad \text{(see Fig. 6.4)}$$

$$= F_X(u_2) + F_Y(v_2) - \mathsf{C}\big(F_X(u_2), F_Y(v_2)\big) \qquad \text{(integral is sum of CDFs)}$$

$$= \mathsf{C}^{\partial}\big(F_X(u_2), F_Y(v_2)\big) \qquad \text{(from the definition (6.47) of } \mathsf{C}^{\partial})$$

$$\leqslant \mathsf{W}^{\partial}\big(F_X(u_2), F_Y(v_2)\big) \qquad \text{(from Example (6.12)).}$$

Finally, observe that $\tau_{\mathsf{W},\circ}$ is the greatest value of $\mathsf{W}\big(F_X(u_1), F_Y(v_1)\big)$ as (u_1, v_1) lies on the curve $u \circ v = x$, and $\rho_{\mathsf{W},\circ}$ is the smallest value of $\mathsf{W}^{\partial}\big(F_X(u_2), F_Y(v_2)\big)$ as (u_2, v_2) lies on the curve $u \circ v = x$. □

Remark 6.7. Notice that some authors (e.g., in [227,272,273]) define CDFs as $F_X(x) \triangleq \mathbb{P}(X < x)$.

Remark 6.8. The bounds in Theorem 6.7 can be proved to be pointwise the best possible [272, §3.3.2], [273, p. 101].

Remark 6.9. The bounds above assume that no information is available about the dependence of X and Y. This is reflected by the use of copula W, since for any copula C_{XY} we have $\mathsf{W} \leqslant \mathsf{C}_{XY}$. Should a better bound $\underline{\mathsf{C}}_{XY}$ to C_{XY} be available, this can be used instead of W in (6.67) to tighten the p-box $[\underline{F}_{X \circ Y}, \overline{F}_{X \circ Y}]$ [272, §3.3.2], [273, p. 100].

6.2.1 Special operations

The special cases $\circ \in \{+, -, \times, \div\}$ yield the following results, valid on the nonnegative real line $\mathbb{R}^+$ [272, pp. 77–78], [273, p. 101]:

$$\underline{F}_{X+Y}(z) = \sup_{x+y=z} \max[F_X(x) + F_Y(y) - 1, 0],$$

$$\overline{F}_{X+Y}(z) = \inf_{x+y=z} \min[F_X(x) + F_Y(y), 1],$$

$$\underline{F}_{X-Y}(z) = \sup_{x+y=z} \max[F_X(x) - F_Y(-y), 0],$$

$$\overline{F}_{X-Y}(z) = 1 + \inf_{x+y=z} \min[F_X(x) - F_Y(-y), 0],$$

$$(6.68)$$

$$\underline{F}_{X \times Y}(z) = \sup_{xy=z} \max[F_X(x) + F_Y(y) - 1, 0],$$

$$\overline{F}_{X \times Y}(z) = \inf_{xy=z} \min[F_X(x) + F_Y(y), 1],$$

$$\underline{F}_{X \div Y}(z) = \sup_{xy=z} \max[F_X(x) - F_Y(1/y), 0],$$

$$\overline{F}_{X \div Y}(z) = 1 + \inf_{xy=z} \min[F_X(x) - F_Y(1/y), 0].$$

To prove (6.68), observe that the results for sum and product follow directly from Theorem 6.7. The cases of difference and quotient can be treated by observing that these two binary operations, which are increasing in their first argument and decreasing in the second, can be converted to sum and product, respectively, as follows [273, p. 101], [272, p. 78]: Let $Y' \triangleq -Y$, and use the results for the sum in (6.68) after observing that $X - Y = X + Y'$ and $F_{Y'}(y) = 1 - F_Y(-y)$. Similarly, let $Y' \triangleq 1/Y$, and use the results for the product after observing that $X \div Y = X \times Y'$ and $F_{Y'}(y) = 1 - F_Y(1/y)$.

Remark 6.10. The results in (6.68) referring to sum and difference hold generally in $\mathbb{R}$, and not only in $\mathbb{R}^+$ [272, p. 78], [207, p. 6].

Example 6.14. Figs. 6.5–6.6 show two p-boxes generated by the dependence bounds on sum and product of two uniformly distributed RVs X and Y. Fig. 6.7 shows the dependence bounds on $F_Z(z)$, with $Z = X + Y$, $X \sim \mathcal{U}[0, 1]$, and $Y \sim \mathcal{U}[0, 1]$, and the actual $F_Z(z)$ under the conditions $Y = X$ (comonotone RVs), X and Y independent RVs, and $Y = 1 - X$ (countermonotone RVs). $\diamond$

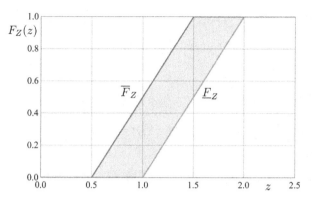

Figure 6.5 Dependence bounds on $F_Z(z)$, with $Z = X + Y$, $X \sim \mathcal{U}[0, 1]$, and $Y \sim \mathcal{U}[0.5, 1]$.

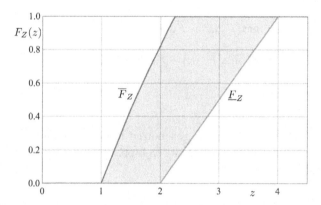

Figure 6.6 Dependence bounds on $F_Z(z)$, with $Z = X \times Y$, $X \sim \mathcal{U}[1, 2]$, and $Y \sim \mathcal{U}[1, 2]$.

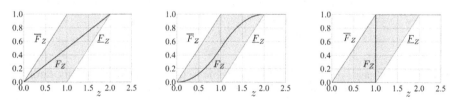

Figure 6.7 Dependence bounds on $F_Z(z)$, with $Z = X + Y$, $X \sim \mathcal{U}[0, 1]$, and $Y \sim \mathcal{U}[0, 1]$, compared with actual $F_Z(z)$ with $Y = X$ (comonotone RVs), X and Y independent RVs, and $Y = 1 - X$ (countermonotone RVs).

6.2.2   Using p-boxes of F_X and F_Y

If F_X, F_Y are CDFs circumscribed by p-boxes $[\overline{F}_X, \underline{F}_X]$ and $[\overline{F}_Y, \underline{F}_Y]$, then the previous results about upper and lower bounds on $F_{X \circ Y}$ can be generalized to determine the p-box $[\overline{F}_{X \circ Y}, \underline{F}_{X \circ Y}]$ that circumscribes $F_{X \circ Y}$. We have [272, §3.3.4], [273, p. 112]

$$\underline{F}_{X+Y}(z) = \sup_{x+y=z} \max[\underline{F}_X(x) + \underline{F}_Y(y) - 1, 0],$$

$$\overline{F}_{X+Y}(z) = \inf_{x+y=z} \min[\overline{F}_X(x) + \overline{F}_Y(y), 1],$$

$$\underline{F}_{X-Y}(z) = \sup_{x-y=z} \max[\underline{F}_X(x) - \overline{F}_Y(-y), 0],$$

$$\overline{F}_{X-Y}(z) = 1 + \inf_{x-y=z} \min[\overline{F}_X(x) - \underline{F}_Y(-y), 0],$$

$$\underline{F}_{X\times Y}(z) = \sup_{xy=z} \max[\underline{F}_X(x) + \underline{F}_Y(y) - 1, 0],$$ (6.69)

$$\overline{F}_{X\times Y}(z) = \inf_{xy=z} \min[\overline{F}_X(x) + \overline{F}_Y(y), 1],$$

$$\underline{F}_{X\div Y}(z) = \sup_{xy=z} \max[\underline{F}_X(x) - \overline{F}_Y(1/y), 0],$$

$$\overline{F}_{X\div Y}(z) = 1 + \inf_{xy=z} \min[\overline{F}_X(x) - \underline{F}_Y(1/y), 0].$$

6.2.3   Operations on d RVs

The results in Section 6.2.1 can be generalized to operations involving more than two RVs by exploiting their associativity [89, pp. 210–211]. For example, explicit equations for the sum of RVs are [63]:

$$\underline{F}(z) = \sup_{x_1+\cdots+x_d=z} \max\left(\sum_{i=1}^{d} F_i(x_i) - (d-1), 0\right),$$

$$\overline{F}(z) = \inf_{x_1+\cdots+x_d=z} \min\left(\sum_{i=1}^{d} F_i(x_i), 1\right).$$ (6.70)

Since in most cases (6.70) are hardly amenable to closed-form expressions, iterative algorithms can be used based on (6.69). For example, assuming $d = 3$, with obvious notations:

$$\overline{F}_{(3)}(z) = \inf_{x_1+x_2+x_3=z} \min\left(F_1(x_1) + F_2(x_2) + F_3(x_3), 1\right)$$

$$= \inf_{x_1} \min\left(F_1(x_1) + \overline{F}_{(2)}(z - x_1), 1\right),$$ (6.71)

where

$$\overline{F}_{(2)}(w) \triangleq \inf_{x_2+x_3=w} \min\left(F_2(x_2) + F_3(x_3), 1\right).$$ (6.72)

6.2.4 Dependence bounds with order statistics

We examine first the case of two RVs X, Y with joint CDF $F_{XY}(x, y)$ and order statistics $\min(X, Y)$ and $\max(X, Y)$. Since we have, with obvious notations [189, p. 141],

$$F_{\max}(x) = F_{XY}(x, x),\tag{6.73}$$

using Fréchet–Hoeffding bound (6.44) with $x = y$, we obtain

$$\max\left(F_X(x) + F_Y(x) - 1, 0\right) \leqslant F_{\max}(x) \leqslant \min\left(F_{X_1}(x), F_{X_2}(x)\right).\tag{6.74}$$

To derive $F_{\min}(x)$, observe first the general relationship

$$F_{\max}(x) + F_{\min}(x) = F_X(x) + F_Y(x),\tag{6.75}$$

which yields, using (6.74),

$$F_X + F_Y - \min\left(F_X, F_Y\right) \leqslant F_{\min} \leqslant F_X + F_Y - \max\left(F_X + F_Y - 1, 0\right)\tag{6.76}$$

and, finally,

$$\max\left(F_{X_1}(x), F_{X_2}(x)\right) \leqslant F_{\min}(x) \leqslant \min\left(F_{X_1}(x) + F_{X_2}(x), 1\right).\tag{6.77}$$

Results (6.74)–(6.77) can be generalized to the extremes of d RVs and to more general order statistics [216,217]. However, while $F_{\max}$ attains the upper bound $\min\left(F_{X_1}, \ldots, F_{X_d}\right)$ and $F_{\min}$ attains the lower bound $\max\left(F_{X_1}, \ldots, F_{X_d}\right)$ for $X_i = F_{X_i}^{\leftarrow}(U)$ with $U \sim \mathcal{U}(0, 1)$, the construction of random variables achieving the other two bounds is less obvious.

6.3 Some examples of application

Example 6.15. Here we examine the impact on the performance of diversity combining in fading channels caused by different joint distributions having the same marginals. We assume two-branch diversity with Rayleigh fading gains R_1 and R_2 on each branch, and maximal-ratio combining [23, Chapter 4]. The equivalent fading $\check{R}$ generated by the combination is given by the square root of $R_1^2 + R_2^2$. With R_1 independent of R_2, the CDF of $\check{R}^2$ is $F_{\check{R}^2}(z) = 1 - e^{-z}(1 + z), z \geqslant 0$. With unknown dependence, one can use the bounds of the first two equations of (6.68), with marginals $F_{R_i^2}(x_i) = 1 - e^{-x_i}, x_i \geqslant 0, i = 1, 2$, which yields, for $z \geqslant 0$,

$$\begin{aligned}\overline{F}(z) &= 1 - e^{-z},\\ \underline{F}(z) &= \max\left(1 - 2e^{-z/2}, 0\right).\end{aligned}\tag{6.78}$$

Fig. 6.8 shows the p-box of $R_1^2 + R_2^2$. Related calculations were performed in [210],

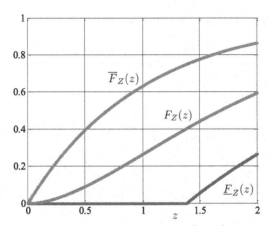

Figure 6.8 Probability box $[\overline{F}_Z, \underline{F}_Z]$ generated by $Z \triangleq R_1^2 + R_2^2$, with Rayleigh-distributed R_1, R_2. The inner CDF is exact, and refers to independent R_1, R_2; $\underline{F}_Z(z)$ is zero for $z \leq 2\log 2 \approx 1.39$. Reproduced with permission from [25].

using a parametric family of copulas (the *Clayton copula*) rather than dependence bounds. The case of branches with different average fading powers can be examined using results in [89]. ⬦

Example 6.16. Consider the RV $Z \triangleq X_1^2 + X_2^2$, with X_1, X_2 two dependent Nakagami-m RVs. The square of a Nakagami-m RV is Gamma distributed, which we write $X_i^2 \sim \mathcal{G}(\alpha, \beta_i)$, $i = 1, 2$. The Gamma pdf is [237]

$$f(x) = \frac{1}{\Gamma(\alpha)} \beta^\alpha x^{\alpha-1} e^{-\beta x}, \qquad x > 0, \ \alpha, \beta > 0, \tag{6.79}$$

and its CDF is

$$F(x) = P(\alpha, \beta x), \tag{6.80}$$

where $P(\cdot, \cdot)$ denotes the regularized Gamma function [2, §6.5.1]. We assume that the correlation coefficient ρ of X_1^2 and X_2^2 is unknown. As derived in [117], Z has a *type-I McKay* distribution. The special case $\alpha = 1$ yields

$$F_{X_i^2}(x) = 1 - \exp(-x/\beta_i), \qquad i = 1, 2, \tag{6.81}$$

and

$$F_Z(z) = \frac{1}{2} \left((c-1)e^{-(c+1)z/b} - (c+1)e^{-(c-1)z/b} + 2 \right), \tag{6.82}$$

where

$$b = \frac{2\beta_1\beta_2}{\sqrt{(\beta_1 + \beta_2)^2 - 4\beta_1\beta_2(1 - \rho)}},$$

$$c = \frac{\beta_1 + \beta_2}{\sqrt{(\beta_1 + \beta_2)^2 - 4\beta_1\beta_2(1 - \rho)}}.$$

(6.83)

Fig. 6.9 shows the p-box of F_Z. $\diamond$

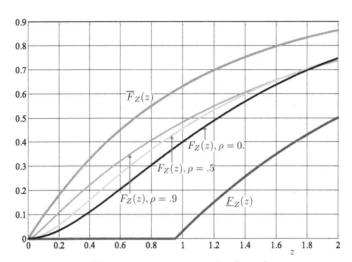

Figure 6.9 Probability box $[\overline{F}_Z, \underline{F}_Z]$ generated by $Z \triangleq X_1^2 + X_2^2$, with X_1, X_2 Nakagami-m RVs and hence $X_i^2 \sim \mathcal{G}(\alpha, \beta_i)$, $i = 1, 2$. The inner CDFs are exact, and refer to X_1^2, X_2^2 having correlation coefficient ρ.

Example 6.17. Let $Z \triangleq X_1^2 + X_2^2$, where $X_1, X_2 \sim \mathcal{N}(0, 1)$ are Gaussian RVs. Suppose first that X_1, X_2 are jointly Gaussian with correlation coefficient ρ. Using a standard procedure, we rewrite Z in the form $Z = \lambda_1 Y_1^2 + \lambda_2 Y_2^2$, where Y_1, Y_2 are independent $\sim \mathcal{N}(0, 1)$, and $\lambda_1 = 1 - \rho$, $\lambda_2 = 1 + \rho$ are the eigenvalues of the covariance matrix of X_1 and X_2. Thus, Z is the sum of two chi-square-distributed RVs with one degree of freedom [237]. The CDF $F_Z(z)$ can be determined by using [237, Eq. (5.8)]. Fig. 6.10 shows $F_Z(z)$ for correlation values $\rho = 0$, $\rho = 0.5$, and $\rho = 1$ (the latter value is obtained for $X_1 = X_2$). If the joint CDF is unknown, the p-box of F_Z, also shown in Fig. 6.10, is obtained from the first two equations of (6.68). The marginals here are, for $i = 1, 2$,

$$F_{X_i^2}(x_i) = \mathrm{erf}\left(\sqrt{x_i/2}\right), \qquad x_i \geqslant 0. \qquad (6.84)$$

Simple calculations using Lagrange multipliers, or the sheer observation of Fig. 6.11, lead to the analytic expressions, valid for $z \geqslant 0$:

$$\underline{F}(z) = \max\left[2\,\mathrm{erf}\left(\sqrt{z/4}\right) - 1, 0\right], \qquad (6.85)$$

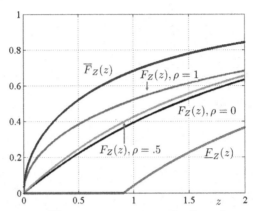

Figure 6.10 Probability box $[\overline{F}_Z, \underline{F}_Z]$ generated by $Z \triangleq X_1^2 + X_2^2$, with $X_1, X_2 \sim \mathcal{N}(0, 1)$. The inner CDFs are exact, and refer to jointly Gaussian X_1, X_2 with correlation coefficient ρ; $\underline{F}(z)$ is zero for $z \leqslant 4(\text{erf}^{-1}(1/2))^2 \sim 0.91$. Reproduced with permission from [25].

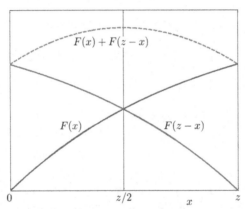

Figure 6.11 Behavior of $F(x)$, $F(z - x)$, and their sum. The maximum value of the sum is achieved at $x = z/2$, and its minimum at $x = 0$ and $x = z$.

$$\overline{F}(z) = \text{erf}\left(\sqrt{z/2}\right). \tag{6.86}$$

The p-box $[\overline{F}(z), \underline{F}(z)]$ is wider than the envelope of the CDFs corresponding to the whole set of values of ρ, because Gaussian marginals do not imply jointly Gaussian RVs (see, e.g., [189, p. 128]). Thus, the gap among the three innermost curves of Fig. 6.10 reflects the uncertainty in the knowledge of the value of ρ for jointly normal X_1, X_2, while the wider gap reflects the uncertainty about their joint distribution. ◇

Example 6.18. Consider the computation of upper and lower dependence bounds on an outage probability of the form

$$p = \mathbb{P}_{X,Y}\left[\log_2(1 + \text{snr}\, XY) < R\right], \tag{6.87}$$

where X and Y have the same exponential CDF

$$F(z) = 1 - e^{-z}, \qquad z \geq 0 \tag{6.88}$$

(see [20] for the description of a communication system leading to (6.87)). We have

$$\begin{aligned} \mathsf{p} &= \mathbb{P}_{X,Y}[XY \leq (2^R - 1)/\mathsf{snr}] \\ &= F_{X \times Y}\big((2^R - 1)/\mathsf{snr}\big). \end{aligned} \tag{6.89}$$

With independent X, Y, the exact value of p is [20]

$$\mathsf{p} = 1 - 2\alpha K_1(2\alpha), \tag{6.90}$$

where

$$\alpha \triangleq \sqrt{(2^R - 1)/\mathsf{snr}} \tag{6.91}$$

and $K_1(\cdot)$ denotes the modified Bessel function of second kind and order 1. For example, using (6.69) with $\mathsf{snr} = 0$ dB and $R = 0.01$, we obtain $\mathsf{p} = 0.0034$, while with no independence assumption we have $\underline{\mathsf{p}} = 6.6 \cdot 10^{-4}$ and $\overline{\mathsf{p}} = 0.16$. $\diamond$

Example 6.19. We compute upper and lower dependence bounds on the error probability of binary antipodal transmission over a channel with fading envelope $R = \sqrt{Z} = \sqrt{X_1^2 + X_2^2}$, X_1, X_2 Gaussian RVs, and $\mathbb{E}R^2 = 1$. The reference value is obtained under the assumption of independent X_1, X_2, which yields a Rayleigh-distributed envelope, and hence error probability [23, p. 89]

$$\mathsf{p} = \frac{1}{2}\left(1 - \sqrt{\frac{\mathsf{snr}}{1 + \mathsf{snr}}}\right), \tag{6.92}$$

where snr denotes the signal-to-noise ratio.

The upper dependence bound is

$$\begin{aligned} \overline{\mathsf{p}} &= \sqrt{\frac{\mathsf{snr}}{4\pi}} \int_0^\infty \mathrm{erf}\left(\sqrt{z/2}\right) \exp(-\mathsf{snr}\, z) z^{-1/2}\, dz \\ &= \frac{1}{\pi} \arctan\left(\frac{1}{\sqrt{2\mathsf{snr}}}\right), \end{aligned} \tag{6.93}$$

while the integral for $\underline{\mathsf{p}}$ does not seem amenable to a closed-form expression. The comparison between $\underline{\mathsf{p}}$ and $\overline{\mathsf{p}}$ is shown in Fig. 6.12, while numerical results show that $\underline{\mathsf{p}}$ is way lower than p. $\diamond$

Example 6.20. We determine the error probability of binary antipodal transmission over a fading channel with diversity 2 and maximal-ratio combination of the two diversity branches, which are affected by Rayleigh fading with envelopes R_1, R_2. Usual analyses assume independence of the fading processes affecting diversity branches,

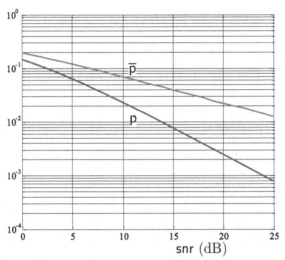

Figure 6.12 Error probability of binary antipodal transmission over a channel with fading envelope $\sqrt{X_1^2 + X_2^2}$, with $X_1, X_2 \sim \mathcal{N}(0, 1)$: (lower curve) X_1 and X_2 are independent; (upper curve) obtained from the upper dependence bound distribution $\overline{F}_Z$, where $Z \triangleq X_1^2 + X_2^2$. Reproduced with permission from [25].

which may be violated in practical systems. With R_1, R_2 independent, we have, using [23, p. 113],

$$p = \left(\frac{1-\mu}{2}\right)^2 (2+\mu), \tag{6.94}$$

with

$$\mu \triangleq \sqrt{\frac{snr}{1+snr}} \tag{6.95}$$

and snr the signal-to-noise ratio. In this case the dependence bounds are, from (5.30),

$$\overline{p} = \sqrt{\frac{snr}{4\pi}} \int_0^\infty \left(1 - e^{-z}\right) \exp(-snr\, z) z^{-1/2}\, dz$$

$$= \frac{1}{2}\left(1 - \sqrt{\frac{snr}{1+snr}}\right) \tag{6.96}$$

and

$$\underline{p} = \sqrt{\frac{snr}{4\pi}} \int_{2\log 2}^\infty \left(1 - 2e^{-z/2}\right) \exp(-snr\, z) z^{-1/2}\, dz \tag{6.97}$$

$$= \frac{1}{2}\mathrm{erfc}\left(\sqrt{2snr\log 2}\right) - \sqrt{\frac{2snr}{1+2snr}}\mathrm{erfc}\left(\sqrt{(1+2snr)\log 2}\right).$$

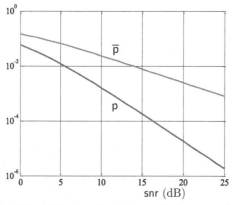

Figure 6.13 Error probability of binary antipodal transmission over a fading channel with diversity 2, Rayleigh fading and maximal-ratio combining: (lower curve) independent fading on the two branches; (upper curve) obtained from the upper dependence bound distribution. Reproduced with permission from [25].

A comparison between p and $\bar{p}$ is shown in Fig. 6.13. The value of p is much lower than $\bar{p}$, which can be explained by observing that it is obtained when the two RVs R_1 and R_2 are countermonotone, i.e., when a small value of one RV is compensated by a large value of the other, thus preventing "deep fading" effects that impair performance.

$\diamond$

Example 6.21. Consider a wireless transmission system with diversity d, Rayleigh fading, and selection combining [23, p. 117]. The Fréchet–Hoeffding dependence bounds can be applied here, with CDFs of the square fading amplitudes given by $F_{X_i}(x) = 1 - e^{-x}$, $i = 1, \ldots, d$. We obtain, for $x \geqslant 0$,

$$\max\left(0, 1 - d e^{-x}\right) \leqslant F_{\max}(x) \leqslant 1 - e^{-x}, \tag{6.98}$$

where the upper bound corresponds to the same fading amplitude in each branch, and hence no diversity. The corresponding error probability for binary antipodal transmission can be obtained as in (5.30), which yields

$$
\begin{aligned}
p &= \sqrt{\frac{\mathsf{snr}}{4\pi}} \int_0^\infty \left(1 - e^{-z}\right)^d \exp(-\mathsf{snr}\, z) z^{-1/2}\, dz \\
&= \frac{1}{2} \sum_{k=0}^d \binom{d}{k} (-1)^k \sqrt{\frac{\mathsf{snr}}{k + \mathsf{snr}}},
\end{aligned}
\tag{6.99}
$$

while

$$\bar{p} = \frac{1}{2}\left(\sqrt{\frac{\mathsf{snr}}{1 + \mathsf{snr}}}\right) \tag{6.100}$$

and

$$\underline{p} = \sqrt{\frac{snr}{4\pi}} \int_{\log d}^{\infty} \left(1 - de^{-z}\right) \exp(-snr\,z)z^{-1/2}\,dz$$

$$= \frac{1}{2}\left[1 - d\sqrt{\frac{snr}{1+snr}} - \mathrm{erf}\left(\sqrt{snr\log d}\right) + d\sqrt{\frac{snr}{1+snr}}\mathrm{erf}\left(\sqrt{(1+snr)\log d}\right)\right].$$

$$(6.101)$$

The values of **p** obtained under the assumption of branch independence and its upper
and lower dependence bounds are shown in Fig. 6.14.   ◇

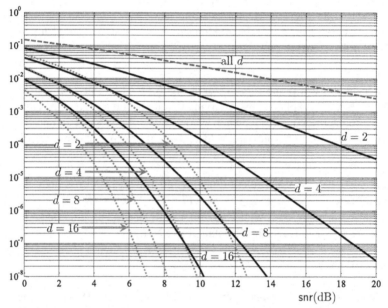

Figure 6.14 Error probability of binary antipodal transmission over a channel with diversity
2, 4, 8, and 16, Rayleigh fading, and selection combining: (dotted curves) lower dependence
bound; (continuous curves) values obtained under the assumption of branch independence;
(dashed curve) upper dependence bound. Reproduced with permission from [29].

Example 6.22. On a nonergodic channel affected by fading with envelope R and additive white Gaussian noise, the *information outage probability*, i.e., the probability
that the transmission rate ρ, measured in bits per dimension pair, exceeds the instantaneous mutual information of the channel, is given, as in Example 2.3, by

$$p_{out} = \mathbb{P}\left[\log_2(1 + R^2 snr) < \rho\right]. \qquad (6.102)$$

From (6.102) we obtain, for a continuous CDF of R^2,

$$\mathsf{p}_{\text{out}} = \mathbb{P}\left[R^2 < (2^\rho - 1)/\mathsf{snr}\right]$$
$$= F_{R^2}\left((2^\rho - 1)/\mathsf{snr}\right). \tag{6.103}$$

In the special case of Rayleigh-distributed fading, we have

$$\mathsf{p}_{\text{out}} = 1 - \exp[-(2^\rho - 1)/2\,\mathsf{snr}],$$

while upper and lower bounds are obtained by replacing the CDFs (6.85)–(6.86) for F_{R^2} in (6.103). Numerical results are shown in Fig. 6.15. $\diamond$

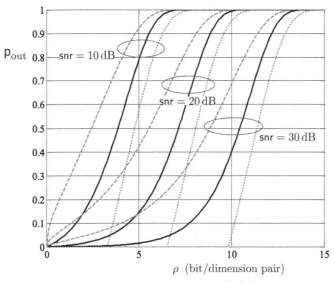

Figure 6.15 Outage probability upper and lower bounds with fading envelope as in Example 6.17: (dotted lines) lower bounds; (continuous lines) baseline values corresponding to Rayleigh fading; (dashed lines) upper bounds.

Example 6.23. Consider a block fading channel with d blocks [23, Section 4.3]. Using independent Gaussian symbols on the d blocks, the outage probability is given by

$$\mathsf{p}_{\text{out}}(\rho) = \mathbb{P}\left(\frac{1}{d}\sum_{i=1}^{d} C(R_i) \leqslant \rho\right), \tag{6.104}$$

where ρ is the average transmission rate, and $C(R_i) \triangleq \log_2(1 + R_i^2\mathsf{snr})$ is the instantaneous mutual information, measured in bits per dimension pair, of the block with fading amplitude R_i and signal-to-noise ratio snr. Eq. (6.104) can be given the form

$$\mathsf{p}_{\text{out}}(\rho) = F_Z(\rho), \tag{6.105}$$

where

$$z \triangleq \sum_{i=1}^{d} X_i \tag{6.106}$$

and

$$X_i \triangleq \frac{1}{d} C(R_i). \tag{6.107}$$

Under the assumption that the RVs R_i have a common Rayleigh distribution, we obtain, for the cdf of X_i,

$$F(x_i) = 1 - \exp\left(-\frac{2^{dx_i} - 1}{\mathsf{snr}}\right). \tag{6.108}$$

To examine the effect on $\mathsf{p}_{\mathrm{out}}$ of the dependence among fading effects across blocks, we derive the dependence p-box of F_Z, viz.,

$$\overline{F}_Z(z) = \inf_{x_1 + \cdots + x_d = z} \min\left[\sum_{i=1}^{d} F(x_i), 1\right],$$

$$\underline{F}_Z(z) = \sup_{x_1 + \cdots + x_d = z} \max\left[\sum_{i=1}^{d} F(x_i) - d + 1, 0\right]. \tag{6.109}$$

To compute $\overline{F}_Z(z)$ and $\underline{F}_Z(z)$, form the Lagrangian functional

$$L(x_1, \ldots, x_d) = \sum_{i=1}^{d} F(x_i) - \sum_{i=1}^{d} \lambda_i x_i - \nu\left(\sum_{i=1}^{d} x_i - x\right), \tag{6.110}$$

which leads to the Karush–Kuhn–Tucker conditions

$$F'(x_i^\star) - \lambda^\star - \nu^\star = 0, \qquad i = 1, \ldots, d, \tag{6.111a}$$

$$x_1^\star + \ldots + x_d^\star - x = 0, \tag{6.111b}$$

$$\lambda_i^\star \geqslant 0, \qquad i = 1, \ldots, d, \tag{6.111c}$$

$$\lambda_i^\star x_i^\star = 0, \qquad i = 1, \ldots, d. \tag{6.111d}$$

Eliminating the "slack" variable $\lambda^\star$, from (6.111a) and (6.111d) we obtain the equation

$$\left(F'(x_i^\star) - \nu^\star\right) x_i^\star = 0, \tag{d'}$$

which shows that, if $x_i^\star > 0 \; \forall i$, then $F'(x_i^\star) = \nu^\star \; \forall i$, which leads to the stationary point

$$x_1^\star = x_2^\star = \cdots = x_d^\star = \frac{x}{d}$$

corresponding the value $dF(x/d)$ of the objective function. Other stationary points are obtained when k of the $x_i^\star$ take value 0, which leads to the value $kF(x/k)$, $0 < k < d$, of the objective function. Assume first that F is concave $\cap$. A form of Jensen inequality (1.44) leads to

$$F\left(\frac{x_1^\star + \cdots + x_d^\star}{d}\right) \geqslant \frac{1}{d}\left[F(x_1^\star) + \cdots + F(x_d^\star)\right].$$ (6.112)

If k of the $x_i^\star$ take value x/k (and the remaining $x_i^\star$ are 0), then (6.112) yields

$$F(x/d) \geqslant \frac{k}{d}F(x/k)$$ (6.113)

and hence, for all $1 \leqslant k \leqslant d$,

$$dF(x/d) \geqslant kF(x/k) \geqslant F(x).$$ (6.114)

In conclusion, if F is concave $\cap$, under the given constraints, the objective function takes maximum value $dF(x/d)$ and minimum value $F(x)$. The opposite occurs if F is convex $\cup$. Thus, for F concave $\cap$, we obtain

$$\begin{aligned} \overline{F}_Z(z) &= F(z), \\ \underline{F}_Z(z) &= \max\left[dF(z/d) - d + 1, 0\right]. \end{aligned}$$ (6.115)

Numerical results are shown in Fig. 6.16. $\diamondsuit$

Example 6.24. While outage probability, as examined in previous examples, is a performance parameter referring to usual point-to-point transmission, *secrecy outage probability* provides a counterpart to it when communications are subject, in addition to reliability constraints, to secrecy requirements. Here, in addition to two "legitimate" users "Alice" and "Bob" of a transmission channel, we consider the presence of a third party "Eve," who tries to steal the information transmitted from Alice to Bob by eavesdropping on the channel. Under the assumptions that both the main channel and the eavesdropper channel are affected by additive white Gaussian noise and block fading, and that Alice and Eve have perfect information about their channel states, the instantaneous mutual information of the main and eavesdropper channels is given by (see also Example 2.3)

$$I_M = \log(1 + \mathsf{snr}_M X),$$ (6.116a)
$$I_E = \log(1 + \mathsf{snr}_E Y),$$ (6.116b)

where snr_M and X (snr_E and Y) denote the signal-to-noise ratio and the squared magnitude of the fading in the main (eavesdropper) channel [21,34]. We have a *secrecy outage* when the difference between I_M and I_E is smaller than the *secrecy rate* ρ_s of the transmission, and hence we define the secrecy outage probability as

$$\mathsf{p}_{\mathsf{s,out}} \triangleq \mathbb{P}\left[\log(1 + \mathsf{snr}_M X) - \log(1 + \mathsf{snr}_E Y) < \rho_s\right].$$ (6.117)

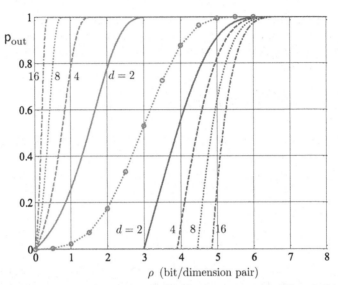

Figure 6.16 Dependence upper and lower bounds for the outage probability of a block fading channel with Rayleigh fading, signal-to-noise ratio $\mathsf{snr} = 10$ dB, and d blocks, $d = 2, 4, 8,$ and 16. The middle curve shows p_{out} for $d = 2$ and independent fading [23, Eq. (4.37)]. The value at which the lower bound equals zero is $\rho_0 = \log_2(1 + \mathsf{snr} \log d)$. Reproduced with permission from [29].

With the definitions $\tilde{X} \triangleq \mathsf{snr}_M X$, $\tilde{Y} \triangleq -\exp(\rho_s)\mathsf{snr}_M Y$, and $s \triangleq \exp(\rho_s) - 1$, (6.117) can be written in the form

$$\mathsf{p}_{s,\text{out}} = \mathbb{P}(\tilde{X} + \tilde{Y} < s). \tag{6.118}$$

If we can assume that $\tilde{X}$ is independent of $\tilde{Y}$, we have, with obvious notations,

$$\begin{aligned}
\mathsf{p}_{s,\text{out}} &= \int_{-\infty}^{0} \int_{0}^{s-\tilde{y}} f_{\tilde{X}}(\tilde{x}) f_{\tilde{Y}}(\tilde{y}) \, d\tilde{x} d\tilde{y} \\
&= \int_{-\infty}^{0} f_{\tilde{Y}}(\tilde{y}) F_{\tilde{X}}(s - \tilde{y}) \, d\tilde{y}.
\end{aligned} \tag{6.119}$$

The special case of Rayleigh fading yields exponentially distributed $\tilde{X}$ and $\tilde{Y}$. Thus, writing

$$\begin{aligned}
F_{\tilde{X}}(\tilde{x}) &= 1 - \exp(-\lambda_X \tilde{x}), && x \in [0, \infty), \\
f_{\tilde{Y}}(\tilde{y}) &= \exp(\lambda_Y \tilde{y}), && y \in (-\infty, 0],
\end{aligned} \tag{6.120}$$

we obtain [21]

$$\mathsf{p}_{s,\text{out}} = 1 - \frac{\lambda_Y}{\lambda_X + \lambda_Y} \exp(-\lambda_X s). \tag{6.121}$$

Should the assumption of independence be unjustified, the dependence bounds (6.68) can be directly applied to yield

$$\sup_{\tilde{x}+\tilde{y}=s} \max\left[F_{\tilde{X}}(\tilde{x})+F_{\tilde{Y}}(\tilde{y})-1,0\right] \leqslant \mathsf{p}_{\mathrm{s,out}} \leqslant \inf_{\tilde{x}+\tilde{y}=s} \min\left[F_{\tilde{X}}(\tilde{x})+F_{\tilde{Y}}(\tilde{y}),1\right]. \quad (6.122)$$

In [21] one can find further details about the calculation of the bounds in (6.122). $\diamond$

6.4 Bivariate dependence bounds on expectations

In Section 5.5 we have seen how upper and lower bounds can be derived for a performance index having the form $\eta = \mathbb{E}_X[h(X)]$ where F_X is known to take values between $\underline{F}_X$ and $\overline{F}_X$. Here we examine the direct derivation of dependence bounds for performance index

$$\eta = \mathbb{E}_{X,Y}[H(X,Y)], \quad (6.123)$$

where the dependence of X and Y is unknown. Specifically, we look for

$$\begin{aligned}\underline{\eta} &= \inf \mathbb{E}_{X,Y}[H(X,Y)],\\\overline{\eta} &= \sup \mathbb{E}_{X,Y}[H(X,Y)],\end{aligned} \quad (6.124)$$

where inf and sup are taken over all joint CDFs $F_{X,Y}$ with known marginals F_X and F_Y.

In the problem examined in Section 5.5 for the one-dimensional case, a simple solution could be found under the condition that the function to be averaged was a.s. strictly increasing or decreasing. To solve the bivariate problem examined here, monotonicity is not sufficient, and must be replaced by *Monge condition* for a right-continuous function $H(x,y)$ [205, p. 108]:

$$H(x,y) - H(x,y_1) \leqslant H(x_1,y) - H(x_1,y_1), \quad \text{for all } x \geqslant x_1, y \geqslant y_1. \quad (6.125)$$

Remark 6.11. If $\partial^2 H(x,y)/\partial x \partial y$ exists, then $H(x,y)$ satisfies Monge condition if and only if $\partial^2 H(x,y)/\partial x \partial y \leqslant 0$ a.e. [43, p. 292].

The following result holds [205, Theorem 3.1.2]:

Theorem 6.8. *Let (X,Y) be a pair of RVs with joint CDF $F_{X,Y}$ and marginal CDFs F_X, F_Y. Moreover, let (X_1,Y_1) a pair of independent RVs with marginal CDFs $F_{X_1} = F_X$ and $F_{Y_1} = F_Y$. If $H(\cdot,\cdot)$ satisfies Monge condition (6.125), and the expectations of $H(X_1,Y_1)$ and $H(X,Y)$ exist and are finite, then*

$$\int_0^1 H\left(F_X^{\leftarrow}(u), F_Y^{\leftarrow}(u)\right) du \leqslant \mathbb{E}_{X,Y}[H(X,Y)] \leqslant \int_0^1 H\left(F_X^{\leftarrow}(u), F_Y^{\leftarrow}(1-u)\right) du.$$

$$(6.126)$$

Remark 6.12. Should $-H$ satisfy Monge condition, Theorem 6.8 could still be applied by exchanging the upper and lower bounds.

Remark 6.13. Theorem 6.8 shows that the bounds are sharp, as they are achieved when X and Y are comonotone and countermonotone RVs.

Example 6.25. Since $-xy$ satisfies Monge condition, if X, Y are $\sim \mathcal{U}[0, 1]$, then [205, p. 108]

$$\int_0^1 F_X^{\leftarrow}(u) F_Y^{\leftarrow}(1-u)\, du \leqslant \mathbb{E}[XY] \leqslant \int_0^1 F_X^{\leftarrow}(u) F_Y^{\leftarrow}(u)\, du. \tag{6.127}$$

Similarly, since the function $(x - y)^2$ satisfies Monge condition, we have

$$\int_0^1 \left(F_X^{\leftarrow}(u) - F_Y^{\leftarrow}(u) \right)^2 du \leqslant \mathbb{E}[(X-Y)^2] \leqslant \int_0^1 \left(F_X^{\leftarrow}(u) - F_Y^{\leftarrow}(1-u) \right)^2 du \tag{6.128}$$

and, more generally, for $H(x, y) = \phi(|x - y|)$ and ϕ convex $\cup$,

$$\int_0^1 \phi\left(|F_X^{\leftarrow}(u) - F_Y^{\leftarrow}(u)| \right) du \leqslant \mathbb{E}\left[\phi(|X - Y|) \right] \leqslant \int_0^1 \phi\left(|F_X^{\leftarrow}(u) - F_Y^{\leftarrow}(1-u)| \right) du. \tag{6.129}$$

$\diamond$

Example 6.26. The function

$$H(x, y) = \frac{x}{1 + y} \tag{6.130}$$

has $\partial^2 H(x, y)/\partial x \partial y \leqslant 0$, and hence satisfies Monge condition. Let X and Y be exponentially distributed with $F_X(x) = 1 - e^{-x}$, $F_Y(y) = 1 - e^{-2y}$, $x \geqslant 0$, $y \geqslant 0$. With $\eta \triangleq \mathbb{E}[H(X, Y)]$, we have $\underline{\eta} = 0.5547$ and $\overline{\eta} = 0.8702$, while independence of X and Y would yield $\eta = 0.7227$. $\diamond$

Example 6.27. The ergodic capacity region for multiple-access fast-fading channel, achieved with successive interference cancellations and fixed decoding order, is determined by the achievable rates [130]

$$\begin{aligned} \rho^{(1)} &= \mathbb{E}_{X,Y} \left[\log_2 \left(1 + \frac{X}{\sigma^2 + Y} \right) \right], \\ \rho^{(2)} &= \mathbb{E}_Y \left[\log_2 \left(1 + \frac{Y}{\sigma^2} \right) \right], \end{aligned} \tag{6.131}$$

where X (resp., Y) denotes the power received by the first (resp., second) decoded user, and σ^2 is the noise variance. The rate of the first decoded user, $\rho^{(1)}$, depends on the joint distribution of X and Y. The joint derivative of $H(x, y) = \log_2\left(1 + x/(\sigma^2 + y)\right)$ is

$$\frac{\partial^2}{\partial x \partial y} H(x, y) = -\frac{1}{(x + y + \sigma^2)^2 \log(2)} \leqslant 0 \qquad (6.132)$$

and hence Monge condition is satisfied. Assuming X and Y to be exponentially distributed with unit second moment, i.e., with the common CDF

$$F(z) = \left(1 - e^{-\sqrt{2}z}\right), \qquad z \geqslant 0, \qquad (6.133)$$

the quasi-inverse CDF is given by

$$F^{\leftarrow}(z) = -\frac{1}{\sqrt{2}} \log(1 - z), \quad 0 \leqslant z < 1. \qquad (6.134)$$

Dependence upper and lower bounds on $\rho^{(1)}$ are plotted in Fig. 6.17 vs. the signal-to-noise ratio $\mathsf{snr} \triangleq 1/\sigma^2$, along with the values obtained under the assumption of independent X, Y.   ◇

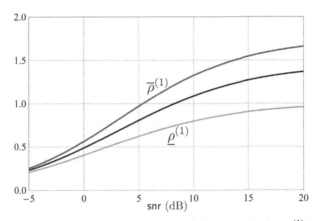

Figure 6.17 Dependence upper and lower bounds for the achievable rate $\rho^{(1)}$ in Example 6.27. The middle curve shows the value of $\rho^{(1)}$ under the assumption of independent X, Y.

Remark 6.14. An extension of Theorem 6.8 to functions H with more than two variables is also possible: see [130, Remark 3] and [205, Remark 3.1.4] for the relevant references.

6.5   Bounds with monotone marginal densities

Previously in this chapter we have seen how wrong independence assumptions may affect the performance even to a considerable extent. It is conceivable to ask how additional information available on a dependence structure can be introduced in the performance bounds to make them tighter. Thus a question arises: what kind of knowledge can be reasonably gathered, under the assumption that it is easy to obtain, and leads to bounds that are significantly tighter while still simple to evaluate? In this section we examine the case in which the marginal pdfs are known to be monotone functions, and we derive the corresponding dependence bounds. An interesting fact is that, while (see *supra*, Section 6.1.6) the lower Fréchet–Hoeffding bound is generally not sharp for $d \geqslant 3$, as in this case $\max[\sum_{i=1}^{d} u_i + 1 - d, 0]$ may not be a copula, under the restriction to *monotone* pdfs sharp dependence bounds become available.

Before proceeding further, two examples will show some properties of random variables with monotone pdf.

Example 6.28. Let X be a continuous RV with monotone pdf f_X and support set $[a, b]$. If $\mathbb{E}[X] = (a + b)/2$, then $X \sim \mathcal{U}[a, b]$. That is, if the mean value of an RV with monotone density is the midpoint of its support set, then the pdf is completely determined [219].   ◇

Example 6.29. Consider a continuous RV X with support set $[a, b]$. If its mean value is known to be μ, applying the Principle of Maximum Entropy as described in Example 3.6, we derive a pdf of the form

$$f_X(x) = e^{-1+\lambda x} = \frac{e^{\lambda x}}{\int_a^b e^{\lambda x}\, dx} = \frac{\lambda e^{\lambda x}}{e^{\lambda b} - e^{\lambda a}}, \qquad x \in [a, b], \tag{6.135}$$

which is monotone. Imposing the constraint of mean value equal to μ, we obtain the equation yielding λ, namely

$$\mu = \frac{b e^{\lambda b} - a e^{\lambda a}}{e^{\lambda b} - e^{\lambda a}} - \frac{1}{\lambda}. \tag{6.136}$$

Fig. 6.18(a) shows λ as a function of μ for an RV with support set $[0, 1]$. Positive values of λ, corresponding to $\mu > 0.5$, yield monotonically increasing pdfs. Negative values of λ, corresponding to $\mu < 0.5$, yield monotonically decreasing pdfs. The mean value $\mu = 0.5$ yields the uniform distribution [219]. Fig. 6.18(b) shows the corresponding pdfs.   ◇

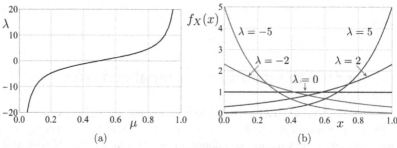

Figure 6.18 (a) Values of parameter λ in maximum-entropy pdf (6.135) vs. the mean value μ. (b) Maximum-entropy pdf (6.135) with parameter λ.

Consider the calculation of

$$\underline{F}^{\star}(z) \triangleq \inf \mathbb{P}\left[\sum_{i=1}^{d} X_i < z\right], \tag{6.137}$$

where the inf is evaluated over the joint distributions of the RVs $X_1, \ldots, X_d$ whose marginal CDFs $F_1, \ldots, F_d$ are known and continuous, and whose pdfs are monotone. The following result holds [19, Theorem 1], [265, Theorem 2.6]:

Theorem 6.9.

$$\underline{F}^{\star}(z) \geqslant \Phi^{-1}(z), \tag{6.138}$$

where

$$\Phi(s) \triangleq \sum_{i=1}^{d} \mathbb{E}\big[X_i \mid X_i \geqslant F_i^{\leftarrow}(s)\big], \qquad s \in (0, 1), \tag{6.139}$$

and $F_i^{\leftarrow}$ denotes the quasi-inverse of F_i as in (6.14).

A tighter bound can be obtained by adding the constraint of equal marginal CDFs, i.e., $F_1 = \cdots = F_d = F$ [20, Theorem 2], [265, Theorem 3.4]:

Theorem 6.10.

$$\underline{F}^{\star}(z) = \phi^{-1}(z). \tag{6.140}$$

The definition of function ϕ depends on the sign of the derivative of the pdf:

   (i) *For decreasing pdfs, we have*

$$\phi(s) \triangleq \begin{cases} H_s\big(c_d(s)\big), & \text{if } c_d(s) > 0, \\ d\,\mathbb{E}[X \mid X > F^{\leftarrow}(s)], & \text{if } c_d(s) = 0, \end{cases} \tag{6.141}$$

   where

$$H_s(x) \triangleq (d-1)F^{\leftarrow}\big(s + (d-1)x\big) + F^{\leftarrow}(1 - x), \qquad s \in (0, 1), \tag{6.142}$$

   and

$$c_d(s) \triangleq \min\left\{c \in [0, (1-s)/d] \,\middle|\, \int_c^{(1-s)/d} H_s(t)\,dt \geqslant \big((1-s)/d - c\big)H_s(c)\right\}. \tag{6.143}$$

   (ii) *For increasing pdfs, we have*

$$\phi(s) \triangleq \begin{cases} H_s(0), & \text{if } c_d(s) > 0, \\ d\,\mathbb{E}[X \mid X > F^{\leftarrow}(s)], & \text{if } c_d(s) = 0, \end{cases} \tag{6.144}$$

where

$$H_s(x) \triangleq (d-1)F^{\leftarrow}(s+x) + (d-1)F^{\leftarrow}\big(1-(d-1)x\big), \qquad s \in (0,1),$$
$$\tag{6.145}$$

and

$$c_n(s) \triangleq \min\left\{c \in [0,(1-s)/d] \,\middle|\, \int_s^{(1-s)/d} H_s(t)\,dt \leqslant \big((1-s)/d - c\big)H_s(c)\right\}.$$
$$\tag{6.146}$$

Theorem 6.11 ([267, Theorem 4.3]). *If the RVs X_i have common CDF, support set $[0,1]$, decreasing pdfs, mean value μ, and satisfy the "regularity condition"*

$$\mathbb{E}[X \mid X \geqslant s] \geqslant s + \frac{1-s}{d}, \tag{6.147}$$

then, with the definition

$$\psi(s) \triangleq \mathbb{E}[X \mid X \geqslant F^{\leftarrow}(s)] \geqslant s + (1-s)/d, \qquad s \in [0,1), \tag{6.148}$$

we have

$$\underline{F}^{\star}(s) = \begin{cases} 0, & s \leqslant d\mu, \\ \psi^{-1}(s/d), & d\mu < s < d, \\ 1, & s \geqslant d. \end{cases} \tag{6.149}$$

Remark 6.15. As observed in [267, p. 1355], the regularity condition (6.147), which prevents the conditional mean of X from being too close to one side, is commonly satisfied by bounded distributions with monotone pdf for d not too small.

Example 6.30. Assume $X \sim \mathcal{U}[0,1]$. Then $F(x) = x$, $F^{\leftarrow}(s) = s$ with $s \in [0,1]$. Moreover, $\phi(s) = d\,\mathbb{E}[X \mid X > s] = d(1+s)/2$ for $s \in [0,1]$ and $c_d(s) = 0$ for $s \in [0,1]$. Thus [265, Example 3.9]

$$F^{\star}(s) = \phi^{-1}(s) = \min[1, \max[2s/d - 1, 0]]. \tag{6.150}$$

$$\diamond$$

Example 6.31. For the decreasing exponential pdf $f_X(x) = \lambda \exp(-\lambda x)$, $x \geqslant 0$, we have [19, Table I]:

$$F_X(x) = 1 - \exp(-\lambda x),$$
$$F_X^{\leftarrow}(x) = -\lambda^{-1}\log(1-x),$$

$$H_s(x) = -\lambda^{-1}\big[(d-1)[\log\big(1-s-(d-1)x\big)] + \log(x)\big],$$

$$\mathbb{E}_X[X \mid X > F_X^\leftarrow(s)] = \lambda^{-1} + F^\leftarrow(s), \tag{6.151}$$

$$\phi(s) = \begin{cases} H_s\big(c_d(s)\big) & \text{if } c_d(s) > 0, \\ \lambda^{-1}d\big(1 - \log(1-s)\big) & \text{if } c_d(s) = 0. \end{cases} \qquad \Diamond$$

Bound (6.138) is actually sharp, unless X does not have a compact support set [265, p. 401]. That a distribution exists achieving the bound is proved in [267, §3] (see also [19]). There, a d-variate copula $Q_d^F(c)$ is derived for $d \geqslant 2$ and some $c \in [0, 1/d]$ such that, if $(U_1, \ldots, U_d)$ has d-variate joint distribution $Q_d^F(c)$, then

   (i) For each i, $i = 1, \ldots, d$, the joint pdf of $(U_1, \ldots, U_d)$, given U_i with support set $[0, c]$, is uniformly supported on line segments $u_j = 1 - (d-1)u_i$ for all $j \neq i$ and $u_i \in [0, c]$.
   (ii) $F^\leftarrow(u_1) + \cdots + F^\leftarrow(u_d)$ is a constant when U_i has support set $\big(c, 1-(d-1)c\big)$ for any $i = 1, \ldots, d$.

Copula $Q_d^F(c)$ may not exist for all values of c, and, if it exists, it may not be unique [267, p. 1350]. When $d = 2$, if it exists, it is exactly the Fréchet–Hoeffding lower-bound copula $W(u, v) \triangleq \max(u + v - 1, 0)$. The problem of deriving an explicit expression for $Q_d^F(c)$ seems to be unsolved yet [267, p. 1352].

The copula $Q_d^F(c)$ solves the following minimization problem [19, Theorem 3], [267, Theorem 3.4]:

Theorem 6.12. *If the RVs $X_1, \ldots, X_d$ share the same marginal CDF F, and their pdf is monotonically increasing, then for any convex function $g(x)$ the following holds:*

$$\min_F \mathbb{E}[g(X_1 + \cdots + X_d)] = \mathbb{E}_{Q_d^F}[g(X_1 + \cdots + X_d)]. \tag{6.152}$$

Remark 6.16. Consider minimizing the variance of the sum $\sum_{i=1}^d X_i$ of RVs with assigned equal marginal CDFs F_X. For $d = 2$, the solution is given by the counter-monotone RVs $X_1 = F^\leftarrow(U)$, $X_2 = F_X^\leftarrow(1 - U)$, with $U \sim \mathcal{U}[0, 1]$. For $d \geqslant 3$, the problem is more difficult, unless monotone marginal pdfs are assumed [267, p. 1344].

Remark 6.17. If the common pdf of $X_1, \ldots, X_d$ is monotonically decreasing, then (6.152) can be applied by observing that $g(-x)$ is also convex, and replacing X_i by $-X_i$ to obtain a monotonically increasing pdf.

Remark 6.18. The minimization of $\mathbb{E}[g(X_1 + \cdots + X_d)]$, with g a convex function and F_X the common CDF of X_i, $1 \leqslant i \leqslant d$, having a monotone pdf, is equivalent to the minimization of $\mathbb{E}\big[g\big(F_X^\leftarrow(X_1) + \cdots + F_X^\leftarrow(X_d)\big)\big]$, where now the RVs $F_X^\leftarrow(X_i)$ are uniformly distributed, and hence their joint CDF is a d-dimensional copula. Moreover, if F_X corresponds to an increasing (resp., decreasing) pdf, then $F^\leftarrow$ is continuous and concave (resp., convex) and increasing [267, p. 1349].

Example 6.32. The ε-outage capacity of a transmission channel is the maximum rate that can be supported with a given outage probability [23, p. 105 ff.]. In a slow-fading channel with signal-to-noise ratio snr, receiver diversity with d branches, maximal-ratio combining, perfect channel-state information at receiver, and no channel-state information at transmitter, the ε-outage capacity is given by

$$\rho_\varepsilon \triangleq \sup\left\{\rho \in \mathbb{R}^+ \;\middle|\; \mathbb{P}\left[\sum_{i=1}^{d} |h_i|^2 < \frac{2^\rho - 1}{\text{snr}}\right] < \varepsilon\right\}, \tag{6.153}$$

where h_i denotes the complex fading gain of the ith diversity branch. For example, with independent Rayleigh-distributed fading gains the ε-outage capacity is obtained by observing that the sum of independent, exponentially distributed RVs with the same mean value has a Gamma distribution [237], and hence [19]

$$\rho_\varepsilon = \log_2\left(1 + \text{snr}P^{-1}(d, \varepsilon)\right), \tag{6.154}$$

where $P(a, x)$ denotes the CDF (6.80) of a Gamma-distributed RV.

Assuming identical distributions and monotone pdfs for the RVs $|h_i|^2$, we compute the two bounds $\underline{\rho}_\epsilon$ and $\overline{\rho}_\epsilon$ on the outage capacity. These can be given the form

$$\underline{\rho}_\varepsilon = \log_2\left(1 - \text{snr}\phi_-(1 - \varepsilon)\right), \tag{6.155a}$$

$$\overline{\rho}_\varepsilon = \log_2\left(1 + \text{snr}\phi(\varepsilon)\right). \tag{6.155b}$$

To prove (6.155a)–(6.155b), consider the lower bound first, as obtained by computing the worst-case CDF

$$\overline{F}(s) \triangleq \sup \mathbb{P}(S < s), \tag{6.156}$$

where $s \triangleq (2^\rho - 1)/\text{snr}$, $S \triangleq \sum_{i=1}^{d} X_i$, and $X_i \triangleq |h_i|^2$. We have [19]

$$\begin{aligned}
s^\star\left(\underline{\rho}_\varepsilon\right) &\triangleq \sup_{s \geq 0}\{s \mid \overline{F}(s) < s\} \\
&= \sup_{s \geq 0}\{s \mid 1 - \inf\left[\mathbb{P}(-S < -s) < \varepsilon\right]\} \\
&= \sup_{s \geq 0}\{s \mid 1 - \phi_-^{-1}(-s) < \varepsilon\},
\end{aligned} \tag{6.157}$$

where ϕ_-^{-1} denotes the inverse of the function ϕ defined either in (6.144) or (6.141) with F the CDF of $-X$. The sup is achieved for $s = s^\star$, with

$$1 - \phi_-^{-1}(-s^\star) = \varepsilon, \tag{6.158}$$

which yields

$$s^\star = -\phi_-(1 - \varepsilon) = \frac{1}{\text{snr}}\left(2^{\underline{\rho}_\varepsilon} - 1\right) \tag{6.159}$$

and hence $\underline{\rho}_\epsilon$ in (6.155a).

The upper bound is derived in a similar way. With

$$\underline{F}(s) \triangleq \inf \mathbb{P}(S \leqslant s), \tag{6.160}$$

we obtain

$$
\begin{aligned}
s^\star\left(\overline{\rho}_\epsilon\right) &\triangleq \sup_{s \geq 0}\{s \mid \underline{F}(s) < s\} \\
&= \sup_{s \geq 0}\{s \mid \phi^{-1}(s) < \varepsilon\} \\
&= \phi(\varepsilon) \\
&= \frac{1}{\mathsf{snr}}\left(2^{\overline{\rho}_\epsilon} - 1\right),
\end{aligned}
\tag{6.161}
$$

which yields $\overline{\rho}_\epsilon$ in (6.155b). Due to the equality (6.140), the bounds (6.155) are sharp.

The special case of Rayleigh fading yields RVs $|h_i|^2$ that are exponentially distributed, and hence have a pdf which is monotonically decreasing. Numerical results showing $\overline{\rho}_\varepsilon$ and $\underline{\rho}_\varepsilon$ are derived in [19]. $\qquad \diamond$

6.5.1 Tail-monotone marginal densities

A pdf $f(x)$ is said to be *tail-monotone* if, for some $b \in \mathbb{R}$, it is decreasing for $x > b$ or is increasing for $x < b$.

Example 6.33. The Gamma pdf (6.79) $f_{\alpha,\beta}(x)$ is tail-monotone, but not monotone, for $\alpha > 1$. $\qquad \diamond$

We have the following result [265, Theorem 3.6]:

Theorem 6.13. *Given an RV with CDF $F(x)$ and pdf $f(x)$ decreasing for $x \in [b, \infty)$, for $z > \phi\big(F(b)\big)$ we have*

$$\underline{F}^\star(z) = \phi^{-1}(z), \tag{6.162}$$

where $\phi(\cdot)$ is defined in (6.141).

6.6 Deriving tighter dependence bounds

In this section we examine how to tighten the dependence bounds by evaluating parameters aimed at restricting the set of multivariate CDFs among which the bounds are sought. Three possible options are:

① Choose a parametric family of copulas based on physical measurements or system analysis, and look for bounds restricted to this family. Some parametric copulas were introduced in Examples 6.3–6.7 and 6.9. This approach was taken in [210], where, by using computer simulation, the error probability of a fading channel was derived for a 2-branch diversity receiver with maximal-ratio combining under the assumption of a channel dependence described by a specific copula. The copula was chosen within the parametric family of Clayton copulas of Example 6.6 *supra* to describe fading distributions with strong dependence in the lower tail (corresponding to deep fading), and near-independence elsewhere. The choice of a copula is generally based on the observation that certain copulas are more appropriate to model dependence either in the upper or in the lower tail of a CDF (see also [224, pp. 17–18]). However, as observed in [273, p. 134], in general the physical implications of the choice of a single copula family are not immediately apparent.

② Use measured parameters that reflect the structure of dependence. For example, in [182, p. 1157] two-dimensional bounds assume knowledge of the medians of X and Y. A simpler parameter to derive, and one that leads to useful results even when its knowledge is confined to a range rather than a single value, is a *measure of dependence* [131,182,183]. We shall describe this approach in the balance of this chapter.

③ Introduce assumptions, based on theoretical considerations, which restrict the range of dependence structures but are not relying on the explicit choice of a copula family. In Section 6.6.5 *infra* we shall describe the effect of the assumption of *positive quadrant/orthant dependence*, which expresses the notion that "large" or "small" values of the RVs tend to occur together [181, p. 187].

6.6.1 Degree of dependence

In this section we introduce a single parameter which, in the words of the authors of [108, p. 249], allows one "to explore different dependence relationships between evidence when the exact nature of dependence is uncertain." Following [9, p. 169], [57], we define the *degree of dependence* of two events A and B as

$$r(A, B) \triangleq \frac{\mathbb{P}(A \cap B)}{\min(\mathbb{P}(A), \mathbb{P}(B))}. \tag{6.163}$$

From definition (6.163) and the results in Section 6.1.4 *supra*, it follows that

(i) $0 \leqslant r(A, B) \leqslant 1$,

(ii) $r(A, B) = 0$ if A and B are mutually exclusive,

(iii) $r(A, B) = 1$ if either $A \subseteq B$ or $B \subseteq A$,

(iv)

$$r(A, B) \geqslant \frac{\max(\mathbb{P}(A) + \mathbb{P}(B) - 1, 0)}{\min(\mathbb{P}(A), \mathbb{P}(B))}, \tag{6.164}$$

(v) $r(A, B) = \max(\mathbb{P}(A), \mathbb{P}(B))$ when A is independent of B.

The conditional probability of B given A can be expressed by

$$\mathbb{P}(B \mid A) = \frac{\mathbb{P}(A \cap B)}{\mathbb{P}(A)}$$

$$= r(A, B) \frac{\min(\mathbb{P}(A), \mathbb{P}(B))}{\mathbb{P}(A)}. \qquad (6.165)$$

It follows that

$$\mathbb{P}(A) \leqslant \mathbb{P}(B) \quad \Rightarrow \quad r(A, B) = \mathbb{P}(B \mid A),$$

$$\mathbb{P}(A) \geqslant \mathbb{P}(B) \quad \Rightarrow \quad r(A, B) = \mathbb{P}(B \mid A)\frac{\mathbb{P}(A)}{\mathbb{P}(B)}. \qquad (6.166)$$

Based on the value taken by $r(A, B)$, we may classify three degrees of dependence between events A and B:

(i) (*Minimum dependence*) This occurs when $r(A, B)$ takes on its smallest value (6.164). This implies that [57, p. 185]

$$\mathbb{P}(A \cap B) = \max(\mathbb{P}(A) + \mathbb{P}(B) - 1, 0),$$

$$\mathbb{P}(A \cup B) = \mathbb{P}(A) + \mathbb{P}(B) - \max(\mathbb{P}(A) + \mathbb{P}(B) - 1, 0)$$

$$= \min(\mathbb{P}(A) + \mathbb{P}(B), 1).$$

(ii) (*Maximum dependence*) This occurs when $r(A, B)$ takes on its maximum value 1. In this case

$$\mathbb{P}(A \cap B) = \min(\mathbb{P}(A), \mathbb{P}(B)).$$

$$\mathbb{P}(A \cup B) = \max(\mathbb{P}(A), \mathbb{P}(B)). \qquad (6.167)$$

(iii) (*Independence*) With $\mathbb{P}(A \cap B) = \mathbb{P}(A)\mathbb{P}(B)$, we have $r(A, B) = \max(\mathbb{P}(A), \mathbb{P}(B))$ and

$$\mathbb{P}(A \cup B) = \mathbb{P}(A) + \mathbb{P}(B) - \mathbb{P}(A) \cdot \mathbb{P}(B). \qquad (6.168)$$

Figs. 6.19–6.20 illustrate these concepts.

Remark 6.19. In [82, p. 18], minimum and maximum dependence are called "opposite" and "perfect" dependence, respectively.

6.6.1.1 Using qualitative knowledge about dependence

The extreme cases of minimum and maximum dependence provide bounds that are useful when no knowledge about dependence is available. In some cases it is possible that, although the exact value of the degree of dependence is unknown, on can exploit some partial knowledge about it. The idea here is to replace one of the bounds with the value achieved under the assumption of independence. Specifically, if events A and B

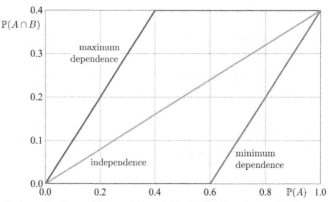

Figure 6.19 Range of values of $\mathbb{P}(A \cap B)$ vs. $\mathbb{P}(A)$ for $\mathbb{P}(B) = 0.4$ under the assumption of minimum dependence, maximum dependence, and independence.

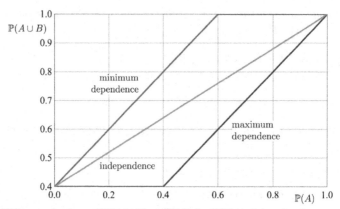

Figure 6.20 Range of values of $\mathbb{P}(A \cup B)$ vs. $\mathbb{P}(A)$ for $\mathbb{P}(B) = 0.4$ under the assumption of minimum dependence, maximum dependence, and independence.

are positively dependent, i.e., no less strongly dependent than independent, then one may think it reasonable to tighten Fréchet inequalities (6.45) to obtain [82, §2.3]

$$\mathbb{P}(A) \cdot \mathbb{P}(B) \leqslant \mathbb{P}(A \cap B) \leqslant \min\big(\mathbb{P}(A), \mathbb{P}(B)\big)),$$
$$\max\big(\mathbb{P}(A), \mathbb{P}(B)\big) \leqslant \mathbb{P}(A \cup B) \leqslant \mathbb{P}(A) + \mathbb{P}(B) - \mathbb{P}(A) \cdot \mathbb{P}(B). \tag{6.169}$$

Vice versa, if A and B are negatively dependent, then

$$\max\big(\mathbb{P}(A) + \mathbb{P}(B) - 1, 0\big) \leqslant \mathbb{P}(A \cap B) \leqslant \mathbb{P}(A) \cdot \mathbb{P}(B),$$
$$\mathbb{P}(A) + \mathbb{P}(B) - \mathbb{P}(A) \cdot \mathbb{P}(B) \leqslant \mathbb{P}(A \cup B) \leqslant \min\big(\mathbb{P}(A) + \mathbb{P}(B), 1\big). \tag{6.170}$$

We see that positive (negative) dependence results in a value of $\mathbb{P}(A \cap B)$ greater (smaller) than the value it would take for A independent of B. This helps understanding the conditions of negative/positive dependence: the assumption of positive dependence means that $\mathbb{P}(A \cap B)$ is expected to be larger than $\mathbb{P}(A) \cdot \mathbb{P}(B)$, while the assumption of negative dependence means that $\mathbb{P}(A \cap B)$ is expected to be smaller than $\mathbb{P}(A) \cdot \mathbb{P}(B)$, which is the value resulting from statistical independence.

Example 6.34. Assume A and B positively dependent, with $\mathbb{P}(A) = \mathbb{P}(B) = 0.1$. Then the minimum value that can be taken by $\mathbb{P}(A \cap B)$ is $\mathbb{P}(A) \cdot \mathbb{P}(B) = 0.01$, and its maximum value is $\min(\mathbb{P}(A), \mathbb{P}(B)) = 0.1$. If A and B are negatively dependent, then 0.01 is the maximum value that can be taken by $\mathbb{P}(A \cap B)$. $\diamond$

6.6.1.2 Degree of dependence as an interval number

Following [57], we may express the degree of dependence as an interval number, writing $r = [\underline{r}, \overline{r}]$. We have, generalizing (6.163),

$$\mathbb{P}(A \cap B) = \left[\underline{r} \min(\underline{\mathbb{P}}(A), \underline{\mathbb{P}}(B)), \, \overline{r} \min(\overline{\mathbb{P}}(A), \overline{\mathbb{P}}(B)) \right]. \tag{6.171}$$

Additional relations can be obtained by using the vertex method described in Section 4.5.1: to compute $\mathbb{P}(A \cup B)$, the vertices of the 3-dimensional parallelepiped are

$$
\begin{aligned}
v_1 &= \left(\underline{\mathbb{P}}(A), \underline{\mathbb{P}}(B), \underline{r} \right), & v_2 &= \left(\underline{\mathbb{P}}(A), \underline{\mathbb{P}}(B), \overline{r} \right), \\
v_3 &= \left(\underline{\mathbb{P}}(A), \overline{\mathbb{P}}(B), \underline{r} \right), & v_4 &= \left(\underline{\mathbb{P}}(A), \overline{\mathbb{P}}(B), \overline{r} \right), \\
v_5 &= \left(\overline{\mathbb{P}}(A), \underline{\mathbb{P}}(B), \underline{r} \right), & v_6 &= \left(\overline{\mathbb{P}}(A), \underline{\mathbb{P}}(B), \overline{r} \right), \\
v_7 &= \left(\overline{\mathbb{P}}(A), \overline{\mathbb{P}}(B), \underline{r} \right), & v_8 &= \left(\overline{\mathbb{P}}(A), \overline{\mathbb{P}}(B), \overline{r} \right).
\end{aligned}
\tag{6.172}
$$

Substituting the coordinates of these vertices into

$$\mathbb{P}(A \cup B) = \mathbb{P}(A) + \mathbb{P}(B) - \mathbb{P}(A \cap B), \tag{6.173}$$

we obtain the following equalities [57, p. 187]:

$$\underline{\mathbb{P}}(A \cup B) = \underline{\mathbb{P}}(A) + \underline{\mathbb{P}}(B) - \overline{r} \min(\underline{\mathbb{P}}(A), \underline{\mathbb{P}}(B)), \tag{6.174a}$$

$$\overline{\mathbb{P}}(A \cup B) = \overline{\mathbb{P}}(A) + \overline{\mathbb{P}}(B) - \underline{r} \min(\overline{\mathbb{P}}(A), \overline{\mathbb{P}}(B)), \tag{6.174b}$$

and, using the same procedure followed to obtain (6.174),

$$\underline{\mathbb{P}}(A^c \cap B) = \underline{\mathbb{P}}(B) - \overline{r} \min(\overline{\mathbb{P}}(A), \underline{\mathbb{P}}(B)), \tag{6.175a}$$

$$\overline{\mathbb{P}}(A^c \cap B) = \overline{\mathbb{P}}(B) - \underline{r} \min(\underline{\mathbb{P}}(A), \overline{\mathbb{P}}(B)), \tag{6.175b}$$

$$\underline{\mathbb{P}}(A \cap B^c) = \underline{\mathbb{P}}(A) - \overline{r} \min(\underline{\mathbb{P}}(A), \overline{\mathbb{P}}(B)), \tag{6.175c}$$

$$\overline{\mathbb{P}}(A \cap B^c) = \overline{\mathbb{P}}(A) - \underline{r} \min(\overline{\mathbb{P}}(A), \underline{\mathbb{P}}(B)), \tag{6.175d}$$

$$\underline{\mathbb{P}}(A^c \cap B^c) = 1 - \overline{\mathbb{P}}(A) - \overline{\mathbb{P}}(B) + \underline{r} \min(\overline{\mathbb{P}}(A), \overline{\mathbb{P}}(B)), \tag{6.175e}$$

$$\overline{\mathbb{P}}(A^c \cap B^c) = 1 - \underline{\mathbb{P}}(A) - \underline{\mathbb{P}}(B) + \overline{r} \min(\underline{\mathbb{P}}(A), \underline{\mathbb{P}}(B)), \tag{6.175f}$$

where the superscript ^{c} denotes set complement.

Some special cases are as follows [57, pp. 188–189]:

(i) *(Minimum dependence)* In this case r is equal to the lower bound in (6.164), and, using the vertex method, we can obtain

$$\underline{r} = \frac{\max(\underline{\mathbb{P}}(A) + \underline{\mathbb{P}}(B) - 1, 0)}{\min(\underline{\mathbb{P}}(A), \underline{\mathbb{P}}(B))}, \tag{6.176a}$$

$$\overline{r} = \frac{\max(\overline{\mathbb{P}}(A) + \overline{\mathbb{P}}(B) - 1, 0)}{\min(\overline{\mathbb{P}}(A), \overline{\mathbb{P}}(B))}, \tag{6.176b}$$

which leads to

$$\underline{\mathbb{P}}(A \cap B) = \max(\underline{\mathbb{P}}(A) + \underline{\mathbb{P}}(B) - 1, 0), \tag{6.177a}$$

$$\overline{\mathbb{P}}(A \cap B) = \max(\underline{\mathbb{P}}(A) + \underline{\mathbb{P}}(B) - 1, 0), \tag{6.177b}$$

$$\underline{\mathbb{P}}(A \cup B) = \underline{\mathbb{P}}(A) + \underline{\mathbb{P}}(B) - \overline{r} \min(\underline{\mathbb{P}}(A), \underline{\mathbb{P}}(B)), \tag{6.177c}$$

$$\overline{\mathbb{P}}(A \cup B) = \overline{\mathbb{P}}(A) + \overline{\mathbb{P}}(B) - \underline{r} \min(\overline{\mathbb{P}}(A), \overline{\mathbb{P}}(B)). \tag{6.177d}$$

(ii) *(Independence)* In this case

$$r = \left[\max(\underline{\mathbb{P}}(A), \underline{\mathbb{P}}(B)), \ \max(\overline{\mathbb{P}}(A), \overline{\mathbb{P}}(B)) \right] \tag{6.178}$$

and

$$\mathbb{P}(A \cap B) = \left[\underline{\mathbb{P}}(A)\underline{\mathbb{P}}(B), \overline{\mathbb{P}}(A)\overline{\mathbb{P}}(B) \right], \tag{6.179a}$$

$$\underline{\mathbb{P}}(A \cup B) = \underline{\mathbb{P}}(A) + \underline{\mathbb{P}}(B) - \max(\overline{\mathbb{P}}(A), \overline{\mathbb{P}}(B)) \cdot \min(\underline{\mathbb{P}}(A) \cap \underline{\mathbb{P}}(B)), \tag{6.179b}$$

$$\overline{\mathbb{P}}(A \cup B) = \overline{\mathbb{P}}(A) + \overline{\mathbb{P}}(B) - \max(\underline{\mathbb{P}}(A), \underline{\mathbb{P}}(B)) \cdot \min(\overline{\mathbb{P}}(A) \cap \overline{\mathbb{P}}(B)). \tag{6.179c}$$

(iii) *(Maximum dependence)* In this case

$$r = [1, 1] \tag{6.180}$$

and

$$\mathbb{P}(A \cap B) = \left[\min(\underline{\mathbb{P}}(A), \underline{\mathbb{P}}(B)), \ \min(\overline{\mathbb{P}}(A), \overline{\mathbb{P}}(B)) \right], \tag{6.181a}$$

$$\mathbb{P}(A \cup B) = \left[\max(\underline{\mathbb{P}}(A), \underline{\mathbb{P}}(B)), \ \max(\overline{\mathbb{P}}(A), \overline{\mathbb{P}}(B)) \right]. \tag{6.181b}$$

(iv) *(Unknown dependence)* In this case

$$r = [0, 1] \tag{6.182}$$

and

$$\mathbb{P}(A \cap B) = \left[0, \min(\overline{\mathbb{P}}(A), \overline{\mathbb{P}}(B)) \right], \tag{6.183a}$$

$$\mathbb{P}(A \cup B) = \left[\max(\underline{\mathbb{P}}(A), \underline{\mathbb{P}}(B)), \min(\overline{\mathbb{P}}(A) + \overline{\mathbb{P}}(B), 1) \right]. \tag{6.183b}$$

6.6.2   Linear correlation

We can gather some information about the dependence of two RVs from the computation of the *linear* correlation coefficient $\rho_{X,Y}$ defined in (3.257). This takes values in $[-1, 1]$ and is actually a measure of dependence, in the sense that $\rho_{X,Y} > 0$ indicates that large values of X and Y tend to occur together, and conversely, while $\rho_{X,Y} < 0$ means that if $X > \mathbb{E}[X]$, then Y tends to be smaller than $\mathbb{E}[Y]$ [59, p. 69]. However, $\rho_{X,Y}$ suffers from a few limitations which makes it play a role that is central essentially only for Gaussian RVs [172, p. 201 ff.]. In fact, given the marginal CDFs F_X and F_Y and a linear correlation coefficient value $\rho_{X,Y} \in [-1, 1]$, it may not be possible to construct a unique F_{XY} with marginals F_X, F_Y and correlation $\rho_{X,Y}$, with the exception of a few cases including jointly Gaussian X and Y. For example, if X and Y are independent RVs, then $\rho_{X,Y} = 0$, but, for the converse to be true, X and Y must be jointly Gaussian, while values of $|\rho_{X,Y}|$ close to zero do not indicate approximate independence, just lack of a linear relation between X and Y. If $|\rho_{X,Y}| = 1$, then X and Y are *perfectly linearly dependent*, i.e., $Y = a + bX$ a.s. with $a \in \mathbb{R}$ and $b \neq 0$, while a value of $|\rho_{X,Y}|$ close to 1 indicates strong linear dependence [59, p. 32]. Moreover, for $W = a + bX$ and $Z = a' + b'Y$ with $b, b' > 0$, we have $\rho_{W,Z} = \rho_{X,Y}$ [172, p. 202], which shows how linear correlation is invariant with respect to strictly increasing *linear* transformations.

Example 6.35. Consider the two RVs $X \sim \mathcal{N}(0, 1)$ and $Y = X^2$. Then

$$\text{cov}(X, Y) = \mathbb{E}[X^3] = 0 \tag{6.184}$$

and hence $\rho_{X,Y} = 0$. Thus, for X and Y zero linear correlation does not translate into independence. $\diamond$

The following general property holds [172, Theorem 5.25]:

Theorem 6.14. *Let X and Y denote two RVs with marginal CDFs F_X and F_Y and finite positive variances. The minimum value of $\rho_{X,Y}$ is attained if and only if X and Y are countermonotone, and its maximum value is attained if and only if X and Y are comonotone.*

Remark 6.20. Since linear correlation is invariant under linear transformations, but not under general transformations, even if strictly increasing, it is not copula-based. For example, the correlation of two lognormal RVs differs from that of the underlying Gaussian RVs.

6.6.3   Rank correlation: Kendall tau, Spearman rho, and Blomqvist beta

Rather than concentrating on the data themselves to derive information about correlation, one might examine the *ranks* of the data, which leads to the concept of *rank correlations*. These are matched to copula theory, because, unlike linear correlations, they depend only on the copula of bivariate CDFs and not on marginal distributions.

The reason of their name follows from the fact that they only depend on how the observed values taken by a set of RVs are ranked, and not on their actual values.

Qualitatively speaking, a pair of random variables are said to be *concordant* if large values of one tend to be associated with large values of the other, and small values of one with small values of the other [181, p. 157]. More precisely, let (x_i, y_i) and (x_j, y_j) denote two realizations of the continuous random pair (X, Y). Assume for simplicity that all the values taken by X and Y are unique. We say that the pairs (x_i, y_i) and (x_j, y_j) are *concordant* if we have simultaneously $x_i < x_j$ and $y_i < y_j$, or $x_i > x_j$ and $y_i > y_j$. Similarly, we say that (x_i, y_i) and (x_j, y_j) are *discordant* if $x_i < x_j$ and $y_i > y_j$, or $x_i > x_j$ and $y_i < y_j$. Equivalently, we may say that (x_i, y_i) and (x_j, y_j) are concordant if $(x_i - x_j)(y_i - y_j) > 0$, and discordant if $(x_i - x_j)(y_i - y_j) < 0$.

We can derive measures $\rho_\kappa(X, Y)$ of concordance between the continuous RVs X and Y connected by copula C. The following properties are sought [181, pp. 168–169]:

(i) $-1 \leqslant \rho_\kappa(X, Y) \leqslant 1$, $\rho_\kappa(X, X) = 1$, $\rho_\kappa(X, -X) = -1$.

(ii) $\rho_\kappa(X, Y) = \rho_\kappa(Y, X)$.

(iii)   If X, Y are independent, then $\rho_\kappa(X, Y) = 0$.

(iv) $\rho_\kappa(-X, Y) = \rho_\kappa(X, -Y) = -\rho_\kappa(X, Y)$.

(v)   If Y is an a. s. increasing function of X, then $\rho_\kappa(X, Y) = 1$, and if Y is an a.s. decreasing function of X, then $\rho_\kappa(X, Y) = -1$.

(vi)   If α and β are both a.s. strictly increasing functions, then

$$\rho_\kappa\big(\alpha(X), \beta(Y)\big) = \rho_\kappa(X, Y).$$

In terms of copulas joining X and Y, we have

(i)   If, for all $(u, v) \in \mathbb{I}^2$, we have $\mathsf{C}_1(u, v) \leqslant \mathsf{C}_2(u, v)$, then $\rho_\kappa(\mathsf{C}_1) \leqslant \rho_\kappa(\mathsf{C}_2)$.

(ii) $\rho(\mathsf{M}) = 1$, $\rho(\sqcap) = 0$, and $\rho(\mathsf{W}) = -1$.

Three concordance measures satisfying these properties are Spearman *rho*, Kendall *tau*, and Blomqvist *beta*. These are defined as follows [172, p. 207]:

1. Spearman *rho* is the linear correlation (3.257) of $F_X(X)$ and $F_Y(Y)$,

$$\rho_s \triangleq \rho_{F_X(X), F_Y(Y)}. \tag{6.185}$$

2. Kendall *tau* is defined as

$$\rho_\tau \triangleq \mathbb{E}\left[\mathsf{sgn}\left((X - \hat{X})(Y - \hat{Y})\right)\right], \tag{6.186}$$

where the pair $(\hat{X}, \hat{Y})$ is independent of (X, Y) and has the same joint CDF F_{XY}.

3. Let m_X and m_Y denote the medians of X and Y. Blomqvist medial correlation coefficient, or Blomqvist *beta* index, is defined as

$$\rho_\beta \triangleq \mathbb{P}\big[(X - m_X)(Y - m_Y) < 0\big] - \mathbb{P}\big[(X - m_X)(Y - m_Y) > 0\big]. \tag{6.187}$$

We have $-1 \leqslant \rho_\beta \leqslant 1$, and the bounds are sharp [183]. From its definition, it follows that

$$\rho_\beta = 4F_{XY}(m_X, m_Y) - 1. \tag{6.188}$$

Kendall *tau* and Spearman *rho* can also be explicitly expressed by using the concepts of concordance and discordance of pairs of RVs. Specifically, let (X_1, Y_1) and (X_2, Y_2) be independent and identically distributed random pairs with the same joint CDF. Kendall *tau* is the difference between their probabilities of concordance and discordance [181, p. 158],

$$\rho_\tau \triangleq \underbrace{\mathbb{P}\big[(X_1 - X_2)(Y_1 - Y_2) > 0\big]}_{\text{prob. of concordance}} - \underbrace{\mathbb{P}\big[(X_1 - X_2)(Y_1 - Y_2) < 0\big]}_{\text{prob. of discordance}}. \qquad (6.189)$$

Similarly, Spearman *rho* can be defined by using two independent random pairs (X_1, Y_1) and (X_2, Y_3) with the same marginals but possibly different joint CDFs, as X_2 and Y_3 are statistically independent. We have [181, p. 167]

$$\rho_s \triangleq 3 \left\{ \underbrace{\mathbb{P}\big[(X_1 - X_2)(Y_1 - Y_3) > 0\big]}_{\text{prob. of concordance}} - \underbrace{\mathbb{P}\big[(X_1 - X_2)(Y_1 - Y_3) < 0\big]}_{\text{prob. of discordance}} \right\}. \qquad (6.190)$$

A *concordance function* may be defined as the difference between the probabilities of concordance and discordance of two independent pairs (X_1, Y_1) and (X_2, Y_2) of continuous RVs with joint CDFs F_{X_1, Y_1}, F_{X_2, Y_2} and the same marginals $F_X = F_{X_1} = F_{X_2}$ and $F_Y = F_{Y_1} = F_{Y_2}$,

$$Q \triangleq \mathbb{P}\big[(X_1 - X_2)(Y_1 - Y_2) > 0\big] - \mathbb{P}\big[(X_1 - X_2)(Y_1 - Y_2) < 0\big]. \qquad (6.191)$$

The function Q depends on the CDFs of (X_1, Y_1) and (X_2, Y_2) only through their copulas. Specifically, if C_1 and C_2 denote the copulas of (X_1, Y_1) and (X_2, Y_2), respectively, then [181, Theorem 5.1.1]

$$Q = Q(C_1, C_2) = 4 \iint_{\mathbb{I}^2} C_2(u, v) \, dC_1(u, v) - 1 \qquad (6.192)$$

(see [181, Corollary 5.1.2] for useful properties of this function).

Example 6.36. The function Q can be easily evaluated for pairs of the basic copulas M, $\sqcap$, and W [181, Example 5.1]:

$$Q(M, M) = 1, \quad Q(\sqcap, \sqcap) = 0, \quad Q(W, W) = -1,$$
$$Q(M, \sqcap) = 1/3, \quad Q(M, W) = 0, \quad Q(W, \sqcap) = -1/3. \qquad (6.193)$$

Moreover, for an arbitrary copula C,

$$Q(C, M) \in [0, 1], \quad Q(C, W) \in [-1, 0], \quad Q(C, \sqcap) \in [-1/3, 1/3]. \qquad (6.194)$$

$\diamond$

Kendall *tau*, Spearman *rho*, and Blomqvist *beta* measures of concordance of a copula C may be expressed in terms of the concordance function Q as follows [181, pp. 161 ff.], [172, p. 207]:

$$\rho_\tau(\mathsf{C}) = 4 \iint_{\mathbb{I}^2} \mathsf{C}(u, v) \, d\mathsf{C}(u, v) - 1 = Q(\mathsf{C}, \mathsf{C}), \tag{6.195a}$$

$$\rho_s(\mathsf{C}) = 12 \iint_{\mathbb{I}^2} \mathsf{C}(u, v) \, du \, dv - 3 = 3 \, Q(\mathsf{C}, \Pi), \tag{6.195b}$$

$$\rho_\beta(\mathsf{C}) = 4\mathsf{C}(1/2, 1/2) - 1. \tag{6.195c}$$

Example 6.37. For a bivariate Gaussian copula with linear correlation coefficient ρ, we have [172, p. 215], [173]

$$\rho_s = \frac{6}{\pi} \arcsin \frac{\rho}{2},$$

$$\rho_\tau = \frac{2}{\pi} \arcsin \rho, \tag{6.196}$$

$$\rho_\beta = \frac{2}{\pi} \arcsin \rho. \qquad \qquad \diamond$$

Unlike linear correlation, rank correlations allow one to specify a bivariate CDF F_{XY} by giving only two marginals F_X, F_Y, and a value of rank correlation in $[-1, 1]$. To do this, one may choose the convex combination [172, p. 208]

$$F_{XY}(x, y) = \lambda \, \mathsf{W}\big(F_X(x), F_Y(y)\big) + (1 - \lambda) \, \mathsf{M}\big(F_X(x), F_Y(y)\big) \tag{6.197}$$

with $\lambda \in [0, 1]$. The resulting random pair has rank correlations

$$\rho_s(X, Y) = \rho_\tau(X, Y) = 1 - 2\lambda, \tag{6.198}$$

which yields any possible value in $[-1, 1]$.

Remark 6.21. Multivariate measures of rank correlation can also be defined: see, e.g., [197] for Spearman *rho*, [236] for Kendall *tau*, and [223] for Blomqvist *beta*.

In the following we describe the application of Fréchet–Hoeffding bounds to pairs of RVs whose Kendall *tau* or Spearman *rho* or Blomqvist *beta* are known. In particular, we compare these bounds with the unconstrained upper bounds $\overline{S}$ and $\underline{S}$ associated with copulas M and W, respectively.

6.6.3.1   Kendall tau

We have the following result [181,182], valid for all $(x, y) \in \mathbb{R}^2$. With X, Y continuous RVs with joint CDF F_{XY}, marginal CDFs F_X, F_Y, and Kendall parameter $\rho_\tau = t$, the tightest possible bounds are [182, Theorem 2]

$$\underline{T}_t\big(F_X(x), F_Y(y)\big) \leqslant F_{XY}(x, y) \leqslant \overline{T}_t\big(F_X(x), F_Y(y)\big), \tag{6.199}$$

where, for $t \in [-1, 1]$ and $(u, v) \in \mathbb{I}^2$,

$$\underline{T}_t(u, v) = \max\left(u + v - 1, \tfrac{1}{2}\left[(u + v) - \sqrt{(u - v)^2 + 1 - t}\right], 0\right), \qquad (6.200a)$$

$$\overline{T}_t(u, v) = \min\left(u, v, \tfrac{1}{2}\left[(u + v - 1) + \sqrt{(u + v - 1)^2 + 1 + t}\right]\right). \qquad (6.200b)$$

Both $\underline{T}_t$ and $\overline{T}_t$ are copulas, and hence the upper and lower bounds in (6.199) are CDFs. In particular, both $\underline{T}_t$ and $\overline{T}_t$ are continuous and nondecreasing in t. Special values are [182, Corollary 3] $\underline{T}_{-1} = \overline{T}_{-1} = \underline{S}$, $\underline{T}_1 = \overline{T}_1 = \overline{S}$, $\underline{T}_0 = \underline{S}$, and $\overline{T}_0 = \overline{S}$. In general, if $t \in [0, 1]$ then $\overline{T}_t = \overline{S}$, and if $t \in [-1, 0]$ then $\underline{T}_t = \underline{S}$.

Example 6.38. To illustrate bounds (6.199) (as well as (6.207) *infra*), consider the largest of two RVs X, Y, whose CDF is [189, p. 141]

$$F_{\max}(x) = F_{XY}(x, x) \qquad (6.201)$$

and determine its upper and lower bounds $\overline{T}_t(x)$ and $\underline{T}_t(x)$ under the assumptions that $F_X = F_Y = F$ and that the value of Kendall *tau* is $\rho_\tau = t$. From (6.201) and (6.199), we have

$$\underline{T}_t(x) = \max\left(2F(x) - 1, \tfrac{1}{2}\left(2F(x) - \sqrt{1 - t}\right), 0\right), \qquad (6.202a)$$

$$\overline{T}_t(x) = \min\left(F(x), \tfrac{1}{2}\left[(2F(x) - 1) + \sqrt{(2F(x) - 1)^2 + 1 + t}\right]\right). \qquad (6.202b)$$

Consider $\overline{T}_t$ first. We have $\overline{T}_t(x) = F(x)$ if

$$F(x) \leqslant \frac{1}{2}\left[(2F(x) - 1) + \sqrt{(2F(x) - 1)^2 + 1 + t}\right],$$

i.e., if $(2F(x) - 1)^2 \geqslant -t$. The last inequality is trivially satisfied if $t \geqslant 0$, while if $t < 0$ it is satisfied for $F(x) \leqslant \tfrac{1}{2}(1 - \sqrt{-t})$ and $F(x) \geqslant \tfrac{1}{2}(1 + \sqrt{-t})$. In conclusion, we have $\overline{T}_t(x) = F(x)$ for $t \geqslant 0$, while for $t < 0$:

$$\overline{T}_t(x) = \begin{cases} F(x), & F(x) \leqslant \tfrac{1}{2}(1 - \sqrt{-t}), \\ \tfrac{1}{2}\left[(2F(x) - 1) + \sqrt{(2F(x) - 1)^2 + 1 + t}\right], \\ & \tfrac{1}{2}(1 - \sqrt{-t}) \leqslant F(x) \leqslant \tfrac{1}{2}(1 + \sqrt{-t}), \\ F(x), & F(x) \geqslant \tfrac{1}{2}(1 + \sqrt{-t}). \end{cases} \qquad (6.203)$$

Consider $\underline{T}_t$ next. After simple calculations, for $t \leqslant 0$ we obtain $\underline{T}_t(x) = \max(2F(x) - 1, 0)$, while for $t > 0$,

$$\underline{T}_t(x) = \begin{cases} 0, & F(x) \leqslant \tfrac{1}{2}\sqrt{1 - t}, \\ F(x) - \tfrac{1}{2}\sqrt{1 - t}, & \tfrac{1}{2}\sqrt{1 - t} \leqslant F(x) \leqslant 1 - \tfrac{1}{2}\sqrt{1 - t}, \\ 2F(x) - 1, & F(x) \geqslant 1 - \tfrac{1}{2}\sqrt{1 - t}. \end{cases} \qquad (6.204)$$

$\diamond$

Example 6.39. Assume the special case $F(x) = 1 - e^{-x}$, and denote by $\underline{S}$ and $\overline{S}$ the dependence bounds obtained without any knowledge of the value t of Kendall *tau*, so that

$$\underline{S}(x) = \max(1 - 2e^{-x}, 0) \quad \text{and} \quad \overline{S}(x) = 1 - e^{-x}. \tag{6.205}$$

With $t = -1$, the upper dependence bound is $\overline{S}$, while the lower dependence bound is $\underline{T}_t = \overline{T}_t = \underline{S}$. With $t = 0$, the dependence bounds are $\overline{T}_t = \overline{S}$ and $\underline{T}_t = \underline{S}$. With $t = 1$, the bounds are $\overline{S} = \underline{T}_t = \overline{T}_t$ and $\underline{S}$. With $t = -0.7$ and $t = 0.7$, the bounds are shown in Figs. 6.21 and 6.22, respectively.

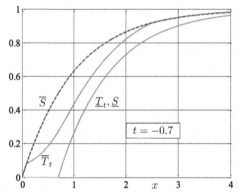

Figure 6.21 Comparison of dependence bounds $\underline{S}, \overline{S}$ on $F_{\max}$ with the bounds $\underline{T}_t, \overline{T}_t$. These are obtained under the assumption that $F(x) = 1 - e^{-x}$, and Kendall *tau* parameter takes value $t = -0.7$. Reproduced with permission from [29].

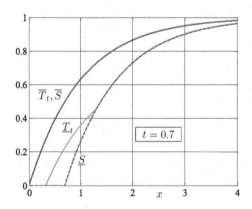

Figure 6.22 As in Fig. 6.21, with $t = 0.7$.

The corresponding bounds on error probability for binary antipodal transmission on a fading channel with diversity 2, Rayleigh fading, and selection combining can be

Table 6.1 Error probability of binary antipodal transmission with diversity 2 with Rayleigh fading and selection combining. Comparison of lower and upper dependence bounds under the assumptions that Kendall *tau* takes value $t = \pm 0.7$ ($\underline{\mathsf{p}}_{0.7}$ and $\overline{\mathsf{p}}_{-0.7}$) or no measure of association is available ($\underline{\mathsf{p}}_0$ and $\overline{\mathsf{p}}_0$). The exact error probability when the two diversity branches are independently faded is denoted $\mathsf{p}_{\text{indep}}$.

snr (dB)	$\underline{\mathsf{p}}_0$	$\underline{\mathsf{p}}_{0.7}$	$\mathsf{p}_{\text{indep.}}$	$\overline{\mathsf{p}}_{-0.7}$	$\overline{\mathsf{p}}_0$
0	0.052	0.069	0.082	0.103	0.146
5	0.004	0.012	0.020	0.040	0.064
10	8e-6	0.0003	0.003	0.018	0.023
15	5e-13	7e-8	0.0003	0.0075	0.0077

evaluated using the general expression (5.30)

$$\mathsf{p} = \sqrt{\frac{\mathsf{snr}}{4\pi}} \int_0^\infty \exp(-\mathsf{snr}\, z) z^{-1/2} T(z)\, dz, \tag{6.206}$$

where T denotes one of the bounds just derived. The results are shown in Table 6.1.

$\diamond$

6.6.3.2 *Spearman* rho

We have the following result [181,182], valid for all $(x, y) \in \mathbb{R}^2$. With X, Y continuous RVs with joint CDF F_{XY}, marginal CDFs F_X, F_Y, and Spearman parameter $\rho_s = t$, the tightest possible bounds are

$$\underline{P}_t (F_X(x), F_Y(y)) \leqslant F_{XY}(x, y) \leqslant \overline{P}_t (F_X(x), F_Y(y)), \tag{6.207}$$

where, for $t \in [-1, 1]$ and $(u, v) \in \mathbb{I}^2$,

$$\underline{P}_t(u, v) = \max\left(0, u + v - 1, \tfrac{1}{2}(u + v) - \phi(u - v, 1 - t)\right), \tag{6.208a}$$

$$\overline{P}_t(u, v) = \min\left(u, v, \tfrac{1}{2}(u + v - 1) + \phi(u + v - 1, 1 + t)\right), \tag{6.208b}$$

with

$$\phi(a, b) \triangleq \frac{1}{6}\left[\left(9b + 3\sqrt{9b^2 - 3a^6}\right)^{1/3} + \left(9b - 3\sqrt{9b^2 - 3a^6}\right)^{1/3}\right]. \tag{6.209}$$

Both $\underline{P}_t$ and $\overline{P}_t$ are copulas, and hence the upper and lower bound in (6.207) are CDFs. In particular, both $\underline{P}_t$ and $\overline{P}_t$ are continuous and nondecreasing in t. Special values are $\underline{P}_{-1} = \overline{P}_{-1} = \underline{S}$ and $\underline{P}_1 = \overline{P}_1 = \overline{S}$. In general, if $t \in [1/2, 1]$ then $\overline{P}_t = \overline{S}$, and if $t \in [-1, -1/2]$ then $\underline{P}_t = \underline{S}$.

Example 6.40. With $F(x) = 1 - e^{-x}$ as in Example 6.39, we have the results plotted in Figs. 6.23, 6.24, and 6.25. The corresponding bounds on error probability are shown in Table 6.2. $\diamond$

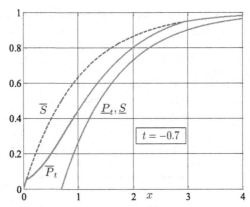

Figure 6.23 Comparison of dependence bounds $\underline{S}$, $\overline{S}$ on $F_{\max}$ with bounds $\underline{P}_t$, $\overline{P}_t$. These are obtained under the assumption that $F(x) = 1 - e^{-x}$, and Spearman *rho* takes value $t = -0.7$.

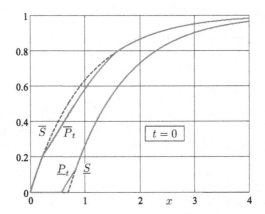

Figure 6.24 As in Fig. 6.23, with $t = 0$.

6.6.3.3 *Blomqvist* beta

Let $\mathcal{B}_t$ denote the set of copulas $\mathbf{C}$ such that $\rho_\beta(\mathbf{C}) = t$, and $\underline{B}_t$, $\overline{B}_t$ the pointwise infimum and supremum of $\mathbf{C}(u, v) \in \mathcal{B}_t$ for $(u, v) \in \mathbb{I}^2$. We have the following result [183, Theorem 1]: If X, Y are continuous RVs with joint CDF F_{XY} and marginal CDFs F_X, F_Y, and such that $\rho_\beta = t$, then the tightest possible bounds are

$$\underline{B}_t\big(F_X(x), F_Y(y)\big) \leqslant F_{XY}(x, y) \leqslant \overline{B}_t\big(F_X(x), F_Y(y)\big) \tag{6.210}$$

for all $(x, y) \in \mathbb{R}^2$, where

$$\begin{aligned}\underline{B}_t(u, v) = \max\big\{0, u + v - 1, (t + 1)/4 - \max\{1/2 - u, 0\} \\ - \max\{1/2 - v, 0\}\big\},\end{aligned} \tag{6.211a}$$

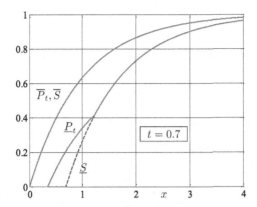

Figure 6.25 As in Fig. 6.23, with $t = 0.7$.

Table 6.2 Error probability of binary antipodal transmission with diversity 2 with Rayleigh fading and selection combining. Comparison of lower and upper dependence bounds under the assumptions that Spearman *beta* takes value $t = \pm 0.7$ ($\underline{p}_{0.7}$ and $\overline{p}_{-0.7}$) or no measure of association is available ($\underline{p}_0$ and $\overline{p}_0$). The exact error probability when the two diversity branches are independently faded is denoted p_{indep}.

snr (dB)	$\underline{p}_0$	$\underline{p}_{0.7}$	p_{indep}	$\overline{p}_{-0.7}$	$\overline{p}_0$
0	0.052	0.066	0.082	0.102	0.146
5	0.004	0.010	0.020	0.037	0.064
10	8e-6	0.0003	0.003	0.016	0.023
15	5e-13	3e-8	0.0003	0.0070	0.0077

$$\overline{B}_t(u, v) = \min\bigl\{u, v, (t + 1)/4 + \max\{u - 1/2, 0\} + \max\{v - 1/2, 0\}\bigr\}.$$
$$(6.211\text{b})$$

Some properties of $\underline{B}_t(u, v)$ and $\overline{B}_t(u, v)$ are listed in [183, Corollary 2]. In particular, $\underline{B}_t = \underline{S}$ if and only if $t = -1$, while $\overline{B}_t(u, v) = \overline{S}$ if and only if $t = +1$. Moreover, $\underline{B}_t(1/2, 1/2) = \overline{B}_t(1/2, 1/2) = (t + 1)/4$. As observed in [183, p. 2305], the six bounds $\underline{T}_t$, $\overline{T}_t$, $\underline{P}_t$, $\overline{P}_t$, $\underline{B}_t$, and $\overline{B}_t$ are copulas, and hence coincide on the boundary of the unit square $\mathbb{I}^2$ (see (6.1)). However, the bounds $\underline{B}_t$, $\overline{B}_t$ are expected to be tighter, as they also coincide at $(1/2, 1/2)$, the center of the square.

Example 6.41. As in Examples 6.39 and 6.40, assume $F_X(x) = F_Y(x) = 1 - e^{-x}$. The corresponding results are shown in Fig. 6.26. $\diamond$

Remark 6.22. The results above show how the knowledge of the value taken by Kendall or Spearman parameter may not result into substantially tightened bounds on the performance parameters. For example, observe that $\underline{T}_t$, $\overline{T}_t$ may improve on $\underline{S}$, $\overline{S}$ only for intermediate values of x, unless Kendall or Spearman parameter is close to -1 or to $+1$, which indicates a considerable amount of knowledge about the dependence structure. An additional observation following the perusal of the results above is that

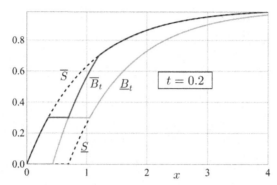

Figure 6.26 Comparison of dependence bounds $\underline{S}, \overline{S}$ on $F_{\max}$ with the bounds $\underline{B}_t, \overline{B}_t$. These are obtained under the assumption that $F(x) = 1 - e^{-x}$, and Blomqvist *beta* takes value $t = 0.2$.

the knowledge of the value taken by ρ_τ or ρ_s may not add a considerable amount of useful information about the actual dependence structure (again, with the exception of cases $t \approx \pm 1$).

6.6.4 Tail dependence

The *coefficients of tail dependence* measure the strength of the dependence in the tails of bivariate CDFs, and depend only on copulas. Consider two RVs $U \sim \mathcal{U}[0, 1]$ and $V \sim \mathcal{U}[0, 1]$. We say that U and V are *upper tail dependent* if large values of V follow from large values of U [224]. Specifically, consider a threshold q and the conditional probability $\mathbb{P}(V > q \mid U > q)$ as $q \to 1$. If this has order smaller than q, we say that U and V have no tail dependence, otherwise they have tail dependence. Formally, for any two RVs X and Y with CFDs F_X and F_Y, we define the *coefficient of upper tail dependence*

$$\lambda_u \triangleq \lim_{q \to 1^-} \mathbb{P}\big(Y > F_Y^{\leftarrow}(q) \mid X > F_X^{\leftarrow}(q)\big) \tag{6.212}$$

under the assumption that the limit exists and $\lambda_u \in [0, 1]$. Similarly, the *coefficient of lower tail dependence* is defined as

$$\lambda_l \triangleq \lim_{q \to 0^+} \mathbb{P}\big(Y \leqslant F_Y^{\leftarrow}(q) \mid X \leqslant F_X^{\leftarrow}(q)\big). \tag{6.213}$$

If $\lambda_u > 0$ we say that X and Y have upper tail dependence. For $\lambda_u = 0$, we say that they are *asymptotically independent in the upper tail*, and analogously for λ_l.

For continuous CDFs, applying Bayes rule, we obtain

$$
\begin{aligned}
\lambda_l &= \lim_{q \to 0^+} \frac{\mathbb{P}\big(Y \leqslant F_Y^{\leftarrow}(q), X \leqslant F_X^{\leftarrow}(q)\big)}{\mathbb{P}\big(X \leqslant F_X^{\leftarrow}(q)\big)} \\
&= \lim_{q \to 0^+} \frac{\mathsf{C}(q, q)}{q}
\end{aligned}
\tag{6.214}
$$

and, using (6.54),

$$
\lambda_u = 2 + \lim_{q \to 0^+} \frac{\hat{\mathsf{C}}(1 - q, 1 - q)}{q},
\tag{6.215}
$$

where $\hat{\mathsf{C}}$ denotes the survival copula (6.53).

Example 6.42. Direct calculations [172, p. 209] show that the Gumbel copula has $\lambda_u = 2 - 2^{1/\theta}$, and hence it has upper tail dependence for $\theta > 1$. The Clayton copula has $\lambda_l = 2^{-1/\theta}$, and hence lower tail dependence for $\theta > 0$, while the Gaussian copula exhibits no tail dependence for $\rho < 1$ [172, p. 211]. $\diamond$

6.6.5 Quadrant/orthant dependence

This type of dependence describes how the values taken by two or more RVs are connected, given that they are simultaneously large or small. The definition of *quadrant dependence* refers to a comparison with independence: the idea here is to examine the probability that two or more RVs be simultaneously small (large), and compare it with the probability that would be achieved if they were independent. Formally, two RVs X and Y are said to be *positive quadrant dependent* (PQD) if the probability that they are simultaneously small (large) is at least as great as it would be were they independent [181, p. 187 ff.]: X and Y are PQD if

$$
F_{XY}(x, y) \geqslant F_X(x) F_Y(y)
\tag{6.216}
$$

or, equivalently,

$$
\tilde{F}_{XY}(x, y) \geqslant \tilde{F}_X(x) \tilde{F}_Y(y),
\tag{6.217}
$$

where $\tilde{F}$ denotes the *survival functions* defined in Section 6.1.5, namely $\tilde{F}_Z(z) \triangleq \mathbb{P}[Z > z] = 1 - F_Z(z)$, $\tilde{F}_{WZ}(w, z) \triangleq \mathbb{P}[W > w, Z > z]$.

In terms of copulas, definition (6.216) is equivalent to

$$
\mathsf{C}(u, v) \geqslant uv = \Pi
\tag{6.218}
$$

for all $(u, v) \in \mathbb{I}^2$.

Remark 6.23. For continuous RVs, quadrant dependence is a property of their copula, and hence is a property invariant under a.s. strictly increasing transformations of the RVs [181, p. 188].

Example 6.43. Assume $Z = X + Y$, where X, Y are PQD. We have the following result [63]:

$$\underline{F}_Z(z) = \sup_{x \in \mathbb{R}} \left[F_X(x) F_Y(z - x) \right],$$

$$\overline{F}_Z(z) = 1 - \sup_{x \in \mathbb{R}} \left[\tilde{F}_X(x) \tilde{F}_Y(z - x) \right]. \tag{6.219}$$

◇

For multivariate RVs, the following definition holds: $X_1, \dots, X_d$ are said to be *positive orthant dependent* (POD) if

$$F_{X_1 \cdots X_d}(x_1, \dots, x_d) \geqslant \prod_{i=1}^{d} F_{X_i}(x_i) \tag{6.220}$$

or, equivalently,

$$\tilde{F}_{X_1 \cdots X_d}(x_1, \dots, x_d) \geqslant \prod_{i=1}^{d} \tilde{F}_{X_i}(x_i). \tag{6.221}$$

Example 6.44. If $Z \triangleq X_1 + \cdots + X_d$, we have [63]:

$$\underline{F}_Z(z) = \sup_{x_1 + \cdots + x_d = z} \left[\prod_{i=1}^{d} F_{X_i}(x_i) \right],$$

$$\overline{F}_Z(z) = 1 - \sup_{x_1 + \cdots + x_d = z} \left[\prod_{i=1}^{d} \tilde{F}_{X_i}(x_i) \right]. \tag{6.222}$$

◇

Example 6.45. Under the same conditions of Example 6.23 and the assumption of POD RVs, we obtain, replicating the calculations of that example,

$$\underline{F}_Z(z) = [F(z/d)]^d,$$

$$\overline{F}_Z(z) = 1 - [1 - F(z/d)]^d. \tag{6.223}$$

Results obtained with the same parameters as in Fig. 6.16 are shown in Fig. 6.27. Comparing the two figures, the increased tightness of the bounds due to the introduction of the POD assumption is apparent. ◇

Sources and parerga

1. In [234, Chapter 1] it is described how in 2008 the rating agencies were charged with assessing the likelihood that trillions of dollars in mortgage-backed securities would go into default, and how they have blown the call. Actually, around

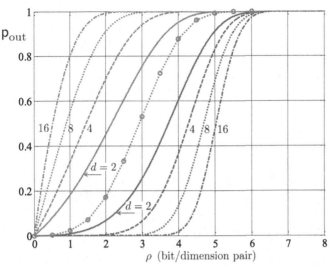

Figure 6.27 Dependence upper and lower bounds for the outage probability of a block fading channel with Rayleigh fading, signal-to-noise ratio $\mathsf{snr} = 10$ dB, and d blocks, $d = 2, 4, 8$, and 16, under the assumption of positive orthant dependence of the diversity branches. The middle curve shows $\mathsf{p}_{\mathrm{out}}$ for $d = 2$ and independent fading [23, Eq. (4.37)].

28% of the AAA-rated collateralized debt obligations, i.e., collections of mortgage debt that are broken into different pools, defaulted, while the prediction of such event was 0.12%. Assuming that we have a pool of five mortgages, each of which has $\mathsf{p} = 0.05$ of defaulting, the probability of all five to default is $\mathsf{p}_{\mathrm{default}} = \mathsf{p}^5 \sim 3 \times 10^{-7}$ if independence is assumed. If instead all mortgages behave exactly alike, then $\mathsf{p}_{\mathrm{default}} = 0.05$, i.e., 160,000 times bigger than with independence. The massive housing bubble occurred in the US made the independence assumption invalid. It was observed in [64, p. 42] that "in the management of large portfolios, the main risk is the joint occurrence of a number of default events or the simultaneous downside evolution of prices. A better knowledge of the dependence between financial assets or claims is crucial to assess the risk of loss clustering." This explains why dependence and copula theory have wide implications in financial theory; see, e.g., [164].

2. The "one-factor Gaussian copula" defined in Example 6.13 was used in the past as an industry standard for the pricing of Collateralized Debt Obligations [164, p. 123]. Around 2008, as financial markets began behaving in ways that exposed its severe weaknesses and made its use inappropriate, it became known as "the formula that killed Wall Street" [218].

3. Dependence is also a common situation in actuarial practice, as, for example, in the assessment of the risk represented by contiguous buildings in fire insurance [63, p. 85].

4. While copulas are the most widely used tool to model the dependence among RVs, dependence can also be described by using the notion of *rearrangement*

of functions [201]. A rearrangement algorithm was introduced in [199] for the numerical derivation of sharp bounds on the CDF of dependent RVs with fixed marginals. The algorithm is computationally efficient, and allows the solution of problems in a large number of dimensions. In [200] it is shown how the same algorithm can be used to find bounds on $\mathbb{E}[H(\mathbf{X})]$ for functions $H : \mathbb{R}^d \mapsto \mathbb{R}$ satisfying the *supermodularity* condition

$$H(\mathbf{x} \wedge \mathbf{y}) + H(\mathbf{x} \vee \mathbf{y}) \geqslant H(\mathbf{x}) + H(\mathbf{y}), \qquad (6.224)$$

where $\wedge$ (resp., $\vee$) denotes componentwise minimum (resp., maximum).

5. The derivations in Section 6.4 are based on the theory of Optimal Transportation (or Transport). Optimal Transportation (see, e.g., [205,259]) was studied in the 18th century by the French mathematician Gaspard Monge to minimize the cost of transporting soil from one location to another predetermined location. Within this theory, $H(x, y)$ denotes the cost of transporting a unit of mass from point x to point y.

6. Functions satisfying Monge condition (6.125) are also called *submodular* or *L-subadditive* [35].

7. In financial mathematics, the quantity $\inf\{s \mid \mathbb{P}[S \leqslant s] \geqslant \alpha\}$ is called *Value-at-Risk* (VaR) at confidence level α [126, §5.2.1], [172, p. 38]. The VaR of a portfolio at level α is the smallest number s such that the probability that the random loss S exceed s is no larger than $(1 - \alpha)$.

8. Although the Gaussian copula is still widely used in financial applications [46, p. 30], it is becoming less and less attractive, especially in case of extreme events [231, Chapter 7]. The Student copula of Example 6.4, along with its multivariate version, is more appropriate to analyze the case of heavy-tailed distributions.

9. Consideration of dependence bounds for tail-monotone pdfs is discussed in [265, p. 407].

10. The problem of finding bounds on the CDF of a sum of nonindependent RVs whose marginal distributions are known is relevant for example in project management. Here each RV models the time duration of an activity needed to complete a project, and hence the sum describes to total project duration [213].

11. Yet another parameter measuring the degree of dependence between random variables is Gini *gamma* index [146].

Beyond probability 7

In this chapter we examine the theories of decision-making in the presence of aleatory and epistemic uncertainties. The most important among these, "Expected Utility Theory" (EUT), is based on the assumption that a rational decision maker (DM) aims at achieving the largest value of an expected "utility." Since the utility actually generated by a choice depends also on contingencies on which the DM has no control (the "State of Nature"), the DM choice is guided by his ideas about the likelihood of the State of Nature at the time of making a decision. This likelihood may be expressed in terms of objective probabilities, which is the case of decisions under risk, or of subjective probabilities, or beliefs, if decisions are made under epistemic uncertainty. An EUT based on axioms of rational decision making was developed by von Neumann and Morgenstern in [184]. Although the expected-utility approach looks reasonable and sound, a number of empirical results exhibit violations of its axioms in the common practice of people making decisions. St. Petersburg, Allais, and Ellsberg paradoxes, among others, have been observed over the years, which questions the descriptive adequacy of EUT. These observations have led to the formulation of a number of questions about the practical validity of EUT. Here we describe EUT, the paradoxes following by its application, and some variations on this theory aimed at solving them.

We also illustrate some extensions of classical probability theory. Such extensions allow epistemic uncertainty to be quantified by distinguishing different degrees of partial knowledge, and avoid the need for the level of precision and internal consistency that is required by probability theory and is higher than that available in many situations.

7.1 Decisions under uncertainty

In decision theory an "agent" (the DM) makes a choice among a number of possible "acts," or "decisions." The consequence, and hence the "utility" of his decision depends on an unknown, or partially known, "State of Nature" (SoN) whose existence is taken into account before a decision is made.

We assume the existence of an exhaustive set S of possible SoNs, with the "true" state belonging to S. The elements of S are mutually exclusive, and only one of them is the true one. It is often convenient, although not necessary, to assume that the set of possible utilities associated with decisions is the set $\mathbb{R}$ of real numbers. A function can be defined which describes mathematically the decision by mapping every State of Nature to a utility value. Since the range of decisions is finite, this function induces a partition of S. We assume finite S (the finiteness restriction can be removed, at the price of some added technicalities) and power set 2^S, the collection of subsets of S.

Dimensions of Uncertainty in Communication Engineering. https://doi.org/10.1016/B978-0-32-399275-6.00015-0

Table 7.1 The "Savage omelet" example illustrating acts, States of Nature, and consequences.

Act	State of Nature	
	Egg is good	*Egg is rotten*
break into bowl	six-egg omelet	no omelet, five good eggs wasted
break into saucer	six-egg omelet, one saucer to wash	five-egg omelet, one saucer to wash
throw away	five-egg omelet, one good egg wasted	five-egg omelet

Remark 7.1. The assumption that the set S is given and known may prove to be a limitation of classical decision theory, as it does not allow the consideration of unforeseen contingencies, i.e., of SoNs that cannot be described or even thought of when the decision is to be made [79, p. 236].

Example 7.1. A gambler places a bet on a soccer game. With a €5 wager on the victory of team A, if A wins he receives €20, otherwise nothing. The set of SoNs includes the three outcomes of the game: A wins, A loses, or the game ends in a draw. Here decisions are bets, and utilities are amounts of money: €15 if team A wins, or €−5 in the other cases.   ◇

Example 7.2. The "Savage omelet" provides a more complex example illustrating a variety of SoNs, acts, and their consequences. Quoting verbatim from [221, p. 13]:

> *Your wife has just broken five good eggs into a bowl when you come in and volunteer to finish making the omelet. A sixth egg, which for some reason must either be used for the omelet or wasted altogether, lies unbroken beside the bowl. You must decide what to do with this unbroken egg. Perhaps it is not too great an oversimplification to say that you must decide among three acts only, namely, to break it into the bowl containing the other five, to break it into a saucer for inspection, or to throw it away without inspection.*

The situation is summarized in Table 7.1. There are two SoNs here: "egg is good," or "egg is rotten," three possible acts, and six consequences. With each consequence, in EUT scheme, the DM should associate a real number reflecting the utility of each act depending on his specific preferences (see, for example, [102, pp. 320–321]).   ◇

In our context, we distinguish between

(i) *Decisions under aleatory uncertainty*, when a probability measure over the set S of States of Nature is available, and

(ii) *Decisions under epistemic uncertainty*, when no such measure is a priori available.

It sounds natural, for a decision under aleatory uncertainty, or "under risk," to choose the act which maximizes the expected utility (or, equivalently, minimizes the expected cost). Assuming a finite set $S = \{x_1, \ldots, x_n\}$ of possible SoNs, each with probability p_i, $i = 1, \ldots, n$, and denoting by $u(x_i)$ the utility of the decision made when the state of nature is x_i, the expected utility criterion chooses the decision which maximizes

$$\mathbb{E}[U] = \sum_i^n p_i u(x_i). \tag{7.1}$$

With EUT, a DM will prefer act "a" to an alternative act "b" if the expected utility of the former is greater that the expected utility of the latter. At the heart of the theory is the utility function u, which explains the behavior of the DM and whose existence is guaranteed if a few axioms are satisfied [184, Chapter 3]. As we shall see, some paradoxes have been observed which show a systematic violation of these axioms, and hence prove that expected utility might not describe peoples' choices in full.

This decision criterion assumes that probabilities and utilities are well defined. Citing verbatim from [102, p. 287], for decisions under risk "expected utility is a simple and natural criterion, because it yields the average utility the decision maker can expect in the long run, taking into account all possible events and their probabilities." For decisions under epistemic uncertainty, a possible option consists of replacing (7.1) with a similar expression in which the probability measure on S is replaced by a *subjective* or *qualitative* measure, as suggested in the "umbrella" Example 7.3 below and described in Section 7.2.2. Or one may look for different "probability-like" measures yielding a satisfactory theory when decisions are made under epistemic uncertainty. Following [261, p. 2], we should ask: (a) what is the best way to model uncertainty in this context? (b) how should we assess, combine and update measures of uncertainty?, and (c) how should we use these measures to make inferences and decisions?

Remark 7.2. The use of (7.1) under epistemic uncertainty implies two a priori unrelated assumptions. The first is that the decision maker has beliefs expressed by a probabilistic structure, and the second is that he accepts to combine them into an expectation, i.e., in a linear manner. As a consequence of the first assumption, since the only restrictions put on a subjective probability measure is that the probability axioms are satisfied, even unreasonable beliefs are in principle acceptable [79, pp. 238–239].

Example 7.3. One wants to decide whether to take an umbrella or not in a walk to the park. If the probability of rain is known because a reliable weather forecast is available, this is a case of decision under risk, where the utility of having an umbrella is high if it rains, and mildly negative (the inconvenience to carry a useless item) if it does not rain. If the probability of rain is unknown, one may resort to assigning probability $1/2$ to both events "rain" and "no rain" (this corresponds to invoking the "Principle of Insufficient Reason," to be described in Section 7.2.1 *infra*). Or one may assign "qualitative probabilities" using terms like "rain is highly probable," and state that the probability of rain lies in the interval $[0.5, 1]$. $\diamond$

Example 7.4 ([260]). Consider a street vendor deciding which merchandise to take along with him tomorrow. Merchandise can be ice cream and hot dogs. The net profits resulting from the merchandise depend on the weather tomorrow, and can be negative

Table 7.2 Utilities obtained from the sale of merchandise and depending on weather.

Merchandise on sale	State of Nature		
	No rain	*Some rain*	*Rain all day*
ice cream (i)	400	100	−400
hot dogs (h)	−400	100	400
neither (n)	0	0	0
both (b)	0	200	0

because goods not sold tomorrow are lost. Assume for simplicity only three SoNs: no rain, some rain, and rain all day. The utility (net profit) in monetary units depends on the weather tomorrow and from the vendor decision, according to Table 7.2. If the probabilities associated with events "no rain," "some rain," and "rain all day" have values 0.4, 0.3, and 0.3, respectively, the vendor expected utilities of carrying along a choice of merchandise are, using (7.1) with obvious notations: $\mathbb{E}[U_i] = 70$, $\mathbb{E}[U_h] = -10$, $\mathbb{E}[U_n] = 0$, and $\mathbb{E}[U_b] = 60$. Thus, a rational vendor using EUT would choose to carry only ice cream. ◇

Remark 7.3. The expected-utility expression (7.1) can be extended, under suitable conditions, to a continuous set $\mathcal{S}$ of States of Nature with pdf $f(x)$. This yields, with obvious notations (see [85, Chapter 3] for mathematical details),

$$\mathbb{E}[U] = \int_{\mathcal{S}} u(x) f(x) \, dx. \tag{7.2}$$

Example 7.5 ([139, pp. 135–136]). Consider the following four acts the DM has to choose from:

A : Earn € 100,000 for sure

B : Earn € 200,000 or € 0 with probability 0.5

C : Earn € 1,000,000 with probability 0.1, or € 0 with probability 0.9

D : Earn € 200,000 with probability 0.9, or lose € 800,000 with probability 0.1

With utility defined as monetary payoff, the expected utility of the four acts is the same, which suggests that it may not provide the most satisfactory criterion for the selection of an act. ◇

Remark 7.4. It may occur that the SoN depends on several occurrences, so that it can be described by a vector $\mathbf{X}$ of RVs with pdf $f(\mathbf{x})$. In principle, one may define a utility function $u(\mathbf{x})$ and look for the decision which maximizes the expected utility as before. This multidimensional utility theory requires the generation of a function such that an alternative is preferred to another if and only if its expected utility is greater, where now expectations must be taken with respect to multivariate probability measures. In practice, the decision problem may become very complex unless simplifying assumptions are introduced, as, for example, the choice of a utility function which is a linear combination of the utilities associated with the individual components x_i

of $\mathbf{x}$, so that $u(\mathbf{x}) = \sum_i w_i u_i(x_i)$. The underlying theory, known under the name of Multiple-Criteria Decision Analysis, can be found described for example in [139], while recent developments are summarized in [103].

7.2 Epistemic vs. aleatory uncertainty

Before proceeding further, we pause for some additional considerations about the mathematical modeling of uncertainty. A key point here is that the mathematical tools used to deal with aleatory and epistemic uncertainty are not necessarily the same. For example, while probability theory is generally used to represent aleatory uncertainty, other theories may conveniently take its place whenever epistemic uncertainty affects the environment in which a decision has to be made.

7.2.1 The Principle of Insufficient Reason

This principle, also called *Principle of Indifference* [140, Chapter IV], states that "in the absence of any relevant evidence, agents should distribute their degrees of belief equally among all the possible States of Nature." In terms of probability, each of the two outcomes of a coin toss is assigned probability $1/2$. Similarly, if a random variable X is only known to take values in $[0, 1]$, the Principle of Insufficient Reason would assign to X a uniform pdf with support set $[0, 1]$ to express a complete lack of knowledge as to the value taken on by X. In J. M. Keynes' own words [140, p. 45],

> The Principle of Indifference asserts that if there is no known reason for predicating of our subject one rather than another of several alternatives, then relatively to such knowledge the assertions of each of these alternatives have an equal probability. Thus equal probabilities must be assigned to each of several arguments, if there is an absence of positive ground for assigning unequal ones.

The Principle of Insufficient Reason, like the principle of maximum entropy that was illustrated in Section 1.6, advocates a way of selecting a probability measure when the prior information about it is insufficient to do so. As criticized in [13, p. 21], when either principle is used, "the prior information is augmented by appeal to a 'Principle' whose validity (it is claimed) is prior to the information in question.[...] A distribution obtained in this manner is not a 'maximum ignorance' distribution since it depends on knowledge of the truth entailed in the Principle which was employed."

Example 7.6 ([90]). The application of the Principle of Insufficient Reason may lead to incongruous results. Take, for example, the information that the exchange rate between Euro and US Dollar is between €1 = \$1 and €1 = \$1.4. The Principle of Insufficient Reason would yield an average exchange rate €1 = \$1.2. An equivalent information is that the exchange rate between US Dollar and Euro is between \$1 = €0.71 and \$1 = €1, which yields the different average €1 = \$1.16. ◇

Example 7.7. This example, illustrated in [13], [14, p. 14], shows another potential pitfall in the application of the Principle of Insufficient Reason. You have to choose between two envelopes, both containing a positive amount of money. You know that one envelope contains twice as much money as the other. After examining the contents, say x, of the envelope you have chosen, you are given the option of keeping the envelope or exchanging it for the other one, which contains either $x/2$ or $2x$. If you use the Principle of Insufficient Reason and assign equal probabilities to these two alternatives, you have that the expected utility of accepting the exchange is $0.5 \times 2x + 0.5 \times x/2 = 1.25x$. Thus, after an envelope is chosen, whatever its content you can increase your reward by 25% if you exchange it for the other envelope (and this holds true even if the first envelope is not opened!).   ◇

7.2.2   Objective vs. subjective probabilities

EUT assumes that the DM can provide a measure of the likelihood of each SoN. This measure may consist of personal or subjective probabilities, yielding a subjective expected utility theory. The difference between two fundamental classes of views on the interpretation of probability is synthetically described by L. J. Savage [221, p. 3] as follows:

> *Objectivistic views hold that some repetitive events, such as tosses of a penny, prove to be in reasonably close agreement with the mathematical concept of independently repeated random events, all with the same probability. [The resulting value of probability] is to be obtained by observation of some repetitions of the event, and from no other source whatsoever.*

> *Personalistic views hold that probability measures the confidence that a particular individual has in the truth of a particular proposition, for example, the proposition that it will rain tomorrow. These views postulate that the individual concerned is in some ways "reasonable," but they do not deny the possibility that two reasonable individuals faced with the same evidence may have different degrees of confidence in the truth of the same proposition.*

What is especially relevant in our context is that objective probabilities can be used only with events that can be repeated (at least in principle), and hence in no way can probability serve as a measure the amount of belief (or of trust) to be put in a proposition. In J. M. Keynes words [140, p. 57],

> *Without compromising the objective character of relations of probability, we must nevertheless admit that there is little likelihood of our discovering a method of recognising particular probabilities, without any assistance whatever from intuition or direct judgment.*

Example 7.8. Consider again the case examined in Example 7.4. In the absence of an objective probability measure for the SoN, the vendor may replace the probability values in (7.1) with values expressing his subjective confidence in tomorrow's weather. A critical point here is that assignments like this must satisfy suitable "rationality" conditions to avoid the occurrence of a situation, known as *Dutch book* [260, Chapter 1], [102, Section 5.3], in which the preferred vendor choices, when combined, yield a worse result than nonpreferred choices. More about this in Example 7.14 *infra*.   ◇

7.2.3   Using interval probabilities

A tool that may be used in a situation of epistemic uncertainty is derived from the theory of "interval probabilities." This is based on the acceptance that the probability of some event may not be stated precisely, but is known to be included in an interval, and hence dealt with as described in Chapters 4 and 5.

Example 7.9 ([143, p. 133]). We assign the value $\mathbb{P}(A) = p$ if we assume that a reasonable agent is willing to pay € p for a ticket returning € 1 if event A occurs, and nothing if A does not occur. Interval probabilities play a role in this context as follows. Assume that the relevant information available to the agent is not sufficient to determine uniquely $\mathbb{P}(A)$. In this situation, one might define a *lower probability* of A as the highest value of p acceptable by the agent for betting on the occurrence of A, and an *upper probability* of A as the highest acceptable ticket price for betting against the occurrence of A. Thus, the agent can assume that $\mathbb{P}(A)$ lies in an interval of values bounded by lower and upper probabilities.   ◇

Although any theory of imprecise probabilities should exhibit the same properties of consistency and coherence as probability, it was observed in [83, p. 109] that "humans do not make decisions according to the theory of subjective expected utilities. One of the reasons is that a traditional probabilist must always be able to discern which of two events is more probable, unless they are equally probable, and which of two options is preferable, unless she (*sic*) is indifferent to the choice between them." The axiom that this is always possible is called *completeness* or *ordering postulate*. A more general theory of uncertainty accepts that one might not be able to compare every two probabilities: Keynes [140] was the first scholar to explicitly emphasize the importance of interval estimates in decision making [150]. Recalling Example 7.9, with imprecise probabilities "the definition of probability can be operationalized as the interval between the highest buying price and the lowest selling price" for a ticket returning € 1 if event A occurs, so that "an agent may elect to neither buy nor sell a gamble if the price is not sufficiently favorable" [83, p. 109]. Interval-valued probabilities were also advocated in [84].

While the use of interval probabilities may sound as a reasonable choice from a purely pragmatic point of view, its mathematical foundations are the object of a good deal of controversy. This is reviewed, for example, in [150], where two main motivations are provided for interval-valued probabilities. The first one argues that interval probabilities can "describe the states of individuals with regard to their degrees of belief more adequately and more realistically than standard subjective Bayesianism can." The second one reflects the idea that, because our degrees of belief ought to be objectively determined, "since our statistical knowledge is never precise, our beliefs are never constrained more precisely than by sets of statistical distributions."

7.3 Lotteries, prospects, and utility functions

An act in decisions under risk in which a finite number n of possible consequences $x_1, \ldots, x_n$ occur with known probabilities $p_1, \ldots, p_n$ is called a "lottery" (or a "bet"), and is denoted by $\mathbf{p} = (p_1, x_1; \ldots; p_n, x_n)$. A limiting case is the "sure" lottery $(1, \alpha)$, yielding outcome α with probability one.

Example 7.10. The carnival prize wheel of Fig. 7.1(a) is divided into 10 equal (and hence equally probable) sectors, each of which is labeled with a monetary amount or utility value: 4 sectors carry label 1, 3 sectors label 5, 2 sectors label 10, and 1 sector label 20. This corresponds to the lottery $\mathbf{p} = (0.4, 1; 0.3, 5; 0.2, 10; 0.1, 20)$. The expected utility of this lottery is

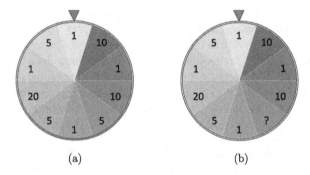

(a)   (b)

Figure 7.1 A carnival wheel.

$$U = 0.4 \cdot 1 + 0.3 \cdot 5 + 0.2 \cdot 10 + 0.1 \cdot 20 = 5.9, \tag{7.3}$$

which means that a rational player should accept to spin the wheel for a fee lower than 5.9 monetary amounts.   ◇

Example 7.11. Given two lotteries $\mathbf{p}$ and $\mathbf{q}$ with probabilities $p_1, \ldots, p_n$ and $q_1, \ldots, q_n$, respectively, and the same set of consequences, their convex combination is the "mixture lottery"

$$\lambda \mathbf{p} + (1 - \lambda)\mathbf{q} = \big(\lambda p_1 + (1 - \lambda q_1), x_1; \ldots; \lambda p_n + (1 - \lambda q_n), x_n\big), \tag{7.4}$$

where $\lambda \in (0, 1)$.   ◇

In a set $\mathcal{L}$ of lotteries we may define a preference relation, denoted $\succcurlyeq$. We say that lottery $\mathbf{p}$ is preferable to $\mathbf{q}$, and write $\mathbf{p} \succcurlyeq \mathbf{q}$, if the expected utility of $\mathbf{p}$ is no smaller than that of $\mathbf{q}$,

$$\sum_i p_i u(x_i) \geqslant \sum_i q_i u(x_i). \tag{7.5}$$

In a similar way, we write $\mathbf{p} \succ \mathbf{q}$ if $\mathbf{p}$ is strictly preferable to $\mathbf{q}$, i.e., (7.5) holds with strict inequality.

Consider the following four assumptions about preferences:

(i) *Completeness.* For any two lotteries $\mathbf{p}$ and $\mathbf{q}$, we may have $\mathbf{q} \succcurlyeq \mathbf{p}$, or $\mathbf{p} \succcurlyeq \mathbf{q}$, or both (when both lotteries are equally desirable).

(ii) *Transitivity.* For any three lotteries $\mathbf{p}$, $\mathbf{q}$, and $\mathbf{r}$, if $\mathbf{p} \succcurlyeq \mathbf{q}$ and $\mathbf{q} \succcurlyeq \mathbf{r}$ then $\mathbf{p} \succcurlyeq \mathbf{r}$.

(iii) *Independence.* For all lotteries $\mathbf{p}, \mathbf{q} \in \mathcal{L}$ such that $\mathbf{p} \succ \mathbf{q}$, and all $\mathbf{r} \in \mathcal{L}$ and $\lambda \in (0, 1)$, the following holds:

$$\lambda \mathbf{p} + (1 - \lambda)\mathbf{r} \succ \lambda \mathbf{q} + (1 - \lambda)\mathbf{r}. \tag{7.6}$$

That is, mixing two lotteries with one and the same lottery does not change the preference order [102, p. 288].

(iv) *Archimedean property.* For all lotteries $\mathbf{p}, \mathbf{q}, \mathbf{r}$ such that $\mathbf{p} \succ \mathbf{q} \succ \mathbf{r}$, there exist a $\lambda \in (0, 1)$ and a $\mu \in (0, 1)$ such that

$$\lambda \mathbf{p} + (1 - \lambda)\mathbf{r} \succ \mathbf{q} \succ \mu \mathbf{p} + (1 - \mu)\mathbf{r}. \tag{7.7}$$

This is a continuity property: no matter how the mixture with $\mathbf{r}$ could change the lottery $\mathbf{p}$, it is always possible to find λ, μ so that the original preference order among $\mathbf{p}$, $\mathbf{q}$, and $\mathbf{r}$ in (7.7) is unchanged [102, p. 288].

Remark 7.5. We shall see *infra* that, although it seems fairly reasonable, the independence assumption need be relaxed in some situations to prevent the occurrence of certain paradoxes.

A key result [102, Theorem 5.6, p. 288] states that, given a preference relation $\succcurlyeq$ satisfying properties (i) to (iv) above, a utility function $u(\cdot)$ exists such that the expected utility "represents" $\succcurlyeq$, that is, for any pair of lotteries $\mathbf{p} = (p_1, x_1; \ldots; p_n, x_n)$ and $\mathbf{p}' = (p_1', x_1'; \ldots; p_n', x_n')$,

$$\mathbf{p} \succcurlyeq \mathbf{p}' \text{ if and only if } \sum_{i=1}^{n} p_i u(x_i) \geq \sum_{i=1}^{n} p_i' u(x_i'). \tag{7.8}$$

Moreover, $u(\cdot)$ is unique up to any transformation which is positive affine, i.e., has the form $\alpha u(\cdot) + \beta$, $\alpha > 0$, $\beta \in \mathbb{R}$. This result justifies the association of the term "rational" with preferences satisfying properties (i)–(iv).

7.3.1 Prospects

When decisions under epistemic uncertainty are involved, the subjective probabilities to be associated with States of Nature may not be immediately apparent. If this is the case, one may be able to derive probabilities from preferences. A *prospect* is a map from SoNs to outcomes (or directly to utilities), and describes courses of action that depend on which SoN is true [260, Chapter 1]. The outcome events are assumed to be exhaustive and mutually exclusive, and hence partition the set of SoNs.

Example 7.12. In the situation described in Example 7.4, there are three States of Nature s_1, s_2, and s_3, corresponding to tomorrow's weather, and four prospects corresponding to the choice of merchandise that the vendor will take with him. With obvious notations, we write the corresponding prospects in the form

$$
\begin{aligned}
\pi_i &= (s_1 : 400, s_2 : 100, s_3 : -400), \\
\pi_h &= (s_1 : -400, s_2 : 100, s_3 : 400), \\
\pi_n &= (s_1 : 0, s_2 : 0, s_3 : 0), \\
\pi_b &= (s_1 : 0, s_2 : 200, s_3 : 0)
\end{aligned}
\tag{7.9}
$$

(notice that $\pi_b = \pi_i + \pi_h$).   ◇

Once a preference relation among prospects is determined, and we can assume that this is based on an expected-utility criterion, then a range of subjective probabilities can be derived from observed preferences by expressing the set of inequalities implied by the preference order for each pair of prospects. For a more accurate computation of subjective probabilities, the actual value of the expected utility should be added.

Example 7.13. Consider the two prospects $\pi_1 = (s_1 : 100, s_2 : 0, s_3 : 0)$ and $\pi_2 = (s_1 : 0, s_2 : 100, s_3 : 0)$. If π_1 is preferred to π_2, and hence the expected utilities satisfy the inequality $\mathbb{P}(s_1) \cdot 100 > \mathbb{P}(s_2) \cdot 100$, any subjective probability assignment over the SoNs satisfying $\mathbb{P}(s_1) > \mathbb{P}(s_2)$ along with $\mathbb{P}(s_1) + \mathbb{P}(s_2) + \mathbb{P}(s_3) = 1$ would provide an appropriate model. One may refine the assignment by observing, for example, that the expected utility of π_1 is twice that of π_2, or, even more precisely, that the utility of π_1 is 50, and that of π_2 is 25. In the latter case, $\mathbb{P}(s_1) \cdot 100 = 50$ and $\mathbb{P}(s_2) \cdot 100 = 25$ yield $\mathbb{P}(s_1) = 1/2$, $\mathbb{P}(s_2) = 1/4$, and hence $\mathbb{P}(s_3) = 1/4$.   ◇

If the decision-maker preferences are not based on expected utilities, then it may happen that the preferred choices give rise to a Dutch book as mentioned at in Example 7.8 *supra* (see [260, p. 28]).

Example 7.14. Consider the following prospects:

$$
\begin{aligned}
\pi_1 &= (s_1 : 0, s_2 : 100, s_3 : 100), \\
\pi_2 &= (s_1 : 100, s_2 : 0, s_3 : 100), \\
\pi_3 &= (s_1 : 100, s_2 : 100, s_3 : 0), \\
\pi_1^\star &= (s_1 : 300, s_2 : 0, s_3 : 0), \\
\pi_2^\star &= (s_1 : 0, s_2 : 300, s_3 : 0), \\
\pi_3^\star &= (s_1 : 0, s_2 : 0, s_3 : 300),
\end{aligned}
\tag{7.10}
$$

and a preference criterion that, between two prospects, favors the one yielding a positive gain for most of the three states of nature. Thus, any of the π_i is preferred to any of the $\pi_j^\star$. This cannot be described by expected value maximization. A Dutch book can be made against this decision maker. In fact, when the preferred prospects are combined, they yield less utility than the nonpreferred prospect for all SoNs:

$$\pi_1 + \pi_2 + \pi_3 = (s_1 : 200, s_2 : 200, s_3 : 200),$$
$$\pi_1^\star + \pi_2^\star + \pi_3^\star = (s_1 : 300, s_2 : 300, s_3 : 300).$$
(7.11)

$\diamond$

7.3.2 St. Petersburg paradox

Consider a game which consists of tossing repeatedly a fair coin. If the first "heads" comes up after n tosses, the agent wins € 2^n. How much would an agent be willing to pay to enter the game? By assuming that the utility of winning € x is x, the expected utility of the game is

$$\mathbb{E}[U] = \sum_{n=1}^{\infty} 2^{-n} \cdot 2^n = \infty$$
(7.12)

and hence a rational player should be willing to pay any finite sum of money to enter the game, even though with very high probability he will win a modest amount. This behavior of the "rational agent" sounds absurd, and gives rise to the so-called *St. Petersburg paradox*. In practice, most people would consider as intuitively fair to pay a finite and rather small amount to play this game [248, p. 50]. A key point stemming from this paradox is that a "linear utility" function may be undesirable (let alone the fact that such a lottery is unrealistic, as no vendor of lotteries would consider selling one whose expected payout is ∞ [161]). To solve this paradox, the Swiss mathematician Daniel Bernoulli [17] started from the observation that

> *The determination of the value of an item must not be based on the price, but rather on the utility it yields [...] There is no doubt that a gain of one thousand ducats is more significant to the pauper than to a rich man though both gain the same amount.*

and introduced the logarithmic utility function $u(x) = \log x$, which causes the expected utility (in Bernoulli's terminology, the "*emolumentum medium*") to be finite and fairly small:

$$\mathbb{E}[U] = \sum_{n=1}^{\infty} 2^{-n} \log \left(2^n \right) = \log 4.$$
(7.13)

This utility function is interpreted as representing a "diminishing marginal utility": the larger the amount of money gained, the lower the rate of increase in utility.

7.3.3 Risk-averse and risk-seeking decisions

Expected Utility Theory assumes implicitly that the preference order between lotteries **p** and **q** is not altered by positive affine transformations. To illustrate this point, consider a lottery **p** in which a fair coin is flipped and the agent receives €1 in any case, and another lottery **q** in which he receives €2 if the result is "heads," and nothing otherwise. This is again a decision under risk. For a rational agent who is *risk neutral*, i.e., whose decisions are based on a utility function equal to the amount of money received, the two lotteries are equivalent. It may also happen that the agent is *risk averse*, i.e., he assumes that a loss is more significant than an equivalent gain, a sure gain is preferable to a random gain, and a random loss is preferable to a sure loss. As risk-averse decisions favor certainty over uncertainty, even when the latter can generate a higher payoff, a risk-averse agent would favor **p** over **q**. The opposite may also occur: we say that the agent is *risk seeking* when **q** is preferred. For example [102, p. 300], risk aversion suggests that earning €1000 one day and then losing the same amount the day after makes an agent feel much worse than with no earning followed by no loss. On the contrary, people tend to be risk-seeking with losses. These risk attitudes can be modeled by a suitable choice of the utility function. Indeed, they can be described by observing the convexity of the utility function $u(\cdot)$. Fig. 7.2 shows the three basic shapes of $u(\cdot)$. Using Jensen inequality (1.44), we have that for a concave

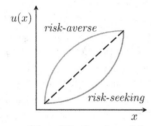

Figure 7.2 The three basic shapes of the utility function: risk-averse, risk-seeking, and neutral (dashed line).

∩ (risk-averse) utility function

$$u\left(\sum_i p_i x_i\right) \geqslant \sum_i p_i u(x_i) = \mathbb{E}[U] \tag{7.14}$$

while the inequality is reversed for convex ∪ (risk-seeking) $u(\cdot)$. A more complex attitude may be reflected by an S-shaped utility function as in Fig. 7.3. This function is concave ∩ in its positive part and convex ∪ for its negative part [133].

7.3.4 Reversing risk aversion: framing effect

Framing-Effect Theory describes predictable shifts of preference occurring when the same lottery is framed in different ways. People, when presented with a lottery, tend sometimes to change their attitude, avoiding risk when the bet is described in a positive frame, but seeking risk when it is described in a negative frame. Decision theory, as we have summarized it, assumes "description invariance," under which it is expected that equivalent formulations of a decision problem lead to the same choices. Contrary to

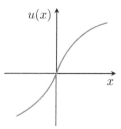

Figure 7.3 S-shaped utility function. The concave ∩ part to the right of the origin reflects risk aversion, while the convex ∪ part to the left reflects a risk-seeking attitude.

this assumption, experimental evidence shows that variations in the framing of options (e.g., in terms of losses rather than of gains) yield predictably different decisions. In fact, it was observed that "the frame that a decision-maker adopts is controlled partly by the formulation of the problem, and partly by the norms, habits, and personal characteristics of the decision-maker" [252, p. 453]. This is a very robust phenomenon, whose function is similar to that of optical illusions [248, p. 110].

Example 7.15. This example is illustrated in [248, p. 103] (other examples can be found in [248, Chapter 9]). A patient has a tumor discovered in his lungs, and a doctor recommends surgery to remove it. The decision the patient makes ("surgery" or "no surgery") is likely to depend strongly on the way the doctor presents the alternatives. One emphasizes survival: "In the past I have performed this surgery on 1,000 patients, and 950 of them survived for more than 5 years." The other emphasizes death: "In the past I have performed this surgery on 1,000 patients, and 50 of them died within 5 years." ◇

To explain choices that vary when their frames are changed, Daniel Kahneman and Amos Tversky [132,253] proposed *Prospect Theory*. This is based on a modified choice criterion whereby the lottery $(p_1, x_1; \ldots; p_n, x_n)$ yields expected utility

$$V \triangleq \sum_{i=1}^{n} \pi(p_i)v(x_i), \tag{7.15}$$

where two functions $\pi(\cdot)$ and $v(\cdot)$ are selected to model framing. The "value function" $v(\cdot)$ is S-shaped, asymmetrical, and passes through a reference point which is defined as the separation between risk-taking and risk-averse behavior, and shifts depending on how the decision problem is framed (see Fig. 7.4). The "decision weight" function $\pi(\cdot)$, which weights the probabilities, captures the concept that agents tend to overreact to rare events and underreact to frequent events (Fig. 7.5).

Example 7.16. This example, presented in [132], illustrates the change in reference point due to framing: The difference in value between a gain of €100 and a gain of €200 is perceived as greater than the difference between a gain of €1,100 and a gain of €1,200. Similarly, the difference between a loss of €100 and a loss of €200

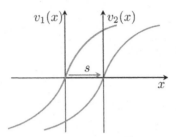

Figure 7.4 Framing effect on an S-shaped utility function: the reference point is shifted by an amount s [248, p. 116].

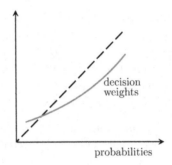

Figure 7.5 Decision weights describe subjective evaluation of probabilities.

appears greater than the difference between a loss of € 1,100 and a loss of € 1,200, unless the larger loss is intolerable.   ◇

Example 7.17. This example [132] justifies the nonlinearity of function $\pi(\cdot)$. The agent, playing Russian roulette, has the opportunity to purchase the removal of a single bullet from the loaded gun. How much would he be willing to pay to reduce the number of bullets? One would expect the agent to be willing to pay much more for a reduction of the probability of death from $1/6$ to zero than for a reduction from $4/6$ to $3/6$.   ◇

7.4   Other paradoxes arising from EUT

In this section we describe two additional paradoxes arising in expected utility theory. These paradoxes, which prove how EUT may not reflect in full the real processes of decision making, can be resolved using suitable extensions of the theory outlined *supra*. *Ellsberg paradox* concerns decision-making under aleatory and epistemic uncertainty, in which probabilities are replaced by subjective beliefs that are available about the States of Nature. *Allais paradox* concerns decision-making under risk, i.e., a situation in which the probability distribution of the states of nature is fully known.

7.4.1   Ellsberg paradox

An urn contains 30 red (r) balls and 60 other balls that are either black (b) or yellow (y) in an unknown proportion. A ball is drawn at random from the urn, and the following bets are proposed [102, §5.3.3], [248, §3.2]:

Bet A: The agent receives €1 if the ball is *red*, €0 otherwise
Bet B: The agent receives €1 if the ball is *black*, €0 otherwise

The following bets are also proposed:

Bet C: The agent receives €1 if the ball is *red or yellow*, €0 otherwise
Bet D: The agent receives €1 if the ball is *black or yellow*, €0 otherwise

This situation reflects both aleatory and epistemic uncertainty, as the percentage of black and yellow balls is not specified. EUT dictates that the agent will prefer Bet A over Bet B if he believes that a red ball is more likely to be drawn than a black ball. Similarly, he will prefer Bet C over Bet D if he believes that a red or yellow ball is more likely to be drawn that a black or yellow ball. Thus, if the agent prefers A to B, he will also prefer C to D (and vice versa). The paradox lies in the fact that, when surveyed, most people prefer A to B (because they are sure that $1/3$ of the balls is red, while there is no certainty about the presence of black balls), but D to C (because $2/3$ of the balls are black or yellow, while there is only certainty that $1/3$ is red or yellow).

Denote by u_1 and u_0 the utilities associated with obtaining €1 and €0, respectively, and assume $u_1 > u_0$ (the agent strictly prefers €1 to nothing) and a subjective probability measure $\mathbb{P}$. Additivity implies that $\mathbb{P}[b \text{ or } y] = \mathbb{P}[b] + \mathbb{P}[y]$ and $\mathbb{P}[r \text{ or } y] = \mathbb{P}[r] + \mathbb{P}[y]$. EUT indicates that the agent prefers A to B if

$$
\begin{aligned}
& u_1 \mathbb{P}[r] + u_0 \big(\mathbb{P}[b] + \mathbb{P}[y]\big) > u_1 \mathbb{P}[b] + u_0 \big(\mathbb{P}[r] + \mathbb{P}[y]\big) \\
& \Leftrightarrow (u_1 - u_0)\,(\mathbb{P}[r] - \mathbb{P}[b]) > 0 \\
& \Leftrightarrow \mathbb{P}[r] > \mathbb{P}[b].
\end{aligned}
\tag{7.16}
$$

Similar calculations show that the agent is expected to prefer D to C if $\mathbb{P}[b] > \mathbb{P}[r]$, hence the contradiction leading to Ellsberg paradox. EUT cannot explain Ellsberg paradox, irrespective of how the subjective probabilities are determined. A lesson that may be learned from this paradox is that most people are not basing their choices on a correct probabilistic analysis, and hence, if we want a mathematical model of their behavior, we may want to abandon one of the axioms of probability theory, in particular additivity [102, p. 309 ff.].

7.4.2   Allais paradox

Let X be a discrete random variable with support set $\mathcal{S} = \{1, 2, \ldots, 100\}$ and values taken with equal probabilities $1/100$. A bet is based on the realization x of X. Consider the following bets [102, pp. 292 ff.]

Bet A: The agent receives €3k for any $x \in \mathcal{S}$
Bet B: The agent receives €4k if $x \leqslant 80$, and nothing otherwise.

The following bets are also proposed:

Bet C: The agent receives €3k if $x \leqslant 5$, and nothing otherwise.
Bet D: The agent receives €4k if $x \leqslant 4$, and nothing otherwise.

Most people prefer A over B, because they prefer a sure gain over a risky bet. Also, more people prefer D over C because the probability of winning €4k with D is about the same as the probability of winning €3k with C. In terms of lotteries, the preferences are

$$(1, €3k) \succ (0.8, €4k; 0.2, €0), \tag{7.17a}$$

$$(0.04, €4k; 0.96, €0) \succ (0.05, €3k; 0.95, €0). \tag{7.17b}$$

Now, if the agent utilities are chosen so that $U(\text{A}) > U(\text{B})$, then we must have

$$\begin{aligned}
U(\text{A}) > U(\text{B}) &\Leftrightarrow u(3k) > 0.2\,u(0) + 0.8\,u(4k) \\
&\Leftrightarrow 0.05\,u(3k) > 0.01\,u(0) + 0.04\,u(4k) \\
&\Leftrightarrow 0.05\,u(3k) + 0.95\,u(0) > 0.96\,u(0) + 0.04\,u(4k) \\
&\Leftrightarrow U(\text{C}) > U(\text{D}).
\end{aligned} \tag{7.18}$$

Hence, an agent choosing A over B should also choose C over D. It was found that most people violate EUT in this experiment, hence Allais paradox.

Remark 7.6. Both Allais and Ellsberg paradoxes may be interpreted as being caused by the empirical invalidity of the independence axiom of EUT, which would call for a revision of this theory [248, Chapter 7] as described in Section 7.9 *infra*.

7.4.3   Entropy-modified expected utility of gambling

It has been recognized that standard EUT does not accommodate the existence of a specific utility of gambling per se. In fact, in EUT the two lotteries $(1, x)$ (pure consequence) and $(0.5, x; 0.5, x)$ (risky action) are categorized as indifferent. Now, it has been suggested that the eagerness or reluctance to bet may be modeled by adding to the utility term an entropy term depending only on the probabilities involved in the lottery [160,161]. The new utility theory combines additively the expected utility

$$\mathbb{E}[U] = \sum_i p_i u(x_i) \tag{7.19}$$

with a "utility of gambling" term given by Shannon entropy

$$H \triangleq -\sum_i p_i \log p_i \tag{7.20}$$

resulting in $A \triangleq \mathbb{E}[U] + \lambda H$. Here the sign of λ indicates whether the utility of gambling is positive or negative, while its magnitude determines the relevance of the utility of gambling (see [160–162] for further details and generalizations).

Allais paradox can be revisited in the light of entropy-modified expected utility. The pair of preferences on which the paradox is based correspond to different probability distributions, so that the entropy term is not the same for both. Choosing a sufficiently large negative value for λ, which corresponds to strong aversion for gambling, makes the two agent choices compatible [160,162] (see also [160, §5.2] for a discussion of Ellsberg paradox based on the version of expected utility induced by subjective probabilities).

7.5 Upper and lower probabilities

In this section we examine how interval probabilities can be formally defined through the use of upper and lower probability measures. Let $\mathcal{P}$ denote a given subset of all probability measures $\mathbb{P}$ on $\mathcal{S}$. This subset, usually assumed to be convex and often referred to as a *credal set*, describes the uncertainty about the probability measure to be used. Define the two set functions $\overline{\mathbb{P}}$ and $\underline{\mathbb{P}}$ as

$$\overline{\mathbb{P}}(A) \triangleq \sup_{\mathbb{P} \in \mathcal{P}} [\mathbb{P}(A)], \tag{7.21a}$$

$$\underline{\mathbb{P}}(A) \triangleq \inf_{\mathbb{P} \in \mathcal{P}} [\mathbb{P}(A)] \tag{7.21b}$$

for $A \in 2^{\mathcal{S}}$. These are called *upper* and *lower* probability measures, respectively, and are *conjugate*, i.e.,

$$\underline{\mathbb{P}}(A) = 1 - \overline{\mathbb{P}}(A^c) \tag{7.22}$$

for all events A.

The following equalities follow immediately from (7.21) (see [143, p. 112 ff.], [186, p. 79 ff.]):

$$\underline{\mathbb{P}}(A) \leqslant \overline{\mathbb{P}}(A), \qquad \text{for all } A \in 2^{\mathcal{S}}, \tag{7.23a}$$

$$\underline{\mathbb{P}}(\emptyset) = \overline{\mathbb{P}}(\emptyset) = 0, \tag{7.23b}$$

$$\underline{\mathbb{P}}(\mathcal{S}) = \overline{\mathbb{P}}(\mathcal{S}) = 1. \tag{7.23c}$$

It should be immediately apparent that the extreme case of full epistemic ignorance yields $[\underline{\mathbb{P}}(A), \overline{\mathbb{P}}(A)] = [0, 1]$, while, if no epistemic uncertainty is present, $\mathcal{P} = \{\mathbb{P}\}$, a known probability measure.

Example 7.18. A flexible, easy-to-work-with, credal set $\mathcal{P}$ capable of modeling uncertainty is provided by the "ε-contamination" class. If we assume that the true probability measure is close to $\mathbb{P}_0$, then we may choose a credal set of the form

$$\mathcal{P} \triangleq \{\mathbb{P} \mid \mathbb{P} = (1 - \varepsilon)\mathbb{P}_0 + \varepsilon\mathbb{P}_1, \mathbb{P}_1 \in \mathcal{Q}\}, \tag{7.24}$$

where $\mathbb{P}_0$ is fixed, $\mathcal{Q}$ is the family of probability measures that describe possible contaminations of $\mathbb{P}_0$, and $\varepsilon \in \mathbb{I}$ reflects how close $\mathbb{P}$ is assumed to be to $\mathbb{P}_0$, and hence the degree of uncertainty to be placed on $\mathbb{P}_0$ (stated otherwise, the DM is only $(1-\varepsilon)\%$ confident that $\mathbb{P}_0$ is the correct probability measure). A possible use of this class is to model a communication channel which is affected by an interference for a fraction ε of the time [195, p. 125]. With this choice, we have

$$
\begin{aligned}
\underline{\mathbb{P}}(A) &= (1-\varepsilon)\mathbb{P}_0(A), \\
\overline{\mathbb{P}}(A) &= (1-\varepsilon)\mathbb{P}_0(A) + \varepsilon.
\end{aligned}
\tag{7.25}
$$

$\diamond$

Example 7.19. If we choose $\mathcal{P}$ to be the set of probability measures generating a given sequence of moments, upper and lower probabilities are generated by moment bounds as described in Chapter 3. $\diamond$

Due to (7.23a), we may define, for every $A \in 2^{\mathcal{S}}$, the closed interval

$$
\left[\underline{\mathbb{P}}(A), \overline{\mathbb{P}}(A)\right],
\tag{7.26}
$$

which can be dealt with by using the tools of interval analysis described in Chapter 4. In particular, if $\underline{\mathbb{P}}(A) = \overline{\mathbb{P}}(A)$ we are reduced to standard (i.e., exact) probabilities.

7.5.1 Conditional upper and lower probabilities

Conditional upper and lower probabilities can be defined as follows [269, §3]. Assume a probability measure $\mathbb{P}$ with support set $\mathcal{S}$, and a partition $\mathcal{B} = \{B_1, \ldots, B_m\}$ of $\mathcal{S}$ with $B_i \neq \mathcal{S}$, $B_i \neq \emptyset$, $B_i \cap B_k = \emptyset$ for $i \neq k$, and $\bigcup_{i=1}^m B_i = \mathcal{S}$. For any $B \in \mathcal{B}$ and $A \subset \mathcal{S}$, we define the conditional interval probabilities $\mathbb{P}(A \mid B) \triangleq [\underline{\mathbb{P}}(A \mid B), \overline{\mathbb{P}}(A \mid B)]$ with

$$
\begin{aligned}
\underline{\mathbb{P}}(A \mid B) &\triangleq \inf \frac{\mathbb{P}(A \cap B)}{\mathbb{P}(B)}, \\
\overline{\mathbb{P}}(A \mid B) &\triangleq \sup \frac{\mathbb{P}(A \cap B)}{\mathbb{P}(B)} = 1 - \inf \frac{\mathbb{P}(A^c \cap B)}{\mathbb{P}(B)},
\end{aligned}
\tag{7.27}
$$

where inf and sup are evaluated with respect to all $\mathbb{P} \in \mathcal{P}$ assigning a nonzero probability to B.

Example 7.20. Consider the set $\mathcal{S} = A_1 \cup A_2 \cup A_3$ and its partition $\mathcal{B} = \{B_1, B_2\}$, with $B_1 = A_1 \cup A_2$ and $B_2 = A_3$. Assuming the following interval probabilities:

$$
\mathbb{P}(A_1) = [0.1, 0.25], \qquad \mathbb{P}(A_2) = [0.2, 0.4], \qquad \mathbb{P}(A_3) = [0.4, 0.6],
$$
$$
\mathbb{P}(A_1 \cup A_2) = [0.4, 0.6], \qquad \mathbb{P}(A_1 \cup A_3) = [0.6, 0.8], \qquad \mathbb{P}(A_2 \cup A_3) = [0.75, 0.9],
$$

we obtain

$$
\begin{aligned}
\mathbb{P}(A_1 \mid B_1) &= \frac{\mathbb{P}(A_1 \cap B_1)}{\mathbb{P}(A_1 \cap B_1) + \mathbb{P}(A_2 \cap B_1)} \\
&= \frac{\mathbb{P}(A_1)}{\mathbb{P}(A_1) + \mathbb{P}(A_2)} \\
&= \frac{[0.1, 0.25]}{[0.1, 0.25] + [0.2, 0.4]} \\
&= [0.2, 0.\overline{5}]
\end{aligned}
\tag{7.28}
$$

and

$$
\begin{aligned}
\mathbb{P}(A_2 \mid B_1) &= \frac{\mathbb{P}(A_2 \cap B_1)}{\mathbb{P}(A_1 \cap B_1) + \mathbb{P}(A_2 \cap B_1)} \\
&= \frac{\mathbb{P}(A_2)}{\mathbb{P}(A_1) + \mathbb{P}(A_2)} \\
&= \frac{[0.2, 0.4]}{[0.1, 0.25] + [0.2, 0.4]} \\
&= [0.\overline{4}, 0.8].
\end{aligned}
\tag{7.29}
$$

Moreover,

$$
\mathbb{P}(A_3 \mid B_1) = [0].
\tag{7.30}
$$

(The final result in (7.28) and (7.29) was computed as in Example 4.13 to avoid interval dependence.)   ◇

Remark 7.7. Further details on definition (7.27), as well as some potential difficulties with it, are discussed in [269].

7.5.2 Inference using upper and lower probabilities

An interval-probability version of Bayes theorem can be derived in a straightforward way from definitions (7.27). We first illustrate this with an example adapted from [269, Example 3.6].

Example 7.21. In cognitive radio with spectrum sensing, let $\mathbb{P}(S) = [0.2, 0.4]$ be the a priori probability that in a given bandwidth interval an information signal be present, and $\mathbb{P}(S^c) = [0.6, 0.8]$ the probability that it be absent. A sensing operation yields the following conditional probabilities of actual presence or absence of the signal, denoted "$+$" and "$-$", respectively:

$$
\mathbb{P}(+ \mid S) = [0.6, 0.8], \qquad \mathbb{P}(- \mid S) = [0.2, 0.4],
$$
$$
\mathbb{P}(+ \mid S^c) = [0.2, 0.3], \qquad \mathbb{P}(- \mid S^c) = [0.7, 0.8].
$$

The a posteriori probabilities are obtained as follows:

$$
\begin{aligned}
\mathbb{P}(S \mid +) &= \frac{\mathbb{P}(+ \mid S)\mathbb{P}(S)}{\mathbb{P}(+ \mid S)\mathbb{P}(S) + \mathbb{P}(+ \mid S^c)\mathbb{P}(S^c)} \\
&= \frac{[0.6, 0.8] \cdot [0.2, 0.4]}{[0.6, 0.8] \cdot [0.2, 0.4] + [0.2, 0.3] \cdot [0.6, 0.8]} \\
&\approx [0.33, 0.73]
\end{aligned}
$$

and, proceeding similarly,

$$
\mathbb{P}(S \mid -) \approx [0.06, 0.28], \quad \mathbb{P}(S^c \mid +) \approx [0.27, 0.67], \quad \mathbb{P}(S^c \mid -) \approx [0.72, 0.94].
$$

$\diamond$

Consider a hypothesis H associated with evidence E. To decide upon H, we want to determine $\mathbb{P}(H)$ based on the knowledge of the conditional probabilities $\mathbb{P}(H \mid E)$ and $\mathbb{P}(H \mid E^c)$. From the "Theorem of Total Probability," we have

$$
\mathbb{P}(H) = \mathbb{P}\big[(H \cap E) \cup (H \cap E^c)\big] \tag{7.31}
$$

or, if the intersection of $H \cap E$ and $H \cap E^c$ is the empty set,

$$
\mathbb{P}(H) = \mathbb{P}(H \mid E)\mathbb{P}(E) + \mathbb{P}(H \mid E^c)\mathbb{P}(E^c) \tag{7.32}
$$

and, equivalently,

$$
\mathbb{P}(H) = \mathbb{P}(H \mid E)\mathbb{P}(E) + \mathbb{P}(H \mid E^c)\big(1 - \mathbb{P}(E)\big). \tag{7.33}
$$

Fig. 7.6 illustrates the possible relationships between E and H. Specifically, when E is *necessary* for H (Fig. 7.6(a)), we have

$$
\mathbb{P}(H \mid E) \subseteq [1, 1], \qquad \mathbb{P}(H \mid E^c) = [0, 0]. \tag{7.34}
$$

When E is *sufficient* for H (Fig. 7.6(b)), we have

$$
\mathbb{P}(H \mid E) = [1, 1], \qquad \mathbb{P}(H \mid E^c) \subseteq [1, 1]. \tag{7.35}
$$

When E is *necessary and sufficient* for H, we have

$$
\mathbb{P}(H \mid E) = [1, 1], \qquad \mathbb{P}(H \mid E^c) = [0, 0]. \tag{7.36}
$$

A weaker condition is that of E being *relevant* to H (Fig. 7.6(c)), which yields

$$
[0, 0] \subset \mathbb{P}(H \mid E) \subseteq [1, 1], \qquad [0, 0] \subseteq \mathbb{P}(H \mid E^c) \subset [1, 1]. \tag{7.37}
$$

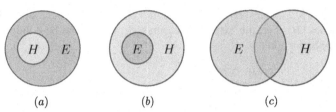

Figure 7.6 (a) E necessary for H; (b) E sufficient for H; (c) E relevant to H (adapted from [108, p. 253]).

It can be proved [108, p. 251 ff.] that

$$\underline{\mathbb{P}}(H) = \begin{cases} \underline{\mathbb{P}}(H \mid E)\,\underline{\mathbb{P}}(E) + \underline{\mathbb{P}}(H \mid E^c)\left(1 - \underline{\mathbb{P}}(E)\right), & \text{if } \underline{\mathbb{P}}(H \mid E) \geqslant \underline{\mathbb{P}}(H \mid E^c), \\ \underline{\mathbb{P}}(H \mid E)\,\overline{\mathbb{P}}(E) + \underline{\mathbb{P}}(H \mid E^c)\left(1 - \overline{\mathbb{P}}(E)\right), & \text{otherwise} \end{cases}$$

(7.38)

and

$$\overline{\mathbb{P}}(H) = \begin{cases} \overline{\mathbb{P}}(H \mid E)\,\overline{\mathbb{P}}(E) + \overline{\mathbb{P}}(H \mid E^c)\left(1 - \overline{\mathbb{P}}(E)\right), & \text{if } \overline{\mathbb{P}}(H \mid E) \geqslant \overline{\mathbb{P}}(H \mid E^c), \\ \overline{\mathbb{P}}(H \mid E)\,\underline{\mathbb{P}}(E) + \overline{\mathbb{P}}(H \mid E^c)\left(1 - \underline{\mathbb{P}}(E)\right), & \text{otherwise.} \end{cases}$$

(7.39)

Remark 7.8. The above can be generalized to a situation where there is more than one item of evidence [108, p. 253 ff.].

7.5.3 The dilation phenomenon

A feature of conditional probability intervals is that they may become uniformly wider after conditioning. Specifically, there may exist an event A and a partition $\mathcal{B} = \{B_1, \ldots, B_m\}$ such that, for all $B \in \mathcal{B}$,

$$\underline{\mathbb{P}}(A \mid B) < \underline{\mathbb{P}}(A) \leqslant \overline{\mathbb{P}}(A) < \overline{\mathbb{P}}(A \mid B) \tag{7.40}$$

so that the interval $[\underline{\mathbb{P}}(A), \overline{\mathbb{P}}(A)]$ is strictly contained in the interval

$$[\underline{\mathbb{P}}(A \mid B), \overline{\mathbb{P}}(A \mid B)].$$

This phenomenon is called *dilation* [228].

Example 7.22. Consider flipping twice a fair coin with nonindependent flips, and denote by H_i, T_i, $i = 1, 2$, the events "heads" and "tails" in flip i. Lower and upper probabilities $\underline{\mathbb{P}}$ and $\overline{\mathbb{P}}$ are modeled as in (7.21) by choosing as $\mathcal{P}$ the set of all probability measures such that $\mathbb{P}(H_1) = \mathbb{P}(H_2) = 1/2$, while the degree of dependence is unspecified. Thus, the two extreme measures in $\mathcal{P}$ are characterized by

$\mathbb{P}_1(H_1 \cap H_2) = 1/2$ and $\mathbb{P}_2(H_1 \cap H_2) = 0$. After observing the outcome of flip 1, represented by set $\{H_1, T_1\}$, the extreme measures $\mathbb{P}_1$ and $\mathbb{P}_2$ can be conditioned using Bayes rule, which yields

$$\mathbb{P}_1(H_2 \mid H_1) = \frac{\mathbb{P}_1(H_1 \cap H_2)}{\mathbb{P}_1(H_1)} = 1,$$

$$\mathbb{P}_2(H_2 \mid H_1) = \frac{\mathbb{P}_2(H_1 \cap H_2)}{\mathbb{P}_2(H_1)} = 0. \tag{7.41}$$

Thus, no matter what the outcome of the first flip was, we have $\mathbb{P}(H_2) = 1/2$, but [228, p. 1140], [262, p. 298]

$$0 = \underline{\mathbb{P}}(H_2 \mid H_1) < \underline{\mathbb{P}}(H_2) = 1/2 = \overline{\mathbb{P}}(H_2) < \overline{\mathbb{P}}(H_2 \mid H_1) = 1,$$

$$0 = \underline{\mathbb{P}}(H_2 \mid T_1) < \underline{\mathbb{P}}(H_2) = 1/2 = \overline{\mathbb{P}}(H_2) < \overline{\mathbb{P}}(H_2 \mid T_1) = 1. \tag{7.42}$$

To interpret this, observe that the probability measures $\mathbb{P}_1$ and $\mathbb{P}_2$ correspond to the outcome of the second flip being completely determined by the outcome of the first: it is either identical (under $\mathbb{P}_1$) or opposite (under $\mathbb{P}_2$). The effect of observing the first flip is the introduction of indeterminacy about the outcome of the second. $\diamond$

Example 7.23. Assume that $\mathcal{P}$ is a ε-contamination class as in Example 7.18. Then

$$\underline{\mathbb{P}}(A \mid B) = \frac{(1 - \varepsilon)\mathbb{P}_0(A \cap B)}{(1 - \varepsilon)\mathbb{P}_0(B) + \varepsilon}. \tag{7.43}$$

Now, if $\mathbb{P}_0(A \cap B) = \mathbb{P}_0(A)\mathbb{P}_0(B)$, with $\varepsilon \neq 0$ we have [228, p. 1144]

$$\begin{aligned}
\underline{\mathbb{P}}(A \mid B) &= (1 - \varepsilon)\mathbb{P}_0(A)\frac{\mathbb{P}_0(B)}{(1 - \varepsilon)\mathbb{P}_0(B) + \varepsilon} \\
&< (1 - \varepsilon)\mathbb{P}_0(A) \\
&= \underline{\mathbb{P}}(A).
\end{aligned} \tag{7.44}$$

$\diamond$

7.6 Expected utility with interval probabilities

The extension of EUT to a case in which the decision-maker knowledge is described using interval probabilities requires a generalization of the concept of expected utility. Let $U(a, S)$ denote the RV expressing the utility of act a when $S \in \mathcal{S}$ is the State of Nature. In general, the DM should choose the act a which yields a maximum of the expected utility $\mathbb{E}[U(a, S)]$, where the expectation is taken with respect to the interval-probability measure over $\mathcal{S}$. A seemingly natural way of extending expectations to probabilities defined over intervals is through interval-valued expectations of the form [7]

$$\mathbb{E}[A] \triangleq \left[\underline{\mathbb{E}}[A], \overline{\mathbb{E}}[A] \right], \tag{7.45}$$

where, consistently with (7.21),

$$\overline{\mathbb{E}}[A] \triangleq \sup_{\mathbb{P} \in \mathcal{P}} [\mathbb{E}(A)], \tag{7.46a}$$

$$\underline{\mathbb{E}}[A] \triangleq \inf_{\mathbb{P} \in \mathcal{P}} [\mathbb{E}(A)], \tag{7.46b}$$

where $\mathcal{P}$ is the family of probability measures $\mathbb{P}$ such that, for all A, $\mathbb{P}(A) \geqslant \underline{\mathbb{P}}(A)$, or, equivalently, $\mathbb{P}(A) \leqslant \overline{\mathbb{P}}(A)$ (recall that $\underline{\mathbb{P}}$ and $\overline{\mathbb{P}}$ are conjugate, i.e., equality (7.22) holds).

Now, given that the generalized expected utility is represented by an interval rather than by a single number, a preference criterion need be defined, yielding a complete order over the set of acts taken by the decision maker. If the expected-utility intervals pertaining to two different acts are disjoint, the preference between the two comes in a natural way, otherwise a possible procedure consists of choosing a real number as a representative of each interval, as examined in Section 4.8. Several simple choices are available, connected with the DM attitude towards epistemic uncertainty. One may accept, for example, to represent $\left[\underline{\mathbb{E}}[U(a, S)], \overline{\mathbb{E}}[U(a, S)]\right]$ using $\underline{\mathbb{E}}[U(a, S)]$, and hence maximize the minimum expected utility ("maximin," or "best of the worst," criterion). In general, the influence of $\underline{\mathbb{E}}$ is higher when the DM is more uncertainty-averse, and hence the choice of $\underline{\mathbb{E}}$ as a representative of the interval corresponds to the limiting case of uncertainty aversion. A more general and flexible choice is obtained when $[\underline{\mathbb{E}}, \overline{\mathbb{E}}]$ is represented by $\eta \underline{\mathbb{E}} + (1 - \eta)\overline{\mathbb{E}}$, where $\eta \in [0, 1]$ is a parameter expressing the caution of the DM. The choice of a value of η close to 0 indicates extreme optimism and risk-prone behavior, while η close to 1 suggests radical pessimism and risk-averse behavior.

A special case of the above yields the *Hurwicz Criterion*, also known as the *optimism–pessimism approach*. Here the DM, having complete ignorance of the SoN, takes simultaneously into account only two utilities, namely the best and worst (see [121] and, for recent developments, [91,122]).

Remark 7.9. If the set $\mathcal{S}$ of SoNs is finite, so that $\mathcal{S} = \{s_1, \ldots, s_n\}$, a probability measure $\mathbb{P}$ on it is identified by a vector $(\mathbb{P}(s_1), \ldots, \mathbb{P}(s_n))$, and every interval-probability set $\mathcal{P}$ is isomorphic to a convex polyhedron [7, p. 11]. In this situation, linear programming can be used to optimize the expected utility.

7.7 Some applications to digital communication

Consider binary communication over a stationary noisy channel. One of two signals x_1 and x_2 is transmitted over the channel, whose behavior is fully described by the conditional pdfs $f_{Z|x_1}(z \mid x_1)$ and $f_{Z|x_2}(z \mid x_2)$ of the transitions between the transmitted signal and the realization z of the random observed signal Z. The combination of transmitted signal and channel form the State of Nature (see Fig. 7.7). The choice on which signal was transmitted is based on the observation of the value taken by Z, and generally results into two possible "terminal" decisions, viz., x_1 was transmitted

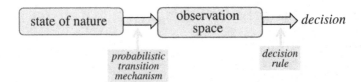

Figure 7.7 A scheme of decision theory in digital communication.

(denoted by $\hat{x}_1$) or x_2 was transmitted (denoted by $\hat{x}_2$), and a "nonterminal" decision (denoted by $\hat{x}_?$), which may entail, for example, a request for retransmission. Decisions are deterministic, and are based on the partition of the observation space into three regions, each corresponding to a specific decision. To simplify our analysis, let us assume than only terminal decisions are made, and define, instead of utilities, the *costs* $c(i \mid j)$ incurred if $\hat{x}_i$ is chosen when x_j was actually transmitted. One simple cost function assumes that any correct decision is costless, while all wrong decisions are equally costly, so that

$$c(i \mid j) = 1 - \delta_{i,j}, \tag{7.47}$$

where $\delta_{i,j}$ denotes Kronecker delta. The expected cost is given by

$$\mathbb{E}[C] = \sum_{j=1}^{2} \sum_{i=1}^{2} \mathbb{P}(\hat{x}_i, x_j) c(i \mid j)$$

$$= \sum_{j=1}^{2} \mathbb{P}(x_j) \sum_{i=1}^{2} \mathbb{P}(\hat{x}_i \mid x_j) c(i \mid j), \tag{7.48}$$

where $\mathbb{P}(\hat{x}_i \mid x_j)$ denotes the probability of choosing $\hat{x}_i$ when x_j was transmitted. If the costs $c(i \mid j)$ are as in (7.47), then

$$\mathbb{E}[C] = \sum_{j=1}^{2} \mathbb{P}(x_j) \sum_{i \neq j} \mathbb{P}(\hat{x}_i \mid x_j) \tag{7.49}$$

is the probability of an incorrect decision, or "average error probability," which we denote by p.

To be able to compute p, we need to know the a priori probabilities $\mathbb{P}(x_i)$, $i = 1, 2$, and to describe the decision (or "demodulation") rule, i.e., the function mapping the conditional densities $f_{Z \mid x_1}(z \mid x_1)$, $f_{Z \mid x_2}(z \mid x_2)$ to $\mathbb{P}(\hat{x}_i \mid x_j)$, $i = 1, 2$. This operation can be interpreted geometrically (Fig. 7.8) by using the partition of the set of observed signals into regions $\mathcal{Z}_1$ and $\mathcal{Z}_2$ associated with terminal decisions $\hat{x}_i$, $i = 1, 2$, so that

$$\mathbb{P}(\hat{x}_i \mid x_j) = \mathbb{P}(Z \in \mathcal{Z}_i \mid x_j), \qquad i \in \{1, 2\}, \ j \in \{1, 2\}. \tag{7.50}$$

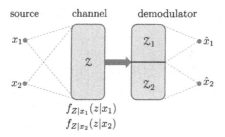

Figure 7.8 Decision procedure for binary communication.

7.7.1 Optimum decisions: Bayes criterion

In a more general situation where there are M transmitted signals and N possible receiver decisions, the expected cost (also called "Bayes risk") is

$$\mathbb{E}[C] = \sum_{j=1}^{M} \mathbb{P}(x_j) \sum_{i=1}^{N} \mathbb{P}(\hat{x}_i \mid x_j) c(i \mid j). \tag{7.51}$$

Again, $\mathbb{P}(\hat{x}_i \mid x_j)$ depends on a function $d(\hat{x}_i \mid z)$ associating a decision with every point in the observation space $\mathcal{Z}$ with partition $\{\mathcal{Z}_1, \ldots, \mathcal{Z}_N\}$, i.e., taking values

$$d(\hat{x}_i \mid z) \triangleq \begin{cases} 1, & z \in \mathcal{Z}_i, \\ 0, & \text{otherwise.} \end{cases} \tag{7.52}$$

We want to minimize $\mathbb{E}[C]$ by seeking the "decision function" $d(\hat{x}_i \mid z)$ so that the expected cost is at a minimum. Observe that, with definition (7.52),

$$\begin{aligned} \mathbb{P}(\hat{x}_i \mid x_j) &= \int_{\mathcal{Z}_i} f_{Z|x_j}(z \mid x_j)\, dz \\ &= \int_{\mathcal{Z}} d(\hat{x}_i \mid z) f_{Z|x_j}(z \mid x_j)\, dz. \end{aligned} \tag{7.53}$$

If $f_Z(z)$ denotes the marginal density of the RV Z,

$$f_Z(z) = \sum_{j=1}^{M} \mathbb{P}(x_j) f_{Z|x_j}(z \mid x_j), \tag{7.54}$$

then the a posteriori probability of x_j given the observation z is, using Bayes rule,

$$\mathbb{P}(x_j \mid z) = \frac{\mathbb{P}(x_j) f_{Z|x_j}(z \mid x_j)}{f_Z(z)}. \tag{7.55}$$

Using (7.51)–(7.53) and (7.55), we may rewrite the expression of the expected cost as

$$\mathbb{E}[C] = \int_{\mathcal{Z}} f_Z(z) \left[\sum_{i=1}^{N} d(\hat{x}_i \mid z) \rho(x_i \mid z) \right] dz, \tag{7.56}$$

where

$$\rho(x_i \mid z) \triangleq \sum_{j=1}^{M} \mathbb{P}(x_j \mid z) c(i \mid j) \tag{7.57}$$

is the conditional (a posteriori) risk, i.e., the average cost of choosing x_i when z is observed. Assuming that $f_Z(z) > 0$ for all $z \in \mathcal{Z}$ (otherwise there would be values z that are never observed), to minimize the risk we minimize the a posteriori risk for all z, which can be obtained using the *Bayes decision* rule

$$z \in \mathcal{Z}_i \quad \text{if} \quad \rho(x_i \mid z) = \min_j \rho(x_j \mid z) \tag{7.58}$$

(should the minimum be achieved by more than one x_j, the choice may be arbitrary among these values).

Example 7.24. Assume $M = N = 2$ and the cost function (7.47), so that the expected cost is error probability. Then we have

$$\rho(x_1 \mid z) = \mathbb{P}(x_2 \mid z) \quad \text{and} \quad \rho(x_2 \mid z) = \mathbb{P}(x_1 \mid z). \tag{7.59}$$

Thus, Bayes decision is $\hat{x}_1$ if $\mathbb{P}(x_1 \mid z) > \mathbb{P}(x_2 \mid z)$, and $\hat{x}_2$ otherwise. Using (7.55), we obtain the decision rule which minimizes error probability, namely

$$\Lambda(z) \underset{\hat{x}_2}{\overset{\hat{x}_1}{\gtrless}} \frac{\mathbb{P}(x_2)}{\mathbb{P}(x_1)}, \tag{7.60}$$

where

$$\Lambda(z) \triangleq \frac{f_{Z|x_1}(z \mid x_1)}{f_{Z|x_2}(z \mid x_2)} \tag{7.61}$$

is the likelihood ratio for the decision between x_1 and x_2.   ◇

Remark 7.10. Observe how Bayes decisions require the knowledge of the a priori probabilities $\mathbb{P}(x_i)$. In the absence of this information, the Principle of Insufficient Reason would suggest the assumption of equal probabilities, unless a theory involving imprecise probabilities is brought into play. If the decision environment does not allow access to the mechanism generating the State of Nature, then Bayes decisions are not acceptable, because the prior distribution is not available. In this situation, an alternative is a criterion that minimizes the maximum of the conditional risks defined in (7.57). This is called *minimax criterion*, and its details can be found for example in [195, Section II.C].

7.7.2   A case in which error probability does not tell the whole story

In cognitive radio (CR), as described in Section 3.7.2, the wireless channels used for transmission can be configured dynamically to minimize interference and congestion. In CR, before accessing a transmission channel, this is sensed, and information signals are transmitted over it according to criteria chosen to allow the existence of more than one concurrent wireless communication in a given spectrum band at one location. Assume that spectrum sensing classifies spectrum spaces as follows [27, p. 150]:

(i)   *White space*, one which is completely empty, except for noise,
(ii) *Black space*, one which is occupied by communication signals, interfering signals, and noise,

and hence faces two States of Nature, as summarized in Table 7.3.

Table 7.3 Decisions, States of Nature, and consequences in cognitive radio.

Decision	State of Nature	
	White	*Black*
transmit in spectrum space	reliable communication	interfered communication
do not transmit in spectrum space	wasted spectrum space	choose a different spectrum space

As for the cost of the decisions possibly made, we have to account for the four possible consequences of Table 7.3. A choice may consist of assigning cost c_1 to transmission in white space or no transmission in black space, and cost c_2 to no transmission in white space or transmission in black space. We may observe that this is by no means the only possible choice, as there is indeed also a cost intrinsic in the transmission in a different spectrum space, which has to be found and tested again for whiteness. The average cost $\mathbb{E}[C]$ resulting from the possible decisions turns out to be a linear combination of four terms.

A common approach consists of defining costs for two specific events, viz., *false alarm* (or *Type I error*), occurring when the decision is "space is black" when it is not, and *missed detection*, or *Type II error*), occurring when the decision is "space is white" when it actually is not. These events, whose probabilities we denote by p_{FA} and p_{MD}, are not independent, and the values they take depend on the detection strategy chosen (more on this was presented in Section 3.7.2). In general, a decision should have both p_{FA} and p_{MD} as small as possible, which turn out to be conflicting objectives in most practical situations. A standard way of expressing the connection between these two probabilities is through a plot called *Receiver Operating Characteristic* (ROC). This illustrates the dependence between p_{FA} and p_{MD} for a given detector and the parameters on which the detection depends, as for example the signal-to-noise ratio snr. A qualitative example of ROC is shown in Fig. 7.9, showing how for different values of snr the tradeoff between p_{FA} and p_{MD} may change. To interpret Fig. 7.9, one may think of the spectrum-sensing strategy described in Section 3.7.2, in which a decision metric Y is evaluated and compared against a discrimination threshold value

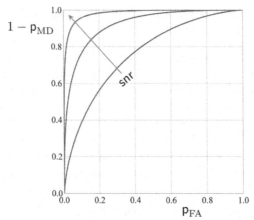

Figure 7.9 Receiver Operating Characteristic for the detection of white space in cognitive radio.

θ. The spectrum space is declared as white if $Y < \theta$, and black otherwise. Let Y have conditional pdf f_0 if the space is actually white, and f_1 otherwise. We have

$$\mathsf{P}_{FA} = \int_{\theta}^{\infty} f_0(y)\,dy \quad \text{and} \quad \mathsf{P}_{MD} = \int_{0}^{\theta} f_1(y)\,dy. \tag{7.62}$$

This strategy is based on the fact that the value of Y in a spectrum space containing only noise is expected to be lower than in a spectrum space in which a communication signal is present. A low threshold entails a high probability of false alarm and correspondingly a small probability of missed detection; the opposite occurs with a high threshold. As snr increases, the ROC curve becomes sharper, indicating more accurate decisions: the best possible decision procedure corresponds to a point located in the upper left corner of the ROC plot, with coordinates $(0, 1)$. In [204], a classification of spectrum sensing strategies is defined, based on their keenness towards opportunistic spectrum usage: *conservative* ($\mathsf{P}_{FA} \geqslant 0.5$, $1 - \mathsf{P}_{MD} > 0.5$), *aggressive* ($\mathsf{P}_{FA} < 0.5$, $1 - \mathsf{P}_{MD} > 0.5$), and *hostile* ($\mathsf{P}_{FA} < 0.5$, $1 - \mathsf{P}_{MD} \leqslant 0.5$).

7.7.3 Neyman–Pearson criterion

Even in the simple situation of $M = N = 2$, selecting the costs of all detection outcomes might be an unrealistic or undesirable task. If this is the case, the *Neyman–Pearson* (NP) criterion [138, p. 89], [255, p. 33 ff.] may be applied. This consists of minimizing P_{MD} subject to a constraint on P_{FA}, for example, a fixed value $\mathsf{P}_{FA} = \alpha$ (α is called the *significance level*, and $1 - \mathsf{P}_{MD}$ the *power* of the decision test). Thus, the Neyman–Pearson criterion aims at finding the most powerful α-level test of the decision between the two hypotheses. We construct the function

$$
\begin{aligned}
L &\triangleq \mathsf{p}_{\mathrm{MD}} + \lambda\big(\mathsf{p}_{\mathrm{FA}} - \alpha\big) \\
&= \int_0^\theta f_1(x)\,dx + \lambda\left[\int_\theta^\infty f_0(x)\,dx - \alpha\right] \\
&= \lambda(1 - \alpha) + \int_0^\theta [f_1(x)\,dx - \lambda f_0(x)]\,dx.
\end{aligned}
\tag{7.63}
$$

To minimize L, a point x is assigned to interval $[0, \theta]$ only when the term in bracket is negative. This is equivalent to the test

$$
\text{choose black space if } \frac{f_1(x)}{f_0(x)} < \lambda, \qquad \text{choose white space otherwise.} \tag{7.64}
$$

The ratio on the LHS of the inequality in (7.64) is recognized as a likelihood ratio. Thus, L is minimized by the LR test

$$
\Lambda(x) \underset{\text{black space}}{\overset{\text{white space}}{\gtrless}} \lambda, \tag{7.65}
$$

where λ is chosen so that the constraint $\mathsf{p}_{\mathrm{FA}} = \alpha$ is satisfied. This is obtained by solving

$$
\mathsf{p}_{\mathrm{FA}} = \int_\lambda^\infty f_0(x)\,dx = \alpha \tag{7.66}
$$

for λ.

Remark 7.11. The environment just described for spectrum sensing can be easily extended to other systems, as, for example, radar detection, where the presence of a target (e.g., an aircraft) should be detected based on radar-signal observations. In this context, a false alarm indicates that a target is not present, but the detector assume that it is, while a missed detection indicates that the actual presence of a target is not detected. The relative costs of the two events depend on the environment in which the radar system is deployed (air traffic control, wartime defense, etc.).

7.7.3.1 Neyman–Pearson criterion with interval probabilities

We examine the NP criterion in the following form, which generalizes the one just described. Assume the observation of a set of values taken by the random variable Y, and the choice between the two hypotheses

$$
\begin{aligned}
H_0 &: Y \sim \mathbb{P}_0, \\
H_1 &: Y \sim \mathbb{P}_1,
\end{aligned}
\tag{7.67}
$$

where $\mathbb{P}_0$ and $\mathbb{P}_1$ are the possible probability measures for Y. Acceptance of H_1 when H_0 is true, the Type-I error, occurs with probability p_{I}, while accepting H_0 when H_1

is true, the Type-II error, occurs with probability p_{II}. With this terminology, the NP criterion consists of minimizing p_{II} while keeping $p_I \leqslant \alpha$.

In the context of interval probabilities, the two hypotheses correspond to Y having a probability measure belonging to either $\mathcal{P}_0$ or $\mathcal{P}_1$, where $\mathcal{P}_0$ and $\mathcal{P}_1$ are two disjoint classes of probability measures. For example, $\mathcal{P}_0$ may include Gaussian measures with zero mean, and $\mathcal{P}_1$ Gaussian measures with known nonzero mean. A possible choice criterion could be the optimization of the worst performance, i.e., the minimization of $\bar{p}_{II}$ subject to $\bar{p}_I \leqslant \alpha$, where

$$
\begin{aligned}
\bar{p}_I &\triangleq \sup_{\mathcal{P}_0} p_I, \\
\bar{p}_{II} &\triangleq \sup_{\mathcal{P}_1} p_{II}.
\end{aligned}
\tag{7.68}
$$

This criterion offers the best worst-case performance, but it could be too conservative, especially when $\mathcal{P}_0$ and $\mathcal{P}_1$ are wide classes. For a discussion of this point, see [195, §III.E.2], and [8, §7.5.2], [119, §10.3] for (rather involved) general results.

7.8 Going beyond probability

The key issue with the selection of a tool to deal with epistemic uncertainty is the choice of the appropriate "probability" measure of the events involved. To classify possible choices, we examine a menagerie of set measures similar to probability, and we discuss their potential applications to decisions under epistemic uncertainty. We start listing some fundamental properties of set functions, among which we can select those whose combination gives rise to the measures we need.

Following Aristotelian logic, it is accepted that an event A and its complement A^c cover the entire event space (the *principium exclusi tertii*), which in probability theory leads to the result $\mathbb{P}(A) + \mathbb{P}(A^c) = 1$. However, when we consider *possibilities* of uncertain outcomes, the situation is different. While A and A^c are indeed incompatible and exhaustive, both of them can be possible. One of the axioms of probability theory assumes additivity: if A and B denote two incompatible events, i.e., such that $A \cap B = \emptyset$, then the probability that either A or B, or both, occur is given by

$$
\mathbb{P}(A \cup B) = \mathbb{P}(A) + \mathbb{P}(B).
\tag{7.69}
$$

This means that A and B do not interact to determine the probability of the joint occurrence of A and B. Now, uncertainty models may arise that cannot be captured by the standard probability axioms (an example is provided by interval probabilities). Thus, we might need measures differing from probability, which we generally denote by $\mathbb{M}$, capable of capturing one of the following situations [143, pp. 103–105]:

(i) Superadditivity,

$$
\mathbb{M}(A \cup B) \geqslant \mathbb{M}(A) + \mathbb{M}(B);
\tag{7.70}
$$

(ii) Subadditivity,

$$\mathbb{M}(A \cup B) \leqslant \mathbb{M}(A) + \mathbb{M}(B). \tag{7.71}$$

Example 7.25. A *coalition* is a group of individuals or entities cooperating to achieve a common goal. Let $\mathbb{M}(A)$ measure the extent to which coalition A is able to achieve its goal. Consider sensors, i.e., devices whose purpose is to detect the state of a system, as, for example, the presence or absence of a signal occupying a given frequency band in cognitive radio [27]. If A denotes a group of cooperating sensors, and A contains a sensor s which is failed, i.e., yields wrong indications, then, if $\mathbb{M}$ measures the quality of system performance, we may have $\mathbb{M}(A - \{s\}) > \mathbb{M}(A)$, which indicates inefficient cooperation. $\diamond$

Example 7.26 (Adapted from [102, p. 29]). In a wine shop, let $\mathbb{M}(\{b\})$ denote the selling price of wine bottle b. In many cases the price is an additive set function, $\mathbb{M}(\{b_i, b_j\}) = \mathbb{M}(\{b_i\}) + \mathbb{M}(\{b_j\})$. However, it is common practice to offer a discount if several bottles are bought together, which yields a subadditive set function, for example, $\mathbb{M}(\{b_1, b_2, b_3\}) < \mathbb{M}(\{b_1\}) + \mathbb{M}(\{b_2\}) + \mathbb{M}(\{b_3\})$. A superadditive set function may model the selling price of a full case of a rare wine, so that $\mathbb{M}(\{b_1, b_2, b_3, b_4, b_5, b_6\}) > \sum_{i=1}^{6} \mathbb{M}(\{b_i\})$. $\diamond$

Example 7.27 ([143, p. 105]). Consider two disjoint adjacent intervals of the real line $A \triangleq [a, b]$ and $B \triangleq (b, c]$, and the outcome of the measurement of a real quantity. We are interested in the computation of $\mathbb{M}(A \cup B)$, interpreted as the evidence that the true value under measure is in $A \cup B$. Assume that observations of values close to b are less reliable than those in the middle of A and B. Thus, if we separately compute $\mathbb{M}(A)$ and $\mathbb{M}(B)$, any measured value close to b is unreliable. On the contrary, if we compute $\mathbb{M}(A \cup B)$, we obtain a more reliable measurement. Thus, $\mathbb{M}(A \cup B) > \mathbb{M}(A) + \mathbb{M}(B)$. $\diamond$

To define set measures making (7.70) or (7.71) admissible under suitable assumptions on A and B, we should consider using a general theory capable of encompassing (7.69), (7.70), and (7.71). One such theory is based on "monotone set functions," to be described in the next section.

7.9   Set functions and their properties

Defining a set function $\mathbb{M} : 2^\mathcal{S} \mapsto \mathbb{R}$, we list the properties that can possibly be shared by $\mathbb{M}$ and are used for classification (see, for example, [102, Chapter 2]). Consider any $A, B \in 2^\mathcal{S}$. Then

Definition 7.1.
   (i) $\mathbb{M}$ is *grounded* if $\mathbb{M}(\emptyset) = 0$.
   (ii) $\mathbb{M}$ is *normalized* if $\mathbb{M}(\mathcal{S}) = 1$.
   (iii) $\mathbb{M}$ is *nonnegative* if $\mathbb{M}(A) \geqslant 0$.

(iv) $\mathbb{M}$ is *additive* if, for $A \cap B = \emptyset$,

$$\mathbb{M}(A \cup B) = \mathbb{M}(A) + \mathbb{M}(B). \tag{7.72}$$

(v) $\mathbb{M}$ is *monotone* if, for $A \subseteq B$,

$$\mathbb{M}(A) \leqslant \mathbb{M}(B). \tag{7.73}$$

(vi) $\mathbb{M}$ is *superadditive* if, for $A \cap B = \emptyset$,

$$\mathbb{M}(A \cup B) \geqslant \mathbb{M}(A) + \mathbb{M}(B). \tag{7.74}$$

(vii) $\mathbb{M}$ is *subadditive* if, for $A \cap B = \emptyset$,

$$\mathbb{M}(A \cup B) \leqslant \mathbb{M}(A) + \mathbb{M}(B). \tag{7.75}$$

(viii) $\mathbb{M}$ is *submodular* if

$$\mathbb{M}(A \cup B) \leqslant \mathbb{M}(A) + \mathbb{M}(B) - \mathbb{M}(A \cap B). \tag{7.76}$$

(ix) $\mathbb{M}$ is *supermodular* if

$$\mathbb{M}(A \cup B) \geqslant \mathbb{M}(A) + \mathbb{M}(B) - \mathbb{M}(A \cap B). \tag{7.77}$$

(x) $\mathbb{M}$ is *k-monotone* if, for a $k \geqslant 2$ and all families $A_1, A_2, \ldots, A_k$ of elements of 2^S, the following inequality holds:

$$\mathbb{M}\left(\bigcup_{i=1}^{k} A_i\right) \geqslant \sum_{\substack{K \subseteq \{1,2,\ldots,k\} \\ K \neq \emptyset}} (-1)^{|K|+1} \mathbb{M}\left(\bigcap_{i \in K} A_i\right). \tag{7.78}$$

(xi) $\mathbb{M}$ is *k-totally monotone* if it is *k*-monotone for any $k \geqslant 2$.

(xii) $\mathbb{M}$ is *k-alternating* if, for a $k \geqslant 2$ and all families $A_1, A_2, \ldots, A_k$ of elements of 2^S, the following inequality holds:

$$\mathbb{M}\left(\bigcap_{i=1}^{k} A_i\right) \leqslant \sum_{\substack{K \subseteq \{1,2,\ldots,k\} \\ K \neq \emptyset}} (-1)^{|K|+1} \mathbb{M}\left(\bigcup_{i \in K} A_i\right). \tag{7.79}$$

(xiii) $\mathbb{M}$ is *k-totally alternating* if it is *k*-alternating for any $k \geqslant 2$.

The following properties of set functions follow immediately from the definitions:

(i) If $\mathbb{M}$ is nonnegative and additive, then it is monotone.

(ii) If $\mathbb{M}$ is additive, then it is grounded: in fact, $\mathbb{M}(\emptyset) + \mathbb{M}(\emptyset) = \mathbb{M}(\emptyset)$ entails $\mathbb{M}(\emptyset) = 0$.

(iii) If $\mathbb{M}$ is monotone, from the inclusions $A \cap B \subseteq A$, $A \cap B \subseteq B$, $A \cup B \supseteq A$, and $A \cup B \supseteq B$, valid for any $A, B \in 2^S$, it follows that

$$\begin{aligned} \mathbb{M}(A \cup B) &\geqslant \max\{\mathbb{M}(A), \mathbb{M}(B)\}, \\ \mathbb{M}(A \cap B) &\leqslant \min\{\mathbb{M}(A), \mathbb{M}(B)\}. \end{aligned} \tag{7.80}$$

Further definitions follow from combinations of some of the properties above:

(i) A grounded $\mathbb{M}$ is called a *game*.
(ii) A *measure* is a nonnegative additive $\mathbb{M}$.
(iii) A *capacity* is a grounded monotone $\mathbb{M}$.
(iv) A normalized capacity satisfying

$$\mathbb{M}(A \cup B) = \max\big(\mathbb{M}(A), \mathbb{M}(B)\big), \qquad A, B \in 2^S \tag{7.81}$$

is called a *possibility measure*.
(v) A normalized capacity satisfying

$$\mathbb{M}(A \cap B) = \min\big(\mathbb{M}(A), \mathbb{M}(B)\big), \qquad A, B \in 2^S \tag{7.82}$$

is called a *necessity measure*.
(vi) A totally monotone normalized capacity is called a *belief measure*.
(vii) A totally alternating normalized capacity is called a *plausibility measure*.
(viii) A grounded normalized measure is called a *probability measure*.

The *conjugate* (or *dual*) $\mathbb{M}^\star$ of a set function $\mathbb{M}$ is defined as $\mathbb{M}^\star(A) \triangleq \mathbb{M}(S) - \mathbb{M}(A^c)$. The following properties can be proved:

(i) If $\mathbb{M}$ is additive, then it is self-conjugate, $\mathbb{M}^\star = \mathbb{M}$.
(ii) The probability measure is self-conjugate.
(iii) The conjugate of a possibility measure is a necessity measure.
(iv) The conjugate of a necessity measure is a possibility measure.
(v) The conjugate of a belief measure is a plausibility measure.
(vi) The conjugate of a plausibility measure is a belief measure.
(vii) Possibility and necessity measures are special cases of belief and plausibility measures.

The classification of set measures is partially summarized in Fig. 7.10.

Example 7.28. The two set functions $\underline{\mathbb{P}}$ and $\overline{\mathbb{P}}$ defined in Section 7.5 are grounded, normalized, and monotone, and may satisfy some stronger conditions. For example, $\underline{\mathbb{P}}$ may be 2-monotone,

$$\underline{\mathbb{P}}(A \cup B) \geqslant \underline{\mathbb{P}}(A) + \underline{\mathbb{P}}(B) - \underline{\mathbb{P}}(A \cap B), \tag{7.83}$$

or $\overline{\mathbb{P}}$ may be 2-alternating,

$$\overline{\mathbb{P}}(A \cap B) \leqslant \overline{\mathbb{P}}(A) + \overline{\mathbb{P}}(B) - \overline{\mathbb{P}}(A \cup B), \tag{7.84}$$

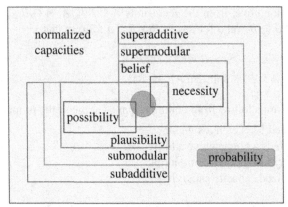

Figure 7.10 Various families of normalized capacities (adapted from [102, Fig. 2.1]). All capacities in the right upper corner are conjugate of those placed symmetrically in the left lower corner, except for subadditive and superadditive capacities.

and $\underline{\mathbb{P}}$ may be totally monotone, or, dually, $\overline{\mathbb{P}}$ may be totally alternating [186, p. 79]. Moreover, lower probability measures are superadditive, while upper probability measures are subadditive. That is, for all A, $B \in 2^S$ with $A \cap B = \emptyset$, we have [143, p. 113]

$$\underline{\mathbb{P}}(A \cup B) \geqslant \underline{\mathbb{P}}(A) + \underline{\mathbb{P}}(B), \tag{7.85a}$$

$$\overline{\mathbb{P}}(A \cup B) \leqslant \overline{\mathbb{P}}(A) + \overline{\mathbb{P}}(B). \tag{7.85b}$$

Upper and lower probabilities are conjugate,

$$\overline{\mathbb{P}}(A) = 1 - \underline{\mathbb{P}}(A^c), \tag{7.86}$$

which shows that one of them is sufficient for calculations.   ◇

Example 7.29. The set function induced by the contaminated probability measures of Example 7.18 can be proved to be an alternating capacity of infinite order [119, p. 259].   ◇

7.10 Infinite sets

The discussions above assume a finite set S of States of Nature, and hence a finite power set 2^S. To include infinite sets, we must ask $\mathbb{M}$ to satisfy some additional properties. To extend measure theory to subsets of an infinite S, we restrict attention to families of subsets that are closed under the union of countably many sets and under the set complement. For example, a measure $\mathbb{M}$ is additive in the sense that its value for a bounded union of a sequence of pairwise disjoints sets is equal to the sum of the values associated with the individual sets.

Specifically, we introduce in 2^S a *σ-algebra* (or *Borel field*) $\mathcal{A}$ of events, that is, a subset of 2^S such that any union of a countable collection of events in $\mathcal{A}$, and the complement A^c of any $A \in \mathcal{A}$, is still in $\mathcal{A}$ [211, p. 16 ff.]. Given a σ-algebra $\mathcal{A}$ of elements of 2^S, a monotone measure is a set function $\mathbb{M} : 2^S \mapsto \mathbb{R}^+$ that satisfies the following properties [143, p. 103 ff.]:

   (i) $\mathbb{M}(\emptyset) = 0$.
   (ii) For all $A, B \in \mathcal{A}$, if $A \subseteq B$, then $\mathbb{M}(A) \leqslant \mathbb{M}(B)$.
   (iii) For any increasing family $A_1 \subseteq A_2 \subseteq \cdots$ of sets in $\mathcal{A}$,

$$\bigcup_{i=1}^{\infty} A_i \in \mathcal{A} \Rightarrow \lim_{i \to \infty} \mathbb{M}(A_i) = \mathbb{M}\left(\bigcup_{i=1}^{\infty} A_i\right). \tag{7.87}$$

   This property is called *continuity from below*.
   (iv) For any decreasing family $A_1 \supseteq A_2 \supseteq \cdots$ of sets in $\mathcal{A}$,

$$\bigcap_{i=1}^{\infty} A_i \in \mathcal{A} \Rightarrow \lim_{i \to \infty} \mathbb{M}(A_i) = \mathbb{M}\left(\bigcap_{i=1}^{\infty} A_i\right). \tag{7.88}$$

   This property is called *continuity from above*.

Notice that, if S is finite, then items (iii) and (iv) are trivial and hence can be disregarded.

Remark 7.12. With some abuse of notation, even when S is infinite, we write 2^S to denote the smallest σ-algebra containing an arbitrary collection of subsets of S.

7.11   Capacities and Choquet integral

Set functions that are monotone (and not necessarily additive) can be used to formalize a wide variety of uncertainty models. Capacities, i.e., set functions which are grounded and monotone, form an important family of these.

Example 7.30. If A denotes an event related to the outcome of a random experiment, then $\mathbb{M}(A)$, the capacity of A, may measure the uncertainty that A contain that outcome, with $\mathbb{M}(A) = 0$ expressing total uncertainty, and $\mathbb{M}(A) = 1$ indicating that there is no uncertainty. ◇

Example 7.31. If a die is thrown, and $S = \{1, 2, 3, 4, 5, 6\}$, then $\mathbb{M}(\{2, 4, 6\})$ may quantify the uncertainty of obtaining an even number. ◇

Example 7.32. As mentioned before, to solve certain paradoxes arising in probability theory, one may resort to measures of uncertainty that are not additive. Normalized capacities provide a large class of such measures [102, p. 309]. Consider, for example, the (supermodular) capacity $\mathbb{M}$ shown in Table 7.4. Using this capacity to model the

Table 7.4 A capacity $\mathbb{M}$ replacing probability and solving Ellsberg paradox.

Event A	r	y	b	r, y	r, b	y, b	r, y, b
$\mathbb{M}(A)$	1/3	0	0	1/3	1/3	2/3	1

uncertainty permeating this example, we obtain the choice criterion

Choose bet A over bet B if and only if $\mathbb{M}(\{r\}) > \mathbb{M}(\{b\})$,

Choose bet D over bet C if and only if $\mathbb{M}(\{y, b\}) > \mathbb{M}(\{r, y\})$,

which avoids Ellsberg paradox. $\diamond$

Remark 7.13. Probability measures are capacities which, for all $k \geqslant 2$, satisfy (7.78) with inequalities replaced by equalities.

Remark 7.14. Since all lower probability measures are superadditive, they are capacities. Thus, capacities are natural representations of lower probability measures [143, p. 114], and as such can be used to formalize a theory of imprecise probabilities.

7.12   Expected values and Choquet integral

"Standard" expected-utility theory is based on a probability measure, which is additive. To obtain the expected value of a utility function u with respect to a more general measure, Lebesgue integrals may be replaced by a functional called *Choquet integral* of u with respect to a capacity $\mathbb{M}$ [102, Chapter 4], [143, §4.5]. If $\mathbb{M}$ is additive, the Choquet integral coincides with the Lebesgue integral, while if u is a nonnegative function it generates a measure that preserves some of the properties of $\mathbb{M}$, namely, monotonicity, subadditivity, superadditivity, and continuity from above/below. Applying the Choquet integral to upper and lower probability measures, lower and upper expected values of a utility function can be obtained.

Definition 7.2. Given a real, bounded, nonnegative function $f : \mathcal{S} \mapsto \mathbb{R}^+$, and a capacity $\mathbb{M}$ on $2^{\mathcal{S}}$, the Choquet integral of f with respect to $\mathbb{M}$ on $A \in 2^{\mathcal{S}}$ is the functional

$$\oint_A f \, d\mathbb{M} \triangleq \int_0^{\infty} G_f(t) \, dt, \tag{7.89}$$

where $G_f(t)$ is the survival function

$$G_f(t) \triangleq \mathbb{M}(\{x \in A \mid f(x) > t\}) \, dt \tag{7.90}$$

and the integral on the RHS of (7.89) is a Riemann integral [102, §4.2].

(A list of properties of Choquet integral can be found, for example, in [143, p. 134].)

To illustrate this definition, consider the special case of a discrete function f taking values $a_1, \ldots, a_n$ with $0 \leqslant a_1 < a_2 < \cdots < a_n$. Consider next the sets $A_i \triangleq \{x \in A \mid f(x) > a_i\}$, $i = 1, \ldots, n$, with $A_1 = A$. With reference to Fig. 7.11, the Choquet integral is the area below the graph of function $G_f(t)$, which can be decomposed either in vertical slices, yielding

$$\oint_A f \, d\mathbb{M} = \sum_{i=1}^{n} (a_i - a_{i-1}) \mathbb{M}(A_i) \tag{7.91}$$

with $a_0 = 0$, or in horizontal slices, so that

$$\oint_A f \, d\mathbb{M} = \sum_{i=1}^{n} a_i \big(\mathbb{M}(A_i) - \mathbb{M}(A_{i+1}) \big) \tag{7.92}$$

with $A_{n+1} = \emptyset$.

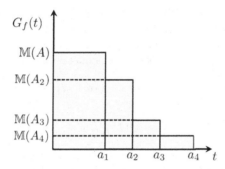

Figure 7.11 Function $G_f(t)$ for a function taking values $a_1 < a_2 < a_3 < a_4$.

Example 7.33 ([102, p. 209]). Let $A = \{x_1, x_2, x_3, x_4\}$ and the function f taking values $f(x_1) = 0.2$, $f(x_2) = 0.1$, $f(x_3) = 0.5$, and $f(x_4) = 0.3$. Use permutation π to arrange f in increasing order: $\pi(1) = 2$, $\pi(2) = 1$, $\pi(3) = 4$, and $\pi(4) = 3$, so that $f(x_{\pi(1)}) < f(x_{\pi(2)}) < f(x_{\pi(3)}) < f(x_{\pi(4)})$. Assuming the following values of capacity $\mathbb{M}$:

$$\mathbb{M}(A) = 1, \ \mathbb{M}(\{x_1, x_3, x_4\}) = 0.8, \ \mathbb{M}(\{x_3, x_4\}) = 0.6, \ \mathbb{M}(\{x_3\}) = 0.2, \tag{7.93}$$

we obtain the integral value

$$\begin{aligned}
\oint_A f \, d\mathbb{M} &= f(x_2)\mathbb{M}(A) + \big(f(x_1) - f(x_2) \big) \mathbb{M}(\{x_1, x_3, x_4\}) \\
&\quad + \big(f(x_4) - f(x_1) \big) \mathbb{M}(\{x_3, x_4\}) + \big(f(x_3) - f(x_4) \big) \mathbb{M}(\{x_3\}) \\
&= 0.1 + 0.1 \times 0.8 + 0.1 \times 0.6 + 0.2 \times 0.2 \\
&= 0.28.
\end{aligned} \tag{7.94}$$

$\diamond$

The example that follows summarizes the calculations of expected utility under some of the decision criteria described in this chapter.

Example 7.34 ([79]). An investor whose wealth is w faces the risk of losing an amount d. The State of Nature can be either L (loss) or L^c (no loss). The result of an act is described by the pair (a, b), where a denotes the amount of money lost if L occurs, and b the amount of money lost if L^c occurs. We examine three possible acts: (a) no insurance bought, which yields the pair $(w - d, w)$; (b) full insurance at a premium r, which yields $(w - r, w - r)$; (c) partial insurance at a premium r' and indemnity s paid in case of loss, which yields $(w - d + s - r', w - r')$. Using the expected-utility criterion, letting $\mathsf{p} \triangleq \mathbb{P}(L)$ and a strictly increasing utility function u, the three values of expected utility are:

$$\mathbb{E}[U_a] = \mathsf{p}\,u(w - d) + (1 - \mathsf{p})u(w),$$
$$\mathbb{E}[U_b] = \mathsf{p}\,u(w - r) + (1 - \mathsf{p})u(w - r) = u(w - r), \tag{7.95}$$
$$\mathbb{E}[U_c] = \mathsf{p}\,u(w - d + s - r') + (1 - \mathsf{p})u(w - r').$$

To use an expected-utility criterion based on expectations computed with Choquet integral, the investor first chooses $1 - \mathbb{M}(L^c)$, the capacity associated with the event "loss occurs" (notice that this capacity may not be equal to $\mathbb{M}(L)$), next makes a decision by comparing the three expected "Choquet utilities"

$$\mathbb{E}_C[U_a] = u(w - d) + \mathbb{M}(L^c)[u(w) - u(w - d)],$$
$$\mathbb{E}_C[U_b] = u(w - r), \tag{7.96}$$
$$\mathbb{E}_C[U_c] = u(w - d + s - r') + \mathbb{M}(L^c)[u(w - r') - u(w - d + s - r')].$$

To use a maximin utility criterion, if the subjective probability of L is assumed to lie in the interval $[\mathsf{p}] \triangleq [\underline{\mathsf{p}}, \overline{\mathsf{p}}]$, the investor act is based on the comparison of these three quantities:

$$U_a = \min_{\mathsf{p}\in[\mathsf{p}]}\left[\mathsf{p}\,u(w - d) + (1 - \mathsf{p})u(w)\right] = \overline{\mathsf{p}}\,u(w - d) + (1 - \overline{\mathsf{p}})u(w),$$
$$U_b = u(w - r),$$
$$U_c = \min_{\mathsf{p}\in[\mathsf{p}]}\left[u(w - d + s - r') + (1 - \mathsf{p})u(w - r')\right] \tag{7.97}$$
$$= \overline{\mathsf{p}}\,u(w - d + s - r') + (1 - \overline{\mathsf{p}})u(w - r'). \qquad \diamond$$

7.13 Dempster–Shafer theory

Dempster–Shafer (DS) theory is at the basis of a technique for modeling reasoning under epistemic uncertainty. It replaces probabilities with beliefs, which allows a clear distinction to be made between epistemic uncertainty and its aleatory counterpart. As a major departure from probability theory, instead of dealing with aleatory experiments resulting in events which may or may not occur, DS theory deals with States of

Nature whose knowledge is blurred by the observer's ignorance. Typically, standard probability theory deals with numbers that reflect how often an event will occur if an experiment is repeated a large number of times. Now, when one performs a random experiment, the probability distribution that governs it is, in the first instance, unknown. If one examines the "belief" in a given hypothesis instead of its probability of being true, two opposite views can be advocated. One assumes that the numerical degree of support of a given proposition is objectively determined by a body of evidence. The competing view claims that the degrees of belief are psychological facts, to be discovered by observing an individual's preferences among alternatives. In DS theory probabilities "must be conceived as a feature of the world," and hence "are not necessarily features of our knowledge or belief" [229, p. 16], while beliefs do not reflect "what Nature chooses, but rather the state of *our knowledge* after making a measurement" [144, p. 28]:

If we know the [probabilities], then we will surely adopt them as our degrees of belief. But if we do not know the [probabilities], then it will be an extraordinary coincidence for our degrees of belief to be equal to them. [229, p. 16]

Since probability theory takes it as given that a statement is either true or false, it "cannot distinguish between lack of belief and disbelief, and does not allow one to withhold belief from a proposition without according that belief to the negation of the proposition" [229, p. 23]. DS theory admits *plausible* propositions, those either supported by evidence or by uncertainty, allows for more nebulous statements such as "I don't know," and can be used in problems in which probabilities are unknown, or only partially known.

Remark 7.15. DS theory avoids the "Bayesian dogma of precision." Citing verbatim from [22, p. 38], this dogma assumes that "the information concerning uncertain statistical parameters, no matter how vague, must be represented by conventional, exactly specified, probability distributions." Specifically, in general "in a situation of ignorance a Bayesian is forced to $(\cdots)$ evenly allocate subjective (additive) probabilities" over the State of Nature.

The formalism of DS theory, introduced in [229, p. 36], assumes that we are concerned with a true hypothesis. Let S denote the set of hypotheses (called a *frame of discernment*), so that all the hypotheses of interest are represented by subsets of S. The propositions of interest are those taking the form "The correct hypothesis is in a subset of S." Now, assign a *belief mass* to all elements of 2^S by defining a set function

$$\mathsf{m} : 2^S \mapsto [0, 1] \tag{7.98}$$

such that

(i) $\mathsf{m}(\emptyset) = 0,$
(ii) $\displaystyle\sum_{A \in 2^S} \mathsf{m}(A) = 1.$

The belief mass $m(A)$ is assigned to A but to no particular subset of A, and measures our confidence that the true hypothesis lies in A but not in any proper subset $B \subset A$ (subsets of A have their own beliefs). We see that there is no belief in the empty set (which might be interpreted as a hypothesis known to be false), while all the assigned belief masses sum to unity. Any $A \in 2^S$ for which $m(A) \neq 0$ is called a *focal element*.

Remark 7.16. Belief masses are sometimes referred to as *basic probability assignment* (see, e.g., [22,94]). We avoid this terminology to prevent any confusion between DS and probability theory.

Two measures of uncertainty are defined for each focal element $A \in 2^S$. The *belief of A* is the degree of belief that directly supports A, and measures our confidence that the true hypothesis lies in A or in any subset $B \subseteq A$. Formally, it is the total mass of all sets implying A, i.e.,

$$\text{bel}(A) \triangleq \sum_{B|B \subseteq A} m(B). \tag{7.99}$$

The relation between belief mass and belief can be expressed as

$$m(A) = \sum_{B|B \subseteq A} (-1)^{|A-B|} \text{bel}(B), \tag{7.100}$$

where (7.100) is usually referred to as the *Möbius Transform* of $\text{bel}(\cdot)$.

The *plausibility of A* is the extent to which we fail to disbelieve A, i.e., the degree of belief that does not directly contradict a specific A. Formally, it is defined as the total mass of all sets not inconsistent with A,

$$\text{pl}(A) \triangleq \sum_{B|B \cap A \neq \emptyset} m(B). \tag{7.101}$$

(In the terminology of [190, p. 41], plausibility is the probability that A is compatible with the available evidence, and hence possible.)

Belief and plausibility are conjugate:

$$\text{bel}(A) = 1 - \text{pl}(A^c), \tag{7.102a}$$

$$\text{pl}(A) = 1 - \text{bel}(A^c) \tag{7.102b}$$

($\text{bel}(A^c)$ may be called the *doubt* about A). Moreover, we have

$$\text{bel}(A) + \text{bel}(A^c) \leqslant 1, \tag{7.103a}$$

$$\text{pl}(A) + \text{pl}(A^c) \geqslant 1. \tag{7.103b}$$

Belief $\text{bel}(A)$ and plausibility $\text{pl}(A)$ yield explicit measures of ignorance about hypothesis A and its complement A^c. The length of the interval $[\text{bel}(A), \text{pl}(A)]$ may also be interpreted as the inaccuracy of our knowledge of the "true probability" $\mathbb{P}(A)$.

Thus, one might think of belief as a lower limit to uncertainty, of plausibility as an upper limit to uncertainty, and of probability as lying somewhere between belief and plausibility (see also [144, p. 26]). (Originally, Dempster referred to bel and pl as lower and upper probability, respectively.) Fig. 7.12 illustrates the relations among belief, plausibility, and probability of an event A.

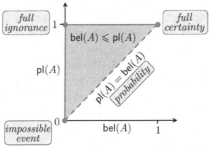

Figure 7.12 Relations among belief, plausibility, and probability of an event A (adapted from [102, p. 416]). The shaded triangular area contains the pairs bel(A) and pl(A).

A further relevant quantity in DS theory is the *commonality* of A [94]. This collects all the belief masses that could be committed to A from all the sets $B \supseteq A$,

$$\mathrm{com}(A) \triangleq \sum_{B \mid B \supseteq A} \mathrm{m}(B). \tag{7.104}$$

Example 7.35. Fig. 7.13 shows an example of cumulative belief and plausibility of the event $\{X \leqslant x\}$, where X is an RV taking values from 0 to 10 (see [114, pp. 612–614]). Cumulative belief and plausibility functions bound below and above the CDF of X.
◇

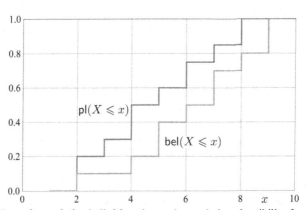

Figure 7.13 Plots of cumulative belief function and cumulative plausibility function of the event $\{X \leqslant x\}$, where X is a RV with support set $\{1, 2, \ldots, 10\}$.

Example 7.36. The belief function

$$
\mathsf{bel}(A) = \begin{cases} 0, & \text{if } A \neq \mathcal{S}, \\ 1, & \text{if } A = \mathcal{S} \end{cases} \tag{7.105}
$$

represents complete ignorance, i.e., no evidence about $\mathcal{S}$.   ◇

Example 7.37. Consider cognitive radio, a given frequency band, and the statement *"a communication source is active in it,"* whose belief is 0.5 and whose plausibility is 0.8. The above means that we have evidence allowing us to state strongly that the source is active with a confidence of 0.5. The evidence contrary to that proposition (*"source is inactive"*) has confidence 0.2. The remaining mass ($0.3 = 0.5 - (1 - 0.8)$) is indeterminate (*"source either active or inactive"*). The plausibility is either supported by evidence or by uncertainty; see Fig. 7.14.   ◇

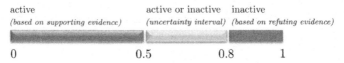

Figure 7.14 An example with belief 0.5 and plausibility 0.8.

Example 7.38 ([229, p. 5]). You have received as a gift a Chinese vase which looks very old. The question is, is it genuine or counterfeit? Let H_1 correspond to the hypothesis "the vase is genuine," and H_0 to the hypothesis "it is counterfeit." Then the set of hypotheses is $\mathcal{S} = \{H_0, H_1\}$, and its power set is

$$
2^{\mathcal{S}} = \{\emptyset, H_0, H_1, \mathcal{S}\}. \tag{7.106}
$$

A belief $b_1 \triangleq \mathsf{bel}(H_1)$ (respectively, $b_0 \triangleq \mathsf{bel}(H_0)$) represents my degree of belief that that vase is genuine (respectively, counterfeit). Different pairs (b_0, b_1) correspond to different weights of evidence associated with both possibilities. Thus, if you have very little evidence on either side, both b_0 and b_1 will take very small values—the limiting case of no evidence will force the choice $b_0 = b_1 = 0$. If there is a strong evidence in favor of H_0, then b_0 will be given a value close to 1, and b_1 close to 0.   ◇

Example 7.39. Anthony, a fine-wine connoisseur, after tasting a glass of barolo wine, is asked to determine the vineyard of its production. Let $\mathcal{S}$ denote the finite set of vineyards allowed to produce barolo. If A denotes the subset of $\mathcal{S}$ including all vineyards in the municipality of Serralunga d'Alba, $\mathsf{bel}(A)$ expresses to what degree Anthony is certain that the vineyard is in that municipality.   ◇

Example 7.40. Vincent, a modest wine connoisseur, is invited to taste a glass of a fairly expensive red wine, and to tell whether the wine is a barolo or a barbaresco. Let us define $\mathcal{S} = \{x_1, x_2\}$ as the set of possible answers to this question, with x_1 corresponding to "a barolo" and x_2 to "a barbaresco." Since this person has a very limited knowledge of wine, he has no specific reason to think that x_1 or x_2 is true. Thus, he allocates to S all the belief mass, thus expressing *total ignorance*:

$$m(\{x_1\}) = m(\{x_2\}) = 0, \qquad m(S) = 1. \tag{7.107}$$

Consequently,

$$\text{bel}(\{x_1\}) = \text{bel}(\{x_2\}) = 0, \tag{7.108a}$$
$$\text{pl}(\{x_1\}) = \text{pl}(\{x_2\}) = 1. \tag{7.108b}$$

Vincent then passes the glass to Anthony, and asks for his opinion. Anthony thinks that there are as many reasons in favor of barolo as there are in favor of barbaresco, and hence allocates to S a zero belief mass, thus expressing *no ignorance*:

$$m(\{x_1\}) = m(\{x_2\}) = 1/2, \qquad m(S) = 0. \tag{7.109}$$

Consequently,

$$\text{bel}(\{x_1\}) = \text{bel}(\{x_2\}) = 1/2, \tag{7.110a}$$
$$\text{pl}(\{x_1\}) = \text{pl}(\{x_2\}) = 1/2. \tag{7.110b}$$

If both persons were willing to use probability theory in this situation, they might invoke the principle of insufficient reason, and assign to x_1 and x_2 the same probability,

$$\mathbb{P}(\{x_1\}) = \mathbb{P}(\{x_2\}) = 1/2, \tag{7.111}$$

without taking in any account their specific wine competence. $\diamond$

Example 7.41. This example, adapted from [22], illustrates potential problems with the choice of the frame of discernment S, and highlights the difference between the representation of epistemic ignorance via probability or via DS theory. Consider the proposition, denoted by α, that "I live in River Road, Smalltown." To construct S among all possible choices admitting a subset A which represents α, we should first gather information on how many roads there are in Smalltown. If there are only two roads, then $S = \{x_1, x_2\}$ and $A = \{x_1\}$. With no additional information, one may want to assign beliefs based on subjective probabilities over S, which is done by using the Principle of Insufficient Reason and hence assuming $\text{bel}(A) = 0.5$. If instead it is estimated that there are about 100 roads in Smalltown, then $S = \{x_1, x_2, \ldots, x_{100}\}$, with $A = \{x_1\}$ and the other x_is representing the other roads. Probability theory would then assume $\text{bel}(A) = 0.01$. This example shows that the amount of belief assigned to A may depend on the choice of S. $\diamond$

Example 7.42. Suppose that we have to choose between two hypotheses H_0 and H_1, so that $S = \{H_0, H_1\}$, and that our decision is based on the observation of the value x taken by an RV X. Under the maximum-likelihood principle, H_i should be chosen against H_j, $j \neq i$, if the following inequality holds between the conditional pdfs of

X, $f(x \mid H_i) > f(x \mid H_j)$. One may define $\mathsf{bel}(H_i \mid x)$ as the degree of support that x provides to H_i as follows [69]:

$$\mathsf{bel}(H_i \mid x) = \begin{cases} 1 - \dfrac{f(x \mid H_j)}{f(x \mid H_i)}, & \text{if } f(x \mid H_j) < f(x \mid H_i), \\ 0, & \text{otherwise.} \end{cases} \tag{7.112}$$

This choice is based on the assumption that the degree to which the evidence provided by x fails to refute $f(x \mid H_j)$ is increasing with $f(x \mid H_j)$. Thus, the degree of support that x provides to H_i is equal to the support provided for the whole of $\mathcal{S}$ minus the degree to which x fails to refute H_j (this is proportional to $f(x \mid H_j)$). $\diamond$

Example 7.43 (Adapted from [279]). Assume that the reception of radio signals in a base station suffers from frequent disruptions, caused either by a hostile transmitter (jammer) or by abrupt variations of the communication medium. Sensors located within the coverage area try to identify the position of the jammer, and the information gathered by n sensors is aggregated to reach a decision about the cause of the disruptions. Assume that sensors $i = 1, \ldots, m$ identify the possible location of the jammer as A_i, where A_i is a subset of the coverage area, while sensors $j = m + 1, \ldots, n$ identify no jammer, or, equivalently, set $A_j = \emptyset$. After aggregation, the question asked is: Is the jammer located in a specific subarea A? The two cases to be considered are: (i) $A_i \subset A$, that is, sensor i deems it *certain* that the jammer is in A; and (ii) $A_i \cap A \neq \emptyset$, that is, sensor i deems it *possible* that the jammer is in A, which is also implied by (i). For aggregation, let the sensor outputs be weighed by assigning belief mass m_i to sensor i, so that $\mathsf{bel}(A)$ and $\mathsf{pl}(A)$ are obtained from their definitions (7.99) and (7.101):

$$\mathsf{bel}(A) = \sum_{i \mid A_i \subseteq A} \mathsf{m}(A_i), \tag{7.113a}$$

$$\mathsf{pl}(A) = \sum_{i \mid A_i \cap A \neq \emptyset} \mathsf{m}(A_i). \tag{7.113b}$$

Should one ask for the probability p that the jammer is in A, the answer would be $\mathsf{p} \in [\mathsf{bel}(A), \mathsf{pl}(A)]$. $\diamond$

Example 7.44. This example, adapted from [143, p. 289 ff.], shows how Expected Utility Theory can be modified when probabilities are replaced by beliefs, which makes utilities take on interval values. Consider again a carnival prize wheel game as in Example 7.10 *supra*. Fig. 7.1(b) shows a wheel in which the value associated with each one of the sectors may take value 1, 5, 10, or 20, but the tag of one sector is hidden from view, and may take any of those values [264, p. 287 ff.]. How much would a rational agent be willing to pay to play this game? To model the uncertainty implicit in this chance game, define a mass distribution with $\mathsf{m}(\emptyset) = 0$ and $\sum \mathsf{m}(A) = 1$. The frame of discernment for this wheel is $\mathcal{S} = \{1, 5, 10, 20\}$. The mass function takes values

$$m(\{1\}) = 0.4,$$
$$m(\{5\}) = 0.2,$$
$$m(\{10\}) = 0.2,$$
$$m(\{20\}) = 0.1,$$
$$m(\{1, 5, 10, 20\}) = 0.1$$

(the last value comes from the observation that the hidden-value sector will be picked with a 0.1 chance). The belief–plausibility intervals are

$$[bel(\{1\}), pl(\{1\})] = [0.4, 0.5],$$
$$[bel(\{5\}), pl(\{5\})] = [0.2, 0.3],$$
$$[bel(\{10\}), pl(\{10\})] = [0.2, 0.3],$$
$$[bel(\{20\}), pl(\{20\})] = [0.1, 0.2].$$

The expected utility of this game is the interval $\left[\underline{\mathbb{E}}[U], \overline{\mathbb{E}}[U]\right]$, where

$$\underline{\mathbb{E}}[U] \triangleq m(\{1\}) \cdot 1 + m(\{5\}) \cdot 5 + m(\{10\}) \cdot 10 + m(\{20\}) \cdot 20$$
$$+ m(\{1, 5, 10, 20\}) \cdot \min(1, 5, 10, 20),$$

$$\overline{\mathbb{E}}[U] \triangleq m(\{1\}) \cdot 1 + m(\{5\}) \cdot 5 + m(\{10\}) \cdot 10 + m(\{20\}) \cdot 20$$
$$+ m(\{1, 5, 10, 20\}) \cdot \max(1, 5, 10, 20),$$

and hence

$$\left[\underline{\mathbb{E}}[U], \overline{\mathbb{E}}[U]\right] = [5.50, 7.40].$$

This result may be compared with that obtained using probabilities along with the Principle of Insufficient Reason. In this case, it is assumed that the four values that can be taken by the hidden sector are equally likely, which yields

$$\mathbb{P}(\{1\}) = 0.4 + 0.1 \cdot 0.25 = 0.425,$$
$$\mathbb{P}(\{5\}) = 0.2 + 0.1 \cdot 0.25 = 0.225,$$
$$\mathbb{P}(\{10\}) = 0.2 + 0.1 \cdot 0.25 = 0.225,$$
$$\mathbb{P}(\{20\}) = 0.1 + 0.1 \cdot 0.25 = 0.125,$$

and consequently

$$\mathbb{E}[U] = 0.425 \cdot 1 + 0.225 \cdot 5 + 0.225 \cdot 10 + 0.125 \cdot 20 = 6.30. \qquad \diamond$$

7.13.1 Bayesian belief functions

Generally, belief is not a probability measure: in fact, $\mathsf{bel}(\emptyset) = 0$, $\mathsf{bel}(2^S) = 1$, and $\mathsf{bel}(A) \leqslant \mathsf{bel}(B)$ if $A \subseteq B$, but

$$\mathsf{bel}(A \cup B) \geqslant \mathsf{bel}(A) + \mathsf{bel}(B), \quad \text{if } A \cap B = \emptyset. \tag{7.114}$$

Now, if $A \cap B = \emptyset$ yields $\mathsf{bel}(A \cup B) = \mathsf{bel}(A) + \mathsf{bel}(B)$, then $\mathsf{bel}(A)$ resembles a probability measure, and is called a *Bayesian belief*. In particular, Bayesian beliefs yield $\mathsf{bel}(A \cup A^c) = 1$, which does not hold for general belief measures: in fact, one might accord to both propositions A and A^c very low degrees of belief, consistently with a form of agnosticism.

Remark 7.17. A belief function is Bayesian if and only if all its focal elements are singletons of 2^S [229, p. 45].

Remark 7.18. Notice that a Bayesian belief function, like probability, does not allow one to distinguish between lack of belief and disbelief: if belief is withheld from a proposition, it is accorded to the negation of the proposition.

7.13.2 Dempster rule of combination

If DS theory is used with information fusion, a fundamental tool is *Dempster rule of combination*, which can be interpreted as "a method for changing prior opinions in the light of new evidence" [229]. Given two belief functions based on distinct and independent bodies of evidence, this rule generates a new belief function obtained by combined evidence. Specifically, given two belief masses m_1 and m_2, Dempster rule yields the following belief masses for their combination

$$\mathsf{m}_{1,2}(\emptyset) = 0, \tag{7.115a}$$

$$\mathsf{m}_{1,2}(A) = \frac{1}{1-K} \sum_{B \cap C = A \neq \emptyset} \mathsf{m}_1(B)\, \mathsf{m}_2(C), \tag{7.115b}$$

where the quantity

$$K \triangleq \sum_{B \cap C = \emptyset} \mathsf{m}_1(B)\, \mathsf{m}_2(C) \tag{7.116}$$

can be interpreted as a measure of the *amount of conflict* between the two mass distributions, and represents the mass which would be assigned to the empty set if masses were not normalized [22, p. 41]. Its value should be taken into account to evaluate the quality of the combination, as the presence of a strong conflict, yielding $K \approx 1$, may produce counterintuitive results: on the other hand, as observed in [163, p. 141], "the fact that two information sources are greatly in conflict indicates that there is something wrong or incomplete in our modeling."

Remark 7.19. In [65] conditions are derived under which Dempster rule of combination reduces to Bayes rule.

Example 7.45. Assume that, with $\mathcal{S} = \{x_1, x_2, x_3\}$, we have two bodies of evidence:

$$
\begin{aligned}
m_1(\{x_1\}) &= 0.99, & m_1(\{x_2\}) &= 0.01, & m_1(\{x_3\}) &= 0.00, \\
m_2(\{x_1\}) &= 0.00, & m_2(\{x_2\}) &= 0.01, & m_2(\{x_3\}) &= 0.99.
\end{aligned}
\tag{7.117}
$$

The two mass distributions are not contradictory, as x_2 is common to both. However, the conflict is high, as shown by the value of K:

$$
\begin{aligned}
K &= m_1(\{x_1\})\big(m_2(\{x_2\}) + m_2(\{x_3\})\big) + m_1(\{x_2\})\big(m_2(\{x_1\}) + m_2(\{x_3\})\big) \\
&\quad + m_1(\{x_3\})\big(m_2(\{x_1\}) + m_2(\{x_2\})\big) \\
&= 0.9999.
\end{aligned}
\tag{7.118}
$$

Dempster rule yields

$$
\begin{aligned}
m_{1,2}(\{x_1\}) &= \frac{1}{1-K}\, m_1(\{x_1\})\, m_2(\{x_1\}) = 0.00, \\
m_{1,2}(\{x_2\}) &= \frac{1}{1-K}\, m_1(\{x_2\})\, m_2(\{x_2\}) = 1.00, \\
m_{1,2}(\{x_3\}) &= \frac{1}{1-K}\, m_1(\{x_3\})\, m_2(\{x_3\}) = 0.00,
\end{aligned}
\tag{7.119}
$$

and hence a choice based on DS theory would select $\{x_2\}$, although the hypothesis supported in evidence 1 is $\{x_1\}$ and that supported in evidence 2 is $\{x_3\}$. The resulting choice is one that both bodies of evidence agree upon, albeit mildly. This result may be counterintuitive, as commented upon in [279], where a case is considered with x_1, x_2, and x_3 corresponding to the following diagnoses of doctors 1 and 2: x_1 = meningitis, x_2 = brain tumor, and x_3 = concussion. So both doctors agree that is highly unlikely that the patient has a brain tumor, and yet Dempster rule leads to the conclusion bel(brain tumor) = 1 (this observation is sometimes referred to as *Zadeh paradox*). On the other hand, the bodies of evidence

$$
\begin{aligned}
m_1(\{x_1\}) &= 1.0, & m_1(\{x_2\}) &= 0.0, & m_1(\{x_3\}) &= 0.0, \\
m_2(\{x_1\}) &= 0.0, & m_2(\{x_2\}) &= 0.0, & m_2(\{x_3\}) &= 1.0
\end{aligned}
\tag{7.120}
$$

would yield the amount of conflict $K = 1$, so that no choice is possible under Dempster's rule of combination. In [163, pp. 140–141], it is argued that it is generally unwise to assign belief masses equal to zero, i.e., to assume that some states of the system are actually impossible. A corrective strategy to avoid Zadeh paradox would be to assign nonzero values, albeit small, to all belief masses. For example, if assignment (7.117) is changed into

$$
\begin{aligned}
m_1(\{x_1\}) &= 0.99, & m_1(\{x_2\}) &= 0.009, & m_1(\{x_3\}) &= 0.001, \\
m_2(\{x_1\}) &= 0.001, & m_2(\{x_2\}) &= 0.009, & m_2(\{x_3\}) &= 0.99,
\end{aligned}
\tag{7.121}
$$

one would obtain $m_{1,2}(\{x_1\}) = m_{1,2}(\{x_3\}) \approx 0.48$ and $m_{1,2}(\{x_2\}) \approx 0.04$, indicating a high state of uncertainty in the combined evidence supporting x_1 and x_3. A different method to manage conflicts in Dempster rule of combination is advocated in [266].

Sources and parerga

1. Imprecise probabilities are intimately related to the theory of Robust Statistics, as described, for example, in [119]. As noted in [8, p. 155], the imprecise probability context emphasizes as an entity the whole "credal set" $\mathcal{P}$ of possible probability measures, while the robust approach focuses on a single central elements in the family, and aims at evaluating the effects of deviations from it.

2. "Behavioral decision theory" studies the psychological aspects of humans making decisions. Takemura's book [248] is entirely devoted to this theory, initiated by the studies of Daniel Kahneman and Amos Tversky and those of H. A. Simon (Nobel Memorial Prize in Economic Sciences, 1978), and investigates the relation between decision-making theory and behavioral decision theory.

3. "Ellsberg paradox" is attributed to Daniel Ellsberg, who introduced it in his doctoral dissertation. Ellsberg is also known as the anti-war activist who released to the media the "Pentagon Papers," a top-secret study of the U.S. government decision-making in relation to the Vietnam War. "Allais paradox" is due to Maurice Félix Charles Allais, who in 1988 won the Nobel Memorial Prize in Economic Sciences "for his pioneering contributions to the theory of markets and efficient utilization of resources."

4. The "framing effect" is an outcome of the research which was carried on by Kahneman and Tversky, and earned the first author a Nobel Memorial Prize in Economic Sciences in 2002. Book [156] is a basic primer on Kahneman and Tversky's research work on heuristics in judgment and decision-making.

5. Framing effects can be used to discuss social implications of choices about deployment of telecommunication infrastructures (especially wireless). In [248, p. 107] it is discussed how public perception is influenced by the presentation of upsides/downsides of the innovation introduced.

6. Prospect theory was proposed initially within the framework of decision-making under risk, and subsequently developed into a theory which can explain decision-making under uncertainty [132,252,253].

7. The term "capacity" in the context of nonadditive measures is due to the French mathematician Gustave Choquet in a challenge to classical measure theory. The same notion was proposed independently by the Japanese mathematician Michio Sugeno under the name of "fuzzy measure."

8. The interpretation of capacity as a measure of the ability of a coalition to achieve a common goal is the core of the theory of *cooperative game theory* [102, Chapter 6].

9. The theory based on certain nonadditive measures and now called DS theory was originated by Arthur P. Dempster around 1967 [62], and, later, fully developed

by Glenn Shafer around 1976 [229]. In [102, p. 378] it is observed that DS theory may be described in several different ways with a high methodological diversity, namely, through upper and lower probabilities, "evidence theory," theory of random sets, and a probabilistic approach as in [148]. Monograph [148], entirely devoted to belief functions, uses an approach based on a compatibility relation between observations and states of a system [102, p. 385].

10. Another theory based upon nonadditive measures, and called *Possibility Theory*, derives from the concept of a fuzzy set, proposed in [278]. Book [109] contains a general treatment of ways of representing uncertainty and of logics of reasoning about it.

11. In [114, p. 606] one can find an extensive set of references for evidence (or Dempster–Shafer) theory, fuzzy set theory, and interval analysis, as well as a discussion of "the strengths and weaknesses of the various mathematical structures for the representation of uncertainty."

12. Other measures in the framework of DS theory were proposed, some of them being categorized in [94].

13. In [190, p. 420 ff.], the value of $m(A)$ is interpreted as a measure of the strength of the argument in favor of hypothesis A, and belief functions as probabilities that a given proposition is provable from a set of other propositions to which probabilities are assigned. The belief of A is called the *degree of assurance* in A [190, p. 418].

14. An early algorithm in Multiple-Criteria Decision Making, called "Moral Algebra," was developed by Benjamin Franklin. As described in [145, p. 5], before making important decisions

[Franklin] would write on one side of a sheet of a paper the arguments in favor of the issue, and on the other side the arguments against it. He would then cross out arguments on each side of the paper of relatively equal importance. When all the arguments on one side were crossed out, the side with any arguments not crossed out was the position on the argument that he felt he should support.

Bibliography

[1] A. Abdi, M. Kaveh, Performance comparison of three different estimators for the Nak-agami m parameter using Monte Carlo simulation, IEEE Communications Letters 4 (4) (April 2000) 119–121.

[2] M. Abramowitz, I.A. Stegun, Handbook of Mathematical Functions with Formulas, Graphs, and Mathematical Tables, Dover, New York, NY, 1965.

[3] H. Akaike, A new look at the statistical model identification, IEEE Transactions on Automatic Control 19 (6) (December 1974) 716–723.

[4] N.I. Akhiezer, The Classical Moment Problem and Some Related Questions in Analysis, Oliver & Boyd, Edinburg and London, UK, 1965.

[5] C. Alsina, M.J. Frank, B. Schweizer, Associative Functions: Triangular Norms and Copulas, World Scientific Publishing Co., Singapore, 2006.

[6] G.A. Anastassiou, Probabilistic Inequalities, World Scientific Publishing Co., Singapore, 2010.

[7] T. Augustin, Expected utility within a generalized concept of probability—a comprehensive framework for decision making under ambiguity, Statistical Papers 43 (2002) 5–22.

[8] T. Augustin, F.P.A. Coolen, G. de Cooman, M.C.M. Troffaes (Eds.), Introduction to Imprecise Probabilities, John Wiley & Sons, Chichester, U.K., 2014.

[9] B.M. Ayyub, G.J. Klir, Uncertainty Modeling and Analysis in Engineering and the Sciences, Chapman & Hall/CRC, Boca Raton, FL, 2006.

[10] C. Baudrit, D. Dubois, Practical representations of incomplete probabilistic knowledge, Computational Statistics & Data Analysis 51 (1) (2006) 86–108.

[11] C. Bayer, C. Teichmann, The proof of Tchakaloff's theorem, Proceedings of the American Mathematical Society 134 (10) (October 2006) 3035–3040.

[12] S. Benedetto, E. Biglieri, Principles of Digital Transmission with Wireless Applications, Kluwer, New York, NY, 1999.

[13] Y. Ben-Haim, Uncertainty, probability and information-gaps, Reliability Engineering and System Safety 85 (1–3) (July–September 2004) 249–266.

[14] Y. Ben-Haim, Info-Gap Decision Theory: Decisions Under Severe Uncertainty, 2nd edition, Academic Press, London, UK, 2006.

[15] C. Berg, The multidimensional moment problem and semigroups, in: H.J. Landau (Ed.), Moments in Mathematics, Proceedings of Symposia in Applied Mathematics, vol. 37, American Mathematical Society, Providence RI, 1987, pp. 110–124.

[16] D. Berleant, K. Villaverde, O.M. Koseheleva, Towards a more realistic representation of uncertainty: an approach motivated by info-gap decision theory, in: 2008 Annual Meeting of the North American Fuzzy Information Processing Society (NAFIPS 2008), New York, NY, May 19–22, 2008.

[17] D. Bernoulli, Exposition of a new theory on the measurement of risk, Econometrica 22 (1) (January 1954) 23–36, Translated from "Specimen Theoriæ Novæ de Mensura Sortis", Commentarii Academiæ Scientiarum Imperialis Petropolitanæ, vol. V, 1738, pp. 175–192 (in Latin).

[18] D. Bertsimas, I. Popescu, Optimal inequalities in probability theory: a convex optimization approach, SIAM Journal on Optimization 15 (3) (2005) 780–804.

[19] K.-L. Besser, E.A. Jorswieck, Reliability bounds for dependent fading wireless channels, IEEE Transactions on Wireless Communications 19 (9) (September 2020) 5833–5845.

[20] K.-L. Besser, E.A. Jorswieck, Copula-based bounds for multi-user communications—Part II: outage performance, IEEE Communications Letters 25 (1) (January 2021) 8–12.

[21] K.-L. Besser, E.A. Jorswieck, Bounds on the secrecy outage probability for dependent fading channels, IEEE Transactions on Communications 69 (1) (January 2021) 443–456.

[22] M. Beynon, B. Curry, P. Morgan, The Dempster–Shafer theory of evidence: an alternative approach to multicriteria decision modelling, Omega—The International Journal of Management Science 28 (2000) 37–50.

[23] E. Biglieri, Coding for Wireless Channels, Springer, New York, NY, 2005.

[24] E. Biglieri, Spectrum sensing under uncertain channel modeling, Journal of Communications and Networks 14 (3) (June 2012) 225–229.

[25] E. Biglieri, Dealing with uncertain models in wireless communications, in: IEEE ICASSP 2016, Shanghai, China, March 2016.

[26] E. Biglieri, R. Calderbank, A. Constantinides, A. Goldsmith, A. Paulraj, H.V. Poor, MIMO Wireless Communications, Cambridge University Press, Cambridge, UK, 2007.

[27] E. Biglieri, A.J. Goldsmith, L.J. Greenstein, N.B. Mandayam, H.V. Poor, Principles of Cognitive Radio, Cambridge University Press, Cambridge, UK, 2012.

[28] E. Biglieri, E. Grossi, M. Lops, Random-set theory and wireless communications, Foundations and Trends® in Communications and Information Theory 7 (4) (2010) 317–462.

[29] E. Biglieri, I-W. Lai, The impact of independence assumptions on wireless communication analysis, in: IEEE ISIT 2016, Barcelona, Spain, July 2016.

[30] E. Biglieri, M. Lops, Linear–quadratic detectors for spectrum sensing, Journal of Communications and Networks 16 (5) (October 2014) 485–492.

[31] E. Biglieri, J. Proakis, S. Shamai (Shitz), Fading channels: information-theoretic and communications aspects, IEEE Transactions on Information Theory 44 (6) (October 1998) 2619–2692, 50th Anniversary Issue.

[32] E. Biglieri, G. Taricco, Bounds on the distribution of a random variable whose first moments are known, Archiv für Elektronik und Übertragungstechnik 41 (6) (1987) 330–336.

[33] E. Biglieri, K. Yao, C.-A. Yang, Fading models from spherically invariant processes, IEEE Transactions on Wireless Communications 14 (10) (October 2015) 5526–5538.

[34] M. Bloch, J. Barros, M.R.D. Rodrigues, S.W. McLaughlin, Wireless information-theoretic security, IEEE Transactions on Information Theory 54 (6) (June 2008) 2515–2534.

[35] H.W. Block, W.S. Griffith, T.H. Savits, L-superadditive structure functions, Advances in Applied Probability 21 (1989) 919–929.

[36] R.P. Boas, Spelling lesson, The College Mathematics Journal 15 (3) (June 1984) 217.

[37] S. Boucheron, G. Lugosi, P. Massart, Concentration Inequalities, Oxford University Press, Oxford, UK, 2013.

[38] S. Boyd, L. Vandenberghe, Convex Optimization, Cambridge University Press, Cambridge, UK, 2008.

[39] H. Brehm, Description of spherically invariant random processes by means of G-functions, in: D. Jungnickel, K. Vedder (Eds.), Combinatorial Theory—Lecture Notes in Mathematics, Springer-Verlag, Berlin, Germany, 1982, pp. 39–73.

[40] P.L. Brockett, S.H. Cox Jr., R.D. MacMinn, B. Shi, Best bounds on measures of risk and probability of ruin for alpha unimodal random variables when there is limited moment information, Applied Mathematics 7 (2016) 765–783.

[41] K.P. Burnham, D.R. Anderson, Model Selection and Multimodel Inference—A Practical Information-Theoretic Approach, 2nd edition, Springer Verlag, New York, NY, 2002.

[42] P.L. Butzer, S. Jansche, A direct approach to the Mellin transform, The Journal of Fourier Analysis and Applications 3 (4) (November 1997) 325–376.

[43] S. Cambanis, G. Simons, W. Stout, Inequalities for $\mathbb{E}k(X,Y)$ when the marginals are fixed, Zeitschrift für Wahrscheinlichkeitstheorie und verwandte Gebiete 36 (1976) 285–294.

[44] J. Cheng, N.C. Beaulieu, Maximum-likelihood based estimation of the Nakagami m parameter, IEEE Communications Letters 5 (3) (March 2001) 101–103.

[45] J. Cheng, N.C. Beaulieu, Generalized moment estimators for the Nakagami fading parameter, IEEE Communications Letters 6 (4) (April 2002) 144–146.

[46] U. Cherubini, F. Gobbi, S. Mulinacci, S. Romagnoli, Dynamic Copula Methods in Finance, John Wiley & Sons, Chichester, UK, 2012.

[47] K. Chowdhari, P. Dupuis, Distinguishing and integrating aleatoric and epistemic variation in uncertainty quantification, ESAIM: Mathematical Modelling and Numerical Analysis 47 (3) (March 2013) 635–662.

[48] B.H. Cheng, L. Vandenberghe, K. Yao, Semidefinite programming bounds on the probability of error of binary communication systems with inexactly known intersymbol interference, IEEE Transactions on Information Theory 51 (8) (August 2005) 2951–2954.

[49] G. Claesken, N.L. Hjort, Model Selection and Model Averaging, Cambridge University Press, Cambridge, UK, 2008.

[50] K. Comanor, L. Vandenberghe, S. Boyd, Semidefinite programming and multivariate Chebyshev bounds, in: Proc. IFAC Symposium on Robust Control Design, Toulouse, France, July 5–7, 2006, pp. 592–596.

[51] R. Cools, Constructing cubature formulae: the science behind the art, Acta Numerica 6 (January 1997) 1–54.

[52] R. Cools, An encyclopaedia of cubature formulas, Journal of Complexity 19 (2003) 445–453.

[53] R. Cools, I.P. Mysovskikh, H.J. Schmidt, Cubature formulae and orthogonal polynomials, Journal of Computational and Applied Mathematics 127 (1-2) (15 January 2001) 121–152. Also in: C. Brezinski, L. Wuytack (Eds.), Numerical Analysis: Historical Developments in the 20th Century, Elsevier, Amsterdam, The Netherlands, 2001, pp. 281–312.

[54] S.L. Cotton, W. Cully, W.G. Scanlon, J. McQuiston, Channel characterization for indoor wearable active RFID at 868 MHz, in: 2011 Loughborough Antennas and Propagation Conference, Loughborough, UK, November 14-15, 2011.

[55] T.M. Cover, J.A. Thomas, Elements of Information Theory, 2nd edition, John Wiley & Sons, New York, NY, 2006.

[56] H. Cramér, Mathematical Methods of Statistics, 19th printing, Princeton University Press, Princeton, NJ, 1999.

[57] W. Cui, D.I. Blockley, Interval probability theory for evidential support, International Journal of Intelligent Systems 5 (1990) 183–192.

[58] H.A. David, H.N. Nagaraja, Order Statistics, 3rd edition, John Wiley & Sons, Hoboken, NJ, 2003.

[59] A.C. Davison, Statistical Models, Cambridge University Press, Cambridge, UK, 2003.

[60] H. Dawood, Theories of Interval Arithmetic: Mathematical Foundations and Applications, Lambert Academic Publishing, Saarbrücken, Germany, 2011.

[61] M. Debbah, R.R. Müller, MIMO channel modeling and the principle of maximum entropy, IEEE Transactions on Information Theory 51 (5) (May 2005) 1667–1690.

[62] A.P. Dempster, Upper and lower probabilities induced by a multivalued mapping, Annals of Mathematical Statistics 38 (2) (1967) 325–339.

[63] M. Denuit, C. Genest, É. Marceau, Stochastic bounds on sums of dependent risks, Insurance: Mathematics and Economics 25 (1999) 85–104.

[64] M. Denuit, O. Scaillet, Nonparametric tests for positive quadrant dependence, Journal of Financial Econometrics 2 (3) (2004) 422–450.

[65] J. Dezert, A. Tchamova, D. Han, J.-M. Tacnet, Why Dempster's fusion rule is not a generalization of Bayes fusion rule, in: F. Smarandache, J. Dezert (Eds.), Advances and Applications of DSmT for Information Fusion. Collected Works, vol. 4, American Research Press, 2015, pp. 195–202.

[66] Y. Dodge, The Concise Encyclopedia of Statistics, Springer-Verlag, 2008.

[67] W. Dong, H.C. Shah, Vertex method for computing fuctions of fuzzy variables, Fuzzy Sets and Systems 24 (1) (October 1987) 65–78.

[68] D.C. Dowson, A. Wragg, Maximum-entropy distributions having prescribed first and second moments, IEEE Transactions on Information Theory 19 (5) (September 1973) 689–693.

[69] E. Drakopoulos, C.-C. Lee, Decision rules for distributed decision networks with uncertainties, IEEE Transactions on Automatic Control 37 (1) (January 1992) 5–14.

[70] M. Dresher, Moment spaces and inequalities, Duke Mathematical Journal 20 (June 1953) 261–271.

[71] M. Dresher, S. Karlin, L. Shapley, Polynomial games, in: H. Bohnenblust, et al. (Eds.), Contributions to the Theory of Games (AM-24). Vol. I, Princeton University Press, Princeton, NJ, 1950, pp. 161–180.

[72] C.F. Dunkl, Y. Xu, Orthogonal Polynomials of Several Variables, Cambridge University Press, Cambridge, UK, 2001.

[73] P. Dupuis, R.S. Ellis, A Weak Convergence Approach to the Theory of Large Deviations, John Wiley & Sons, New York, NY, 2000.

[74] F. Durante, C. Sempi, Principles of Copula Theory, CRC Press, Boca Raton, FL, 2016.

[75] A. Eckberg Jr., Sharp bounds on Laplace-Stieltjes trasforms, with applications to various queueing problems, Mathematics of Operations Research 2 (2) (May 1977) 135–142.

[76] B. Efron, C. Morris, Stein's estimation rule and its competitors–an empirical Bayes approach, Journal of the American Statistical Association 68 (341) (March 1973) 117–130.

[77] B. Efron, C. Morris, Stein's paradox in statistics, Scientific American 236 (5) (1977) 119–127.

[78] A. Erdelyi, Tables of Integral Transforms, McGraw-Hill, New York, NY, 1954.

[79] J. Etner, M. Jeleva, J.-M. Tallon, Decision theory under ambiguity, Journal of Economic Surveys 26 (2) (2012) 234–270.

[80] W. Feller, An Introduction to Probability Theory and Its Applications, vol. 2, John Wiley & Sons, New Yor, NY, 1966.

[81] S. Ferson, V. Kreinovich, L. Ginzburg, D.S. Myers, K. Sentz, Constructing Probability Boxes and Dempster–Shafer Structures, Report SAND 2002-4015, Sandia National Laboratory, Albuquerque, NM, January 2003.

[82] S. Ferson, R.B. Nelsen, J. Hajagos, D.J. Berleant, J. Zhang, W.T. Tucker, L.R. Ginzburg, W.L. Oberkampf, Dependence in Probabilistic Modeling, Dempster–Shafer Theory, and Probability Bounds Analysis, Report SAND 2004-3072, Sandia National Laboratory, Albuquerque, NM, October 2004.

[83] S. Ferson, J. Siegrist, Verified computation with probabilities, in: A.M. Dienstfrey, R.F. Boisvert (Eds.), Uncertainty Quantification in Scientific Computing. Proceedings of 10th IFIP WG 2.5 Working Conference, WoCoUQ 2011, Boulder, CO, 2012, pp. 95–122.

[84] T.L. Fine, Lower probability models for uncertainty and nondeterministic processes, Journal of Statistical Planning and Inference 20 (3) (1988) 389–411.

[85] P.C. Fishburn, The Foundations of Expected Utility, Springer, Dordrecht, Germany, 1982.

[86] G.S. Fishman, Monte Carlo: Concepts, Algorithms, and Applications, Springer, New York, NY, 1999.

[87] A. Fort, C. Desset, Ph. De Doncker, P. Wambacq, L. Van Biesen, An ultra-wideband body area propagation channel model—from statistics to implementation, IEEE Transactions on Microwave Theory and Techniques 54 (4) (April 2006) 1820–1826.

[88] C.R. Fox, G. Ülkümen, Distinguishing two dimensions of uncertainty, in: W. Brun, G.B. Keren, G. Kirkebøen, H. Montgomery (Eds.), Perspectives on Thinking, Judging, and Decision Making: A Tribute to Karl Halvor Teigen, Universitetsforlagets, Oslo, Norway, 2011, pp. 21–35.

[89] M.J. Frank, R.B. Nelsen, B. Schweizer, Best-possible bounds for the distribution of a sum—a problem of Kolmogorov, Probability Theory and Related Fields 74 (2) (June 1987) 199–211.

[90] T. Gajdos, J.-M. Tallon, J.-C. Vergnaud, Decision making with imprecise probabilistic information, Journal of Mathematical Economics 40 (6) (2004) 647–681.

[91] H. Gaspars-Wieloch, Modifications of the Hurwicz's decision rule, Central European Journal of Operations Research 22 (2014) 779–794.

[92] W. Gautschi, On the construction of Gaussian quadrature rules from modified moments, Mathematics of Computation 24 (110) (April 1970) 245–260.

[93] W. Gautschi, Orthogonal Polynomials: Computation and Approximation, Oxford University Press, Oxford, UK, 2004.

[94] T. George, N.R. Pal, Quantification of conflict in Dempster–Shafer framework: a new approach, International Journal of General Systems 24 (4) (1996) 407–423.

[95] B.K. Ghosh, Probability inequalities related to Markov's theorem, The American Statistician 56 (3) (August 2002) 186–190.

[96] H.J. Godwin, On generalizations of Tchebychef's inequality, Journal of the American Statistical Association 50 (271) (September 1955) 923–945.

[97] G.H. Golub, Some modified matrix eigenvalue problems, SIAM Review 15 (1973) 318–334.

[98] G.H. Golub, C.F. Van Loan, Matrix Computations, 3rd edition, Johns Hopkins University Press, Baltimore, MD, 1996.

[99] G.H. Golub, J.H. Welsch, Calculation of Gauss quadrature rules, Mathematics of Computation 23 (April 1969) 221–230.

[100] K. Gourgoulias, M.A. Katsoulakis, L. Rey-Bellet, J. Wang, How biased is your model? Concentration inequalities, information and model bias, IEEE Transactions on Information Theory 66 (5) (May 2020) 3079–3097.

[101] J. Goutsias, R.P.S. Mahler, H.T. Nguyen (Eds.), Random Sets: Theory and Applications, Springer-Verlag, New York, NY, 1997.

[102] M. Grabisch, Set Functions, Games and Capacities in Decision Making, Springer, Switzerland, 2016.

[103] S. Greco, M. Ehrgott, J.R. Figueira (Eds.), Multiple Criteria Decision Analysis: State of the Art Surveys, Springer, New York, NY, 2016.

[104] P.D. Grünwald, The Minimum Description Length Principle, The MIT Press, Cambridge, MA, 2007.

[105] V. Gupta, T. Osogami, On Markov–Krein characterization of the mean waiting time in $M/G/K$ and other queueing systems, Queueing Systems 68 (2011) 339–352.

[106] A. Haegemans, R. Piessens, Construction of cubature formulas of degree seven and nine symmetric planar regions, using orthogonal polynomials, SIAM Journal on Numerical Analysis 14 (3) (June 1977) 492–508.

[107] T.C. Hales, S.P. Ferguson, The Kepler Conjecture: The Hales–Ferguson Proof, Springer, New York, NY, 2011.

[108] J.W. Hall, D.I. Blockley, J.P. Davis, Uncertain inference using interval probability theory, International Journal of Approximate Reasoning 19 (1998) 247–264.

[109] J.Y. Halpern, Reasoning About Uncertainty, 2nd edition, The MIT Press, Cambridge, MA, 2017.

[110] E. Hansen, G.W. Walster, Global Optimization Using Interval Analysis, 2nd edition, Marcel Dekker, New York, NY, 2004.

[111] M.H. Hansen, B. Yu, Model selection and the principle of minimum description length, Journal of the American Statistical Association 96 (454) (June 2001) 746–774.

[112] T.L. Heath, The Works of Archimedes, Dover Publications, Mineola, NY, 2002.

[113] R.W. Heath Jr., A. Lozano, Foundations of MIMO Communication, Cambridge Univesity Press, Cambridge, UK, 2019.

[114] J.C. Helton, J.D. Johnson, W.L. Oberkampf, C.J. Sallaberry, Representation of analysis results involving aleatory and epistemic uncertainty, International Journal of General Systems 39 (6) (August 2010) 605–646.

[115] C.C. Heyde, On a property of the lognormal distribution, Journal of the Royal Statistical Society. Series B (Methodological) 25 (2) (1963) 392–393.

[116] W. Hoeffding, Probability inequalities for sums of bounded random variables, Journal of the American Statistical Association 58 (301) (March 1963) 13–30.

[117] H. Holm, M.-S. Alouini, Sum and difference of two squared correlated Nakagami variates in connection with the McKay distribution, IEEE Transactions on Communications 52 (8) (August 2004) 1367–1376.

[118] C. Hu, R.B. Kearfott, A. de Korvin, V. Kreinovich (Eds.), Knowledge Processing with Interval and Soft Computing, Springer, London, UK, 2008.

[119] P.J. Huber, E.M. Ronchetti, Robust Statistics, 2nd edition, John Wiley & Sons, Hoboken, NJ, 2009.

[120] W. Hürlimann, Extremal moment methods and stochastic orders: application in actuarial science, Chapters I–III, Boletín de la Asociación Matemática Venezolana XV (1) (2008) 5–301.

[121] L. Hurwicz, The Generalized Bayes-Minimax Principle: a Criterion for Decision-Making Under Uncertainty, Cowles Commission Discussion Paper: Statistics No. 355, February 8, 1951.

[122] V.N. Huynh, C. Hu, Y. Nakamori, V. Kreinovich, On decision making under interval uncertainty: a new justification of Hurwicz optimism-pessimism approach and its use in group decision making, in: Proceedings of the 39th International Symposium on Multiple-Valued Logic (ISMVL'2009), Naha, Okinawa, Japan, May 21–23, 2009, pp. 214–220.

[123] I. Ioannou, C.D. Charalambous, S. Loyka, Outage probability under channel distribution uncertainty, IEEE Transactions on Information Theory 58 (11) (November 2012) 6825–6838.

[124] K. Isii, On sharpness of Tchebycheff-type inequalities, Annals of the Institute of Statistical Mathematics 14 (1963) 185–197.

[125] L. Jaulin, M. Kieffer, O. Didrit, É. Walter, Applied Interval Analysis, Springer-Verlag, London, UK, 2001.

[126] P. Jaworski, F. Durante, W.K. Härdle, T. Rychlik (Eds.), Copula Theory and Its Applications, Springer, Berlin, Germany, 2010.

[127] E.T. Jaynes, Information theory and statistical mechanics I, Physical Review 106 (4) (1957) 620–630.
[128] N.L. Johnson, C.A. Rogers, The moment problem for unimodal distributions, The Annals of Mathematical Statistics 22 (3) (September 1951) 433–439.
[129] M.A. Johnson, M.R. Taaffe, Tchebycheff systems for probabilistic analysis, American Journal of Mathematical and Management Sciences 13 (1–2) (1993) 83–111.
[130] E.A. Jorswieck, K-L. Besser, Copula-based bounds for multi-user communications— Part I: average performance, IEEE Communications Letters 25 (1) (January 2021) 3–7.
[131] R. Kaas, R.J.A. Laeven, R.B. Nelsen, Worst VaR scenario with given marginals and measures of association, Insurance: Mathematics and Economics 44 (2009) 146–158.
[132] D. Kahneman, A. Tversky, Prospect theory: an analysis of decision under risk, Econometrica 47 (2) (March 1979) 263–292.
[133] D. Kahneman, A. Tversky, The psychology of preferences, Scientific American 246 (1) (1982) 160–173.
[134] T. Kailath, The divergence and Bhattacharyya distance measures in signal selection, IEEE Transactions on Communications Technology COM-15 (1) (February 1967) 52–60.
[135] S. Karlin, Dynamic inventory policy with varying stochastic demands, Management Science 6 (3) (April 1960) 231–258.
[136] S. Karlin, W.J. Studden, Tchebycheff Systems: With Applications in Analysis and Statistics, John Wiley & Sons, New York, NY, 1966.
[137] S.M. Kay, Fundamentals of Statistical Signal Processing: Estimation Theory, Prentice Hall, Upper Saddle River, NJ, 1993.
[138] S.M. Kay, Fundamentals of Statistical Signal Processing: Detection Theory, Prentice Hall, Upper Saddle River, NJ, 1998.
[139] R.L. Keeney, H. Raiffa, Decisions with Multiple Objectives: Preferences and Value Tradeoffs, Cambridge University Press, Cambridge, UK, 1993.
[140] J.M. Keynes, A Treatise on Probability, MacMillan, London, UK, 1921.
[141] J.H.B. Kemperman, The general moment problem, a geometric approach, Annals of Mathematical Statistics 39 (1968) 93–122.
[142] J.H.B. Kemperman, Geometry of the moment problem, in: H.J. Landau (Ed.), Moments in Mathematics, Proceedings of Symposia in Applied Mathematics, vol. 37, American Mathematical Society, Providence RI, 1987, pp. 16–53.
[143] G.J. Klir, Uncertainty and Information: Foundations of Generalized Information Theory, John Wiley & Sons, Hoboken, NJ, 2006.
[144] D. Koks, S. Challa, An Introduction to Bayesian and Dempster–Shafer Data Fusion, Report DSTO-TR-1436, DSTO Systems Sciences Laboratory, Edinburgh, Australia, 2003.
[145] M. Köksalan, J. Wallenius, S. Zionts, An early history of multiple criteria decision making, in: S. Greco, M. Ehrgott, J.R. Figueira (Eds.), Multiple Criteria Decision Analysis: State of the Art Surveys, Springer, New York, NY, 2016, pp. 3–17.
[146] J. Komorník, M. Komorníková, J. Kalická, Dependence measures for perturbations of copulas, Fuzzy Sets and Systems 324 (1 October 2017) 100–116.
[147] C.J. Kowalski, Non-normal bivariate distributions with normal marginals, The American Statistician 27 (3) (June 1973) 103–106.
[148] I. Kramosil, Probabilistic Analysis of Belief Functions, Springer, New York, NY, 2001.
[149] M.G. Kreĭn, A.A. Nudel'man, The Markov Moment Problem and Extremal Problems, American Mathematical Society, Providence, Rhode Island, 1977 (English Translation of: М.Г. Крейн и А.А. Нудельман, Проблема Моментов Маркова и Экстремальные Задачи. Издательство "Наука", Москва, СССР, 1973 (на русском языке)).

[150] H.E. Kyburg, Interval-valued probabilities, Contribution to the Documentation Section on the website www.sipta.org, 1999, of the Society for Imprecise Probability Theory and Applications (SIPTA).

[151] H.J. Landau (Ed.), Moments in Mathematics, Proceedings of Symposia in Applied Mathematics, vol. 37, American Mathematical Society, Providence, RI, 1987.

[152] A. Lapidoth, P. Narayan, Reliable communication under channel uncertainty, IEEE Transactions on Information Theory 44 (6) (October 1998) 2148–2177.

[153] J.B. Lasserre, Moments, Positive Polynomials and Their Applications, Imperial College Press, London, UK, 2010.

[154] A.J. Laub, Matrix Analysis for Scientists & Engineers, Society for Industrial and Applied Mathematics, Philadelphia, PA, 2005.

[155] M. Ledoux, The Concentration of Measure Phenomenon, American Mathematical Society, Providence, RI, 2001.

[156] M. Lewis, The Undoing Project: A Friendship That Changed Our Minds, Norton, New York, NY, 2017.

[157] J. Li, D. Xiu, Computation of failure probability subject to epistemic uncertainty, SIAM Journal of Scientific Computing 34 (6) (January 2012) A2946–A2964.

[158] G.D. Lin, Recent developments on the moment problem, Journal of Statistical Distributions and Applications 4 (5) (2017) 1–17.

[159] L.J. Lucas, H. Owhadi, M. Ortiz, Rigorous verification, validation, uncertainty quantification and certification through concentration-of-measure inequalities, Computer Methods in Applied Mechanics and Engineering 197 (51-52) (15 October 2008) 4591–4609.

[160] R.D. Luce, A.J. Marley, C.T. Ng, Entropy-related measures of the utility of gambling, in: S. Brams, W.G. Gehrlein, F.S. Roberts (Eds.), The Mathematics of Preference, Choice and Order, Springer, Berlin, Germany, 2009, pp. 5–25.

[161] R.D. Luce, C.T. Ng, A.J. Marley, J. Aczél, Utility of gambling I: entropy modified linear weighted utility, Economic Theory 36 (2008) 1–33.

[162] R.D. Luce, C.T. Ng, A.J. Marley, J. Aczél, Utility of gambling II: risk, paradoxes, and data, Economic Theory 36 (2008) 165–187.

[163] R.P.S. Mahler, Statistical Multisource–Multitarget Information Fusion, Artech House, Norwood, MA, 2007.

[164] J.-F. Mai, M. Scherer, Financial Engineering with Copulas Explained, Palgrave Macmillan, UK, 2014.

[165] C. Mallows, Another comment on O'Cinneide, The American Statistician 45 (3) (August 1991) 257.

[166] C.F. Manski, Partial Identification of Probability Distributions, Springer, New York, NY, 2003.

[167] C.F. Manski, Public Policy in an Uncertain World: Analysis and Decisions, Harvard University Press, Cambridge, MA, 2013.

[168] A.W. Marshall, I. Olkin, Multivariate Chebyshev inequalities, Annals of Mathematical Statistics 31 (4) (1960) 1001–1014.

[169] G. Matheron, Random Sets and Integral Geometry, John Wiley & Sons, New York, NY, 1975.

[170] J.W. Matthews, Sharp error bounds for intersymbol interference, IEEE Transactions on Information Theory IT-19 (4) (July 1973) 440–447.

[171] C. McDiarmid, Concentration, in: M. Habib, C. McDiarmid, J. Ramirez-Alfonsin, B. Reed (Eds.), Probabilistic Methods for Algorithmic Discrete Mathematics, Springer, Berlin, Germany, 1998.

[172] A.J. McNeil, R. Frey, P. Embrechts, Quantitative Risk Management: Concepts, Techniques and Tools, Princeton University Press, Princeton, NJ, 2005.

[173] C. Meyer, The bivariate normal copula, Communications in Statistics—Theory and Methods 42 (13) (2013) 2402–2422.

[174] R. Mitra, A.K. Mishra, T. Choubisa, Maximum likelihood estimate of parameters of Nakagami-m distribution, in: Proceedings of the 2012 International Conference on Communications, Devices and Intelligent Systems, Kolkata, India, December 28-29, 2012, pp. 9–12.

[175] I.S. Molchanov, Theory of Random Sets, Springer Verlag, London, UK, 2005.

[176] R.E. Moore, Methods and Applications of Interval Analysis, Society for Industrial and Applied Mathematics, Philadelphia, PA, 1979.

[177] R.E. Moore, R.B. Kearfott, M.J. Cloud, Introduction to Interval Analysis, Society for Industrial and Applied Mathematics, Philadelphia, PA, 2009.

[178] G.S. Mudholkar, P.S.R.S. Rao, Some sharp multivariate Tchebycheff inequalities, Annals of Mathematical Statistics 38 (2) (1967) 393–400.

[179] I.P. Mysovskih, On the construction of cubature formulas with the smallest number of nodes, Soviet Mathematics Doklady 9 (1968) 277–280. Translated from Doklady Akademii Nauk USSR 178 (1968) 1252–1254.

[180] M. Nakagami, The m-distribution: a general formula of intensity distribution of rapid fading, in: W.C. Hoffman (Ed.), Statistical Methods in Radio Wave Propagation, Pergamon Press, New York, NY, 1960, pp. 3–36.

[181] R.B. Nelsen, An Introduction to Copulas, 2nd edition, Springer, New York, NY, 2006.

[182] R.B. Nelsen, J.J. Quesada-Molina, J.A. Rodríguez-Lallena, M. Úbeda-Flores, Bounds on bivariate distribution functions with given margins and measures of association, Communications in Statistics—Theory and Methods 30 (6) (2001) 1155–1162.

[183] R.B. Nelsen, M. Úbeda-Flores, A comparison of bounds on sets of joint distribution functions derived from various mesures of association, Communications in Statistics—Theory and Methods 33 (10) (2004) 2299–2305.

[184] J. von Neumann, O. Morgenstern, Theory of Games and Economic Behavior, 3rd edition, Princeton University Press, Princeton, NJ, 1953.

[185] I. Newton, The Principia: Mathematical Principles of Natural Philosophy, University of California Press, Oakland, CA, 1999.

[186] H.T. Nguyen, An Introduction to Random Sets, Chapman & Hall, Boca Raton, FL, 2006.

[187] I.E. Ovcharenko, Two-dimensional moment sequences, Ukrainian Mathematical Journal 36 (1) (1984) 46–51.

[188] H. Owhadi, C. Scovel, T.J. Sullivan, M. McKerns, M. Ortiz, Optimal uncertainty quantification, SIAM Review 55 (2) (2013) 271–345.

[189] A. Papoulis, Probability, Random Variables, and Stochastic Processes, 3rd edition, McGraw-Hill, New York, NY, 1991.

[190] J. Pearl, Probabilistic Reasoning in Intelligent Systems: Networks of Plausibile Inference, revised 2nd printing, Morgan Kaufmann, San Francisco, CA, 1988.

[191] L.C. Petersen, On the relation between the multidimensional moment problem and the one-dimensional moment problem, Mathematica Scandinavica 51 (1982) 361–366.

[192] R. Piessens, A. Haegemans, Cubature formulas of degree nine for symmetric planar regions, Mathematics of Computation 29 (1975) 810–815.

[193] A. Pinkus, J.M. Quesada, On Chebyshev–Markov–Krein inequalities, Journal of Approximation Theory 164 (2012) 1262–1282.

[194] D. Pollard, A User's Guide to Measure Theoretic Probability, Cambridge University Press, Cambridge, UK, 2002.

[195] H.V. Poor, An Introduction to Signal Detection and Estimation, 2nd edition, Springer, New York, NY, 1994.

[196] I. Popescu, A semidefinite programming approach to optimal moment bounds for convex classes of distributions, Mathematics of Operations Research 30 (3) (August 2005) 632–657.

[197] M. Preischl, Bounds on integrals with respect to multivariate copulas, Dependence Modeling 4 (2016) 277–287.

[198] W.H. Press, S.A. Teukolsky, W.T. Vetterling, B.P. Flannery, Numerical Recipes: The Art of Scientific Computing, 3rd edition, Cambridge University Press, Cambridge, UK, 2007.

[199] G. Puccetti, L. Rüschendorf, Computation of sharp bounds on the distribution of a function of dependent risks, Journal of Computational and Applied Mathematics 236 (2012) 1833–1840.

[200] G. Puccetti, L. Rüschendorf, Computation of sharp bounds on the expected value of a supermodular function of risks with given marginals, Communication in Statistics–Simulation and Computation 44 (3) (March 2015) 705–718.

[201] G. Puccetti, R. Wang, Extremal dependence concepts, Statistical Science 30 (4) (2015) 485–517.

[202] M. Putinar, A note on Tchakaloff's theorem, Proceedings of the American Mathematical Society 125 (8) (August 1997) 2409–2414.

[203] M. Putinar, C. Scheiderer, Multivariate moment problems: geometry and indeterminateness, Annali della Scuola Normale di Pisa–Classe di Scienze, Series 5 5 (2) (2006) 137–157.

[204] Z. Quan, S. Cui, A.H. Sayed, Optimal linear cooperation for spectrum sensing in cognitive radio networks, IEEE Journal of Selected Topics in Signal Processing 2 (1) (February 2008) 28–40.

[205] S.T. Rachev, L. Rüschendorf, Mass Transportation Problems. Volume I: Theory, Springer-Verlag, New York, NY, 1998.

[206] J. Radon, Zur mechanischen Kubatur, Monatshefte für Mathematik 52 (1948) 286–300.

[207] H.M. Regan, S. Ferson, D. Berleant, Equivalence of methods for uncertainty propagation of real-valued random variables, International Journal of Approximate Reasoning 36 (1) (2004) 1–30.

[208] A.M. Reza, R.L. Kirlin, Maximum entropy estimation of density functions using order statistics, IEEE Transactions on Information Theory 67 (5) (May 2021) 3075–3094.

[209] J. Rissanen, Modeling by shortest data description, Automatica 14 (5) (September 1978) 465–471.

[210] J.A. Ritcey, Copula models for wireless fading and their impact on wireless diversity combining, in: Forty-First Asilomar Conference on Signals, Systems and Computers, Pacific Grove, CA, November 4–7, 2007, pp. 1564–1567.

[211] H.L. Royden, Real Analysis, 2nd edition, Macmillan, New York, NY, 1968.

[212] L. Rüschendorf, Sharpness of Fréchet-bounds, Zeitschrift für Wahrscheinlichkeitstheorie und verwandte Gebiete 57 (2) (1981) 293–302.

[213] L. Rüschendorf, Random variables with maximum sums, Advances in Applied Probability 14 (1982) 623–632.

[214] L. Rüschendorf, Fréchet-bounds and their applications, in: G. Dall'Aglio, S. Kotz, G. Salinetti (Eds.), Advances in Probability Distributions with Given Marginals—Beyond the Copulas, Kluwer, Dordrecht, The Netherlands, 1991.

[215] B. Russell, Introduction to Mathematical Philosophy, 2nd edition, George Allen & Unwin, London, UK, 1920.

[216] T. Rychlik, Bounds for order statistics based on dependent variables with given nonidentical distributions, Statistics & Probability Letters 23 (1995) 351–358.

[217] T. Rychlik, Order statistics of variables with given marginal distributions, in: L. Rüschendorf, B. Schweizer, M.D. Taylor (Eds.), Distributions with Fixed Marginals and Related Topics, in: IMS Lecture Notes, Monograph Series, vol. 28, 1996, pp. 297–306.

[218] F. Salmon, Recipe for disaster: the formula that killed Wall Street, Wired Magazine 17 (3) (February 23, 2009).

[219] W. Sans, E. von Collani, A note on monotonic probability distributions, Economic Quality Control 18 (1) (2003) 67–81.

[220] I.R. Savage, Probability inequalities of the Tchebycheff type, Journal of Research of the National Bureau of Standards—B. Mathematics and Mathematical Physics 65B (3) (July–September 1961) 211–222.

[221] L.J. Savage, The Foundations of Statistics, 2nd edition, Dover, New York, NY, 1972.

[222] J.G. Saw, M.C.K. Yang, T.C. Mo, Chebyshev inequality with estimated mean and variance, The American Statistician 38 (2) (May 1984) 130–132.

[223] F. Schmid, R. Schmidt, Nonparametric inference on multivariate versions of Blomqvist's beta and related measures of tail dependence, Metrika 66 (3) (November 2007) 323–354.

[224] T. Schmidt, Coping with copulas, in: J. Rank (Ed.), Copulas—From Theory to Applications in Finance, Risk Books, London, UK, 2007.

[225] U.G. Schuster, H. Bölcskei, Ultrawideband channel modeling on the basis of information-theoretic criteria, IEEE Transactions on Wireless Communications 6 (7) (July 2007) 2464–2475.

[226] G. Schwarz, Estimating the dimension of a model, Annals of Statistics 6 (2) (March 1978) 461–464.

[227] B. Schweizer, A. Sklar, Probabilistic Metric Spaces, Elsevier, New York, NY, 1983, and Dover, Mineola, NY, 2005.

[228] T. Seidenfeld, L. Wasserman, Dilation for sets of probabilities, The Annals of Statistics 21 (3) (1993) 1139–1154.

[229] G. Shafer, A Mathematical Theory of Evidence, Princeton University Press, Princeton, NJ, 1976.

[230] C.E. Shannon, A mathematical theory of communication, Bell System Technical Journal 27 (October 1948) 623–656.

[231] A. Shemyakin, A. Kniazev, Introduction to Bayesian Estimation and Copula Models of Dependence, John Wiley & Sons, Hoboken, NJ, 2017.

[232] J.A. Shohat, J.D. Tamarkin, The Problem of Moments, AMS Mathematical Surveys, vol. 2, American Mathematical Society, New York, 1943.

[233] J.H. Shore, R.W. Johnson, Axiomatic derivation of the principle of maximum entropy and the principle of minimum cross-entropy, IEEE Transactions on Information Theory IT-26 (1) (January 1980) 26–37.

[234] N. Silver, The Signal and the Noise: The Art and Science of Prediction, The Penguin Press, New York, NY, 2012.

[235] S.D. Silvey, Statistical Inference, Chapman and Hall, London, UK, 1975.

[236] G. Simon, Multivariate generalization of Kendall's tau with application to data reduction, Journal of the American Statistical Association 72 (358) (June 1977) 367–376.

[237] M.K. Simon, Probability Distributions Involving Gaussian Random Variables, Springer, New York, NY, 2006.

[238] M.K. Simon, M.-S. Alouini, Digital Communication over Fading Channels, 2nd edition, John Wiley & Sons, Hoboken, NJ, 2005.

[239] N.J.A. Sloane, Kepler's conjecture confirmed, Nature 395 (1 October 1998) 435–436.

[240] J.E. Smith, Generalized Chebychev inequalities: theory and applications in decision analysis, Operation Research 43 (5) (September-October 1995) 807–825.

[241] S.L. Sobolev, V.L. Vaskevich, The Theory of Cubature Formulas, Springer, Dordrecht, Netherlands, 1997.

[242] F.-X. Socheleau, C. Laot, J.-M. Passerieux, Concise derivation of scattering function from channel entropy maximization, IEEE Transactions on Communications 58 (11) (November 2010) 3098–3103.

[243] M.D. Springer, The Algebra of Random Variables, John Wiley & Sons, New York, NY, 1979.

[244] A.H. Stroud, Approximate Calculation of Multiple Integrals, Prentice-Hall, Englewood Cliffs, NJ, 1971.

[245] A.H. Stroud, D. Secrest, Gaussian Quadrature Formulas, Prentice-Hall, Englewood Cliffs, NJ, 1966.

[246] T.J. Sullivan, Introduction to Uncertainty Quantification, Springer, Cham, Switzerland, 2015.

[247] G. Szegö, Orthogonal Polynomials, 4th edition, Americal Mathematical Society, Providence, RI, 1975.

[248] K. Takemura, Behavioral Decision Theory: Psychological and Mathematical Descriptions of Human Choice Behavior, Springer, Tokyo, Japan, 2014.

[249] M.A. Taneda, J. Takada, K. Araki, The problem of the fading model selection, IEICE Transactions on Communications E84-B (3) (March 2001) 660–666.

[250] M.C.M. Troffaes, T. Basu, A Cantelli-type inequality for constructing non-parametric p-boxes based on exchangeability, Proceedings of Machine Learning Research 103 (2019) 386–393.

[251] A.B. Tsybakov, Introduction to Nonparametric Estimation, Springer, New York, NY, 2009.

[252] A. Tversky, D. Kahneman, The framing of decisions and the psychology of choice, Science 211 (4481) (January 30, 1981) 453–458.

[253] A. Tversky, D. Kahneman, Advances in prospect theory: cumulative representation of uncertainty, Journal of Risk and Uncertainty 5 (1992) 297–323.

[254] L. Vandenberghe, S. Boyd, K. Comanor, Generalized Chebyshev bounds via semidefinite programming, SIAM Review 49 (2007) 52–64.

[255] H.L. Van Trees, Detection, Estimation, and Modulation Theory. Part I: Detection, Estimation, and Linear Modulation Theory, John Wiley & Sons, New York, NY, 2001.

[256] A.F. Veinott Jr., Optimal policy in a dynamic, single product, nonstationary inventory model with several demand classes, Operations Research 13 (5) (September-October 1965) 761–778.

[257] S. Verdú, H.V. Poor, Minimax robust discrete-time matched filters, IEEE Transactions on Communications 31 (2) (February 1983) 208–215.

[258] R. Vershynin, High-Dimensional Probability: An Introduction with Applications in Data Science, Cambridge University Press, Cambridge, UK, 2018.

[259] C. Villani, Optimal Transport: Old and New, Springer-Verlag, Berlin, Germany, 2009.

[260] P.P. Wakker, Prospect Theory for Risk and Ambiguity, Cambridge University Press, Cambridge, UK, 2010.

[261] P. Walley, Measures of uncertainty in expert systems, Artificial Intelligence 83 (1986) 1–58.

[262] P. Walley, Statistical Reasoning with Imprecise Probabilities, Chapman and Hall, London, UK, 1991.

[263] D.J. Walters, G. Ülkümen, C. Erner, D. Tannenbaum, C.R. Fox, Investment behaviors under epistemic versus aleatory uncertainty, Available at https://doi.org/10.2139/ssrn.3695316, September 18, 2020.

[264] Z. Wang, G.J. Klir, Fuzzy Measure Theory, Plenum Press, New York, NY, 1992.

[265] R. Wang, L. Peng, J. Yang, Bounds for the sum of dependent risks and worst Value-at-Risk with monotone marginal densities, Finance Stochastics 17 (2) (April 2013) 395–417.

[266] J. Wang, K. Qiao, Z. Zhang, F. Xiang, A new conflict management method in Dempster–Shafer theory, International Journal of Distributed Sensor Networks 13 (3) (2017) 1–11.

[267] B. Wang, R. Wang, The complete mixability and convex minimization problems with monotone marginal densities, Journal of Multivariate Analysis 102 (10) (November 2011) 1344–1360.

[268] M. Wax, T. Kailath, Detection of signals by information theoretic criteria, IEEE Transactions on Acoustics, Speech, and Signal Processing 33 (2) (April 1985) 387–392.

[269] K. Weichselberger, The theory of interval-probability as a unifying concept for uncertainty, International Journal of Approximate Reasoning 24 (2–3) (May 2000) 149–170.

[270] P. Whittle, A multivariate generalization of Tchebichev's inequality, The Quarterly Journal of Mathematics 9 (1) (1958) 232–240.

[271] H.S. Wilf, Mathematics for the Physical Sciences, John Wiley & Sons, New York, 1962.

[272] R.C. Williamson, Probabilistic Arithmetic, Ph.D. Dissertation, Department of Electrical Engineering, University of Queensland, August 1989.

[273] R.C. Williamson, T. Downs, Probabilistic arithmetic. I. Numerical methods for calculating convolutions and dependency bounds, International Journal of Approximate Reasoning 4 (2) (1990) 89–158.

[274] R.R. Yager, OWA aggregation over a continuous interval argument with applications to decision making, IEEE Transactions on Systems, Man, and Cybernetics—Part B: Cybernetics 34 (5) (October 2004) 1952–1963.

[275] K. Yao, A representation theorem and its application to spherically invariant random processes, IEEE Transactions on Information Theory 19 (5) (September 1973) 600–608.

[276] K. Yao, E. Biglieri, Multidimensional moment error bounds for digital communication systems, IEEE Transactions on Information Theory IT-26 (4) (July 1980) 454–464.

[277] K. Yao, R.M. Tobin, Moment space upper and lower error bounds for digital systems with intersymbol interference, IEEE Transactions on Information Theory IT-22 (1) (January 1976) 65–74.

[278] L.A. Zadeh, Fuzzy sets, Information and Control 8 (1965) 338–353.

[279] L.A. Zadeh, Review of Shafer's 'A mathematical theory of evidence', The AI Magazine 5 (3) (1984) 81–83.

[280] A. Zellner, R.A. Highfield, Calculation of maximum entropy distributions and approximation of marginal posterior distributions, Journal of Econometrics 37 (2) (1988) 195–209.

[281] Q.T. Zhang, A note on the estimation of Nakagami-m fading parameter, IEEE Communications Letters 6 (6) (June 2002) 237–238.

Further reading

[282] A. Abdi, M. Kaveh, On the utility of gamma pdf in modeling shadow fading (slow fading), in: Proceedings of the IEEE 49th Vehicular Technology Conference (VTC Spring '99), Houston, TX, May 1999, pp. 2308–2312.

[283] M. Allais, Le comportement de l'homme rationnel devant le risque: Critique des postulats et axiomes de l'École Americaine, Econometrica 21 (4) (October 1953) 503–546.

[284] G. Apostolakis, S. Kaplan, Pitfalls in risk calculations, Reliability Engineering 2 (1981) 135–145.

[285] S. Benedetto, E. Biglieri, A. Luvison, V. Zingarelli, Moment-based performance evaluation of digital transmission systems, IEE Proceedings-I 139 (3) (June 1992) 258–266.

[286] Y. Ben-Haim, M. Zacksenhouse, R. Eshel, R. Levi, A. Fuerst, W. Bentley, Failure detection with likelihood ratio tests and uncertain probabilities: an info-gap application, Mechanical Systems and Signal Processing 48 (2014) 1–14.

[287] J.O. Berger, Statistical Decision Theory and Bayesian Analysis, 2nd edition, Springer, New York, NY, 1985.

[288] J. Berger, L.M. Berliner, Robust Bayes and empirical Bayes analysis with ε-contaminated priors, The Annals of Statistics 14 (2) (June 1986) 461–486.

[289] K. Beven, Facets of uncertainty: epistemic uncertainty, non-stationarity, likelihood, hypothesis testing, and communication, Hydrological Sciences Journal 61 (9) (2016) 1652–1665.

[290] E. Biglieri, Probability of error for digital systems with inaccurately known interference, IEEE Transactions on Information Theory 30 (2) (March 1984) 443–446.

[291] E. Biglieri, The impact of uncertain channel models on wireless communication, Journal of Communications and Information Networks 1 (1) (2016) 1–13.

[292] E. Biglieri, N. Alrajeh, The robustness of coding and modulation for body-area networks, Journal of Communications and Networks 16 (3) (June 2014) 1–6.

[293] N. Blomqvist, On a measure of dependence between two random variables, Annals of Mathematical Statistics 21 (4) (1950) 593–600.

[294] P.L. Brockett, S.H. Cox Jr., Insurance calculations using incomplete information, Scandinavian Actuarial Journal (April 1985) 94–108.

[295] J. Bukowski, L. Korn, D. Wartenberg, Correlated inputs in quantitative risk assessment: the effects of distributional shape, Risk Analysis 15 (2) (1995) 215–219.

[296] S. Ferson, Model uncertainty in risk analysis, in: Proceedings of the 6th International Workshop of Reliable Engineering Computing: Reliability and Computations of Infrastructures, Chicago, IL, May 25–28, 2014, pp. 27–43.

[297] S. Ferson, W.T. Tucker, Probability boxes as info-gap models, in: 2008 Annual Meeting of the North American Fuzzy Information Processing Society (NAFIPS 2008), New York, NY, May 19–22, 2008.

[298] T.L. Fine, Theories of Probability: An Examination of Foundations, Academic Press, New York, NY, 1973.

[299] P.C. Fishburn, Nonlinear Preference and Utility Theory, The Johns Hopkins University Press, Baltimore, MD, 1988.

[300] G. Fraidenraich, M.D. Yacoub, The α-η-μ and α-κ-μ fading distributions, in: Proceedings of the IEEE Ninth International Symposium on Spread Spectrum Techniques and Applications, Manaus, Brazil, August 28–31, 2006, pp. 16–20.

[301] I. Gilboa, Theory of Decision Under Uncertainty, Cambridge University Press, Cambridge, UK, 2009.

[302] P.R. Halmos, Measure Theory, Springer-Verlag, New York, NY, 1974.

[303] G.H. Hardy, J.E. Littlewood, G. Pólya, Inequalities, 2nd edition, Cambridge University Press, Cambridge, UK, 1952.

[304] K. Isii, Inequalities of the types of Chebyshev and Cramér–Rao and mathematical programming, Annals of the Institute of Statistical Mathematics 16 (1964) 277–293.

[305] J. Jakeman, M. Eldred, D. Xiu, Numerical approach for quantification of epistemic uncertainty, in: PRISM: NNSA Center for Prediction of Reliability, Integrity and Survivability of Microsytems, 2010, Paper 49.

[306] D. Kahneman, Thinking, Fast and Slow, Farrar, Straus and Giroux, New York, NY, 2011.

[307] D. Kahneman, A. Tversky, Variants of uncertainty, Cognition 11 (1982) 143–152.

[308] A.J. Keith, D.K. Ahner, A survey of decision making and optimization under uncertainty, Annals of Operations Research 300 (2021) 319–353.

[309] D. Kokol Bukovšek, T. Košir, B. Mojškerc, M. Omladič, Spearman's footrule and Gini's gamma: local bounds for bivariate copulas and the exact region with respect to Blomqvist's beta, arXiv:2009.06221v2 [math.ST], 5 Jan 2021.

[310] M.G. Krein, The ideas of P.L. Čebyšev and A.A. Markov in the theory of limiting values of integrals and their further developments, Americal Mathematical Society Translations, Ser. 2 12 (1951) 1–121. Translated from Uspekhi Matematičeskih Nauk 6 (4(44)) (1951) 3–120.

[311] В.П. Кузнецов, Интервальные Статистические Модели, Радио и Связь, Москва, Россия, 1991 (на русском языке).

[312] J.B. Lasserre, Bounds on measures satisfying moment conditions, The Annals of Applied Probability 12 (3) (2002) 1114–1137.

[313] J.B. Lasserre, A semidefinite programming approach to the generalized problem of moments, Mathematical Programming, Series B 112 (2008) 65–92.

[314] D.P. Laurie, Computation of Gauss-type quadrature formulas, Journal of Computational and Applied Mathematics 127 (2001) 201–217.

[315] D.V. Lindley, Understanding Uncertainty, John Wiley & Sons, Hoboken, NJ, 2006.

[316] S. Loyka, C.D. Charalambous, I. Ioannou, Ergodic capacity under channel distribution uncertainty, in: Fifty-Second Annual Allerton Conference, October 1–3, Allerton House, UIUC, IL, 2014, pp. 779–784.

[317] R.D. Luce, Where does subjective expected utility fail descriptively?, Journal of Risk and Uncertainty 5 (1992) 5–27.

[318] M.J. Machina, Choice under uncertainty: problems solved and unsolved, The Journal of Economic Perspectives 1 (1) (1987) 121–154, Summer.

[319] R.J. Muirhead, Aspects of Multivariate Statistical Theory, 2nd edition, John Wiley & Sons, Hoboken, NJ, 2005.

[320] A. Saumard, F. Navarro, Finite sample improvement of Akaike's information criterion, IEEE Transactions on Information Theory 67 (10) (October 2021) 6328–6343.

[321] K. Schmüdgen, The Moment Problem, Springer International, Cham, Switzerland, 2017.

[322] G. Shafer, R. Logan, Implementing Dempster's rule for hierarchical evidence, Artificial Intelligence 33 (1987) 271–298.

[323] A.E. Smith, P.B. Ryan, J.S. Evans, The effect of neglecting correlations when propagating uncertainty and estimating the population distribution of risk, Risk Analysis 12 (4) (December 1992) 467–474.

[324] D.D. Stancu, A.H. Stroud, Quadrature formulas with simple Gaussian nodes and multiple fixed nodes, Mathematics of Computation 17 (84) (October 1963) 384–394.

[325] T.M. Strat, Decision analysis using belief functions, International Journal of Approximate Reasoning 4 (5–6) (September–November 1990) 391–417.

[326] M.D. Taylor, Multivariate measures of concordance for copulas and their marginals, Dependence Modeling 4 (2016) 224–236.

[327] A.H. Tchen, Inequalities for distributions with given marginals, The Annals of Probability 8 (4) (1980) 814–827.

[328] H. Thuneberg, H. Salmi, To know or not to know: uncertainty is the answer. Synthesis of six different science exhibition contexts, Journal of Science Communication 17 (2) (2018) A01.

[329] W. Whitt, On approximation for queues, I: extremal distributions, AT&T Bell Laboratories Technical Journal 63 (1984) 115–138.

[330] G. Winkler, Extreme points of moment sets, Mathematics of Operations Research 13 (4) (November 1988) 581–587.

[331] M.D. Yacoub, Foundations of Mobile Radio Engineering, CRC Press, Boca Raton, FL, 1993.

[332] R.R. Yager, L. Liu (Eds.), Classic Works of the Dempster–Shafer Theory of Belief Functions, Springer, New York, NY, 2008.

[333] K. Yao, Error probability of asynchronous spread spectrum multiple access communication systems, IEEE Transactions on Communications COM-25 (8) (August 1977) 803–809.

[334] L.A. Zadeh, A simple view of the Dempster–Shafer theory of evidence and its implication for the rule of combination, The AI Magazine 7 (2) (1986) 85–90.

Index

Printed in the United States
by Baker & Taylor Publisher Services